94.3° UCF

Global Structural Analysis of Buildings

Global Structural Analysis of Buildings

Karoly A. Zalka

London and New York

First published 2000 by E & FN Spon
11 New Fetter Lane, London EC4P 4EE

Simultaneously published in the USA and Canada
by E & FN Spon
29 West 35th Street, New York, NY 10001

E & FN Spon is an imprint of the Taylor & Francis Group

© 2000 Karoly A. Zalka

Printed and bound in Great Britain by TJ International, Padstow, Cornwall

All rights reserved. No part of this book may be reprinted or reproduced or utilised in any form or by any electronic, mechanical, or other means, now known or hereafter invented, including photocopying and recording, or in any information storage or retrieval system, without permission in writing from the publishers.

Publisher's Note

This book has been prepared from camera-ready copy provided by the author.

British Library Cataloguing in Publication Data
A catalogue record for this book is available from the British Library

Library of Congress Cataloging in Publication Data

Zalka, K. A.

   Global structural analysis of buildings / Karoly A. Zalka.

     p. cm.
   Includes bibliographical references and index.
    1.Structural analysis (Engineering) 2. Global analysis (Mathematics)
     I. Title.
TA645 .Z35 2000
690'.21--dc21   00-026451

ISBN 0-415-23483-2

Contents

Preface	xi
Notations	xiii

1 Introduction — 1
  1.1 Background — 1
  1.2 General assumptions — 3
  1.3 The structure of the book — 4

2 Spatial behaviour — 7
  2.1 Basic principles — 7
  2.2 The equivalent column and its characteristics — 8
  2.3 The spatial behaviour of the equivalent column — 16

3 Stability and frequency analyses — 18
  3.1 Stability analysis — 19
    3.1.1 Doubly symmetrical systems – basic critical loads — 20
    3.1.2 Coupling of the basic modes; combined sway-torsional buckling — 26
    3.1.3 Concentrated top load; single-storey buildings — 32
    3.1.4 Shear mode situations — 33
    3.1.5 Soil-structure interaction — 36
    3.1.6 Individual beam-columns — 41
  3.2 Frequency analysis — 43
    3.2.1 Doubly symmetrical systems – basic natural frequencies — 44
    3.2.2 Coupling of the basic modes; combined lateral-torsional vibrations — 52
    3.2.3 Concentrated mass at top level; single-storey buildings — 55
    3.2.4 Soil-structure interaction — 57
    3.2.5 Supplementary remarks — 60

4	**Stress analysis: an elementary approach**	63
	4.1 Horizontal load	64
	4.1.1 Wind	64
	4.1.2 Seismic load	65
	4.1.3 Construction misalignment	67
	4.1.4 Comparisons	68
	4.2 Buildings braced by parallel walls	69
	4.2.1 Basic principles	70
	4.2.2 Load distribution	72
	4.2.3 Deformations	75
	4.3 Buildings braced by perpendicular walls	76
	4.3.1 Load distribution	77
	4.3.2 Deformations	81
	4.4 Buildings braced by frameworks	82
	4.4.1 Frameworks in a symmetrical arrangement	82
	4.4.2 Frameworks in an asymmetrical arrangement	83
	4.5 Maximum bending moments in the bracing elements	86
	4.6 Worked examples	90
	4.6.1 Example 1: building braced by parallel walls	90
	4.6.2 Example 2: building braced by perpendicular walls	91
	4.6.3 Comparison	94
	4.6.4 Example 3: building braced by frameworks and a single wall	94
	4.7 Discussion	96
5	**Stress analysis: an advanced approach**	98
	5.1 The equivalent column and its load	99
	5.2 Deformations of the equivalent column	103
	5.2.1 Horizontal displacements	103
	5.2.2 Rotations	105
	5.3 Deformations of the building	108
	5.4 Load distribution among the bracing elements	110
	5.4.1 Shear forces and bending moments	110
	5.4.2 Torsional moments	121
	5.5 Stresses in the bracing elements	126
	5.6 Concentrated force at top level; single-storey buildings	129
	5.7 Buildings with $I_{xy} = 0$, subjected to uniformly distributed horizontal load	137
	5.8 Worked example: a 6-storey building in London	139
	5.8.1 Model: individual shear walls	140
	5.8.2 Model: built-up shear walls and cores	144

	5.9	Supplementary remarks	149
		5.9.1 Frameworks and coupled shear walls	149
		5.9.2 Bracing systems with shear or a mixture of shear and bending deformations	150
		5.9.3 Special cases – scope for simplification	150
		5.9.4 Second-order effects	151
		5.9.5 Soil-structure interaction	152
6	**Illustrative example; Qualitative and quantitative evaluation**	154
	6.1	Case 1	155
		6.1.1 Critical load	157
		6.1.2 Fundamental frequency	159
		6.1.3 Maximum stresses and deformations	160
	6.2	Case 2	163
		6.2.1 Critical load	165
		6.2.2 Fundamental frequency	166
		6.2.3 Maximum stresses and deformations	167
	6.3	Case 3	169
		6.3.1 Critical load	171
		6.3.2 Fundamental frequency	171
		6.3.3 Maximum stresses and deformations	172
	6.4	Evaluation	174
7	**Global critical load ratio**	176
	7.1	Global critical load ratio – Global safety factor	177
	7.2	Global critical load ratio – Performance indicator	178
	7.3	Further applications	181
8	**Use of frequency measurements for the global analysis**	182
	8.1	Stiffnesses	183
	8.2	Critical loads	184
		8.2.1 Multistorey buildings under uniformly distributed floor load	184
		8.2.2 Concentrated top load; single-storey buildings	186
	8.3	Deformations	186
		8.3.1 Multistorey buildings subjected to horizontal load of trapezoidal distribution	187
		8.3.2 Concentrated force at top level; single-storey buildings	189
		8.3.3 Deformations of the building	191

9 Equivalent wall for frameworks; Buckling analysis of planar structures — 192
   9.1 Introduction — 192
   9.2 Characteristic deformations, stiffnesses and part critical loads — 194
   9.3 Frameworks on fixed supports — 199
      9.3.1 The application of summation theorems — 200
      9.3.2 The continuum model — 201
      9.3.3 The sandwich model — 204
      9.3.4 Design formulae — 208
   9.4 Frameworks on pinned supports — 210
      9.4.1 Frameworks without ground floor beams — 212
      9.4.2 Frameworks with ground floor beams — 213
   9.5 Frameworks with ground floor columns of different height — 214
   9.6 Analysis of coupled shear walls by the frame model — 216
   9.7 Frameworks with cross-bracing — 218
      9.7.1 Shear stiffness and shear critical load — 220
      9.7.2 Critical loads — 224
      9.7.3 Structures with global regularity — 225
   9.8 Infilled frameworks — 226
   9.9 Equivalent wall for 3-dimensional analysis — 228
   9.10 Shear walls — 230
   9.11 Symmetrical cross-wall system buildings — 231
   9.12 Planar bracing elements: a comparison — 233
   9.13 Supplementary remarks — 236

10 Test results and accuracy analysis — 238
   10.1 Description of the models — 238
   10.2 Horizontal load on Model 'M1' — 241
   10.3 Horizontal load on Model 'M2' — 243
   10.4 Comparative analysis of the formulae for horizontal load — 245
   10.5 Dynamic tests — 246
   10.6 Stability tests — 248
      10.6.1 Model 'M1' — 250
      10.6.2 Model 'M2' — 251
      10.6.3 Deformation of the bracing elements — 252

11 Evaluation; design guidelines — 254
   11.1 Spatial behaviour — 255
   11.2 Stability analysis — 255
   11.3 Frequency analysis — 256

11.4 Stresses and deformations	258
11.5 Structural performance of the bracing system	259
11.6 Stability of planar structures	260
11.6.1 Low-rise to medium-rise (4–25-storey) structures	260
11.6.2 Tall (over 25-storey) structures	263
11.6.3 Structural performance of planar bracing elements	263

Appendix A Cross-sectional characteristics for bracing elements 266

Appendix B The generalized power series method for eigenvalue problems 278

Appendix C Mode coupling parameter κ 283

References 315

Further reading 325

Name index 331

Subject index 335

Preface

A shift in emphasis can be seen in the approach to structural design. More often structures are looked at 'globally', as whole structural units, rather than a group of individual elements. The investigation of this global behaviour, also described as 'holistic' or 'whole building' behaviour has been made possible by new theoretical achievements and the spectacular advance in computer technology during the last decades.

The global structural analysis of buildings can be carried out following two routes. First, sophisticated and complex computer packages based on the finite element method offer endless facilities and can handle even huge structures with a great number of elements. Second, analytical methods can also deal with whole structures leading to simple closed-form solutions, with the additional benefit of providing fast checking facilities for the computer-based methods.

This book follows the latter route and, after describing and solving the complex theoretical problems of bracing systems covering many practical cases, intends to achieve the following three objectives:

- To present simple procedures and closed-form formulae which make it possible for the practising structural engineer to carry out a general structural analysis of the bracing system of building structures in minutes.
- To show that the main areas of structural design (stability, stress and frequency analyses) are not independent; indeed they can be linked by the global critical load ratio which can be used to achieve optimum structural solutions with high performance and adequate safety.
- To help to understand global behaviour better and to develop structural engineering common sense through the introduction of the most representative stiffness characteristics for the stability, stress and frequency analyses.

Notations

CAPITAL LETTERS

A	cross-sectional area; area of the plan of the building; floor area
A_a	area of lower flange
A_b	cross-section of beams
A_c	cross-section of columns
A_d	cross-section of diagonal bars in cross-bracing
A_h	cross-section of horizontal bars in cross-bracing
A_f	area of upper flange; contact area between foundation and soil
A_g	area of web
A_i	cross-sectional area of the ith bracing element
A_j	incremental area
A_o	area of closed cross-section defined by the middle line of the walls
A_{ref}	reference area for the force coefficient
B	plan breadth of the building (in direction y)
C	centre of vertical load; centroid
C_1, C_2, C_3, C_4	constants of integration
E	modulus of elasticity
E_b	modulus of elasticity of beams
E_c	modulus of elasticity of columns
E_d	modulus of elasticity of diagonal bars in cross-bracing
E_h	modulus of elasticity of horizontal bars in cross-bracing
F	concentrated load (on top floor level); horizontal force
F_{cr}	critical concentrated load
$F_{cr,X}, F_{cr,Y}$	critical concentrated load in directions X and Y
$F_{cr,\varphi}$	critical concentrated load for pure torsional buckling
F_g	full-height (global) bending critical concentrated load
F_i	vertical load on the ith bracing element/framework; vertical force at x_i, y_i
F_l	local bending critical concentrated load
F_m	total horizontal load due to misalignment
F_x, F_y	components of the resultant of the horizontal load in directions x and y
F_w	global wind force
$F_{w,x}, F_{w,y}$	wind force in directions x and y
F_{wj}	global wind force for height/width > 2
G	modulus of elasticity in shear
H	height of building/frame/coupled shear walls; horizontal force
I	second moment of area
I_b	second moment of area of beams

I_c	second moment of area of columns
I_{cg}	fictitious 'global' second moment of area of a column of storey height
I_e	effective second moment of area
I_f	second moment of area of the foundation
I_g	global second moment of area of the columns of the framework; gross (uncracked) second moment of area
I_o	polar moment of inertia of the ground plan area with respect to the shear centre of the bracing system
I_x, I_y	second moments of area with respect to centroidal axes x and y
I_X, I_Y	second moments of area with respect to principal axes X and Y
I_{xp}, I_{yp}	second moments of area of the plan of the building with respect to centroidal axes x and y
I_{xy}	product of inertia with respect to axes x and y
I_{XY}	product of inertia with respect to principal axes X and Y
I_ω	warping (bending torsional) constant
J	Saint-Venant torsional constant
K	shear stiffness of frameworks; shear critical load; seismic constant
K^*	shear stiffness/shear critical load of coupled shear walls
K_d	shear stiffness representing the effect of the diagonal bars in cross-bracing
K_g	full height (global) shear stiffness; global shear critical load
K_g^*	full height (global) shear stiffness of coupled shear walls; full height (global) shear critical load of coupled shear walls
K_h	shear stiffness representing the effect of the horizontal bars in cross-bracing
K_l	local shear stiffness; local shear critical load
L	width of framework; plan length of building (in direction x); width of equivalent shear wall
M_p	bending moments due to load of intensity p
M_Q	bending moments due to virtual force $Q = 1$
M_t	Saint-Venant torsional moment
$M_{x,i}, M_{y,i}$	bending moment in the ith bracing element in planes xz and yz
M_z	total torque
M_ω	warping torsional moment
M^*	bending moment on the equivalent column
$\overline{M}$	concentrated moments representing the supporting effect of the beams
MSC	Mercalli-Sieberg-Cancani seismic scale
N	total applied uniformly distributed load (measured at ground floor level)
$N_{combined}$	combined sway-torsional critical load for the monosymmetric case
N_{cr}	critical load for the uniformly distributed load
N_{cr}^D	combined (F + N) critical load
N_{cr}^{flex}	critical load of rigid column on flexible support
N_{cr}^{int}	critical load which takes into consideration soil-structure interaction
$N_{cr,X}, N_{cr,Y}$	critical UDL in directions X and Y
$N_{cr,\varphi}$	critical UDL for pure torsional buckling
N_g	full-height (global) bending critical UDL for frameworks
N_l	local bending critical UDL for frameworks

xv

$N(z)$	total vertical load at z
O	shear centre
Q	intensity of uniformly distributed floor load; weight per unit area of top floor; shear force on floor level
R	radius
S, S_x, S_y	seismic force
S'_x, S'_y	first (statical) moments of area about the neutral axis
S_ω	sectorial static moment
T	natural period; shear force at contraflexure point
$T_{x,i}, T_{y,i}$	shear force in the ith bracing element in planes xz and yz
T_X, T_Y	first natural period for lateral vibration
UDL	uniformly distributed load
V	vertical load
V_{CR}	critical load for model structure
W	bimoment; width of structure
X, Y	principal axes
Z_1, Z_2	auxiliary functions defined by formulae (5.18)

SMALL LETTERS

a	length of wall section
a_0, a_1, a_2	coefficients
b	length of wall section; width of perforated wall section
b_w	width of diagonal strip for infill
c	translation; length of wall section; depth of lintel with coupled shear walls; critical load parameter for different end conditions
c_F	translation due to F passing through the shear centre
c_{yF}	translation in direction y due to F passing through the shear centre
c_M	translation due to M acting around the shear centre
c_{yM}	translation in direction y due to M acting around the shear centre
c_{ALT}	altitude factor
c_d	dynamic coefficient
c_{DIR}	wind direction factor
$c_e(z_e)$	exposure coefficient
c_f	force coefficient
c_{fj}	force coefficient for incremental area A_j
c_i	coefficients in a series
c_{TEM}	temporary (seasonal) factor
d	length of wall section; length of diagonal with cross-bracing
dz	length of elementary section
e	perpendicular distance between the line of action of the horizontal load and the shear centre; distance of upper flange from centroid
e^*	distance of lower flange from centroid (with bracing cores)
f	frequency
f_c	frequency when the effect of the compressive force is taken into account
$f_{combined}$	combined lateral-torsional frequency for the monosymmetric case

f_{int}	natural frequency which takes into account soil-structure interaction
f_{flex}	natural frequency of rigid column on flexible support
f_d	frequency when the effect of damping is taken into account
f_X, f_Y	natural frequency of lateral vibration in principal directions X and Y
f_φ^t	pure torsional frequency associated with the Saint-Venant stiffness
f_φ^ω	pure torsional frequency associated with the warping stiffness
f_φ	fundamental frequency of torsional vibration
g	global axis; gravity acceleration
h	height of storey; length of wall section
h^*	equivalent height of storey
$\bar{h}$	height of first storey columns
$h_{i,k}$	length of the kth section of the ith bracing element
i	parameter relating to the number of bracing elements (from 0 to n)
i_p	radius of gyration
k	torsion parameter; translational stiffness; parameter relating to the number of elements in a series; spring constant; stiffness of a framework
k_x, k_y	translational stiffness with respect to axes x and y, respectively
k_s	modified torsion parameter; modulus of subgrade reaction
l	width of bay; distance between shear walls; local axis
l_1, l_2, l_3, l_4	'torsion arm' of shear walls; distance between the wall sections of coupled shear walls; width of bay
l_i	frequency factor
m	distributed moments for the stability analysis; length of section of beam for cross-bracing
m_z	distributed torque
m_{z0}	uniformly distributed part of torque
$\bar{m}$	distributed moments for the stress analysis, representing the supporting effect of the beams
n	number of columns/walls/bracing elements; number of storeys
n_h	number of structural elements on one floor level
p	intensity of the uniformly distributed load on the beams
p_s	intensity of seismic load
q	intensity of the uniformly distributed load on the floors; intensity of the uniformly distributed vertical load; horizontal load
q_0	intensity of the uniformly distributed part of the horizontal load
q_{0x}, q_{0y}	components of the uniformly distributed part of the horizontal load in directions x and y
q_1	intensity of the variable part of the horizontal load at top
q_{1x}, q_{1y}	components of the variable part of the horizontal load in directions x and y
q_{ref}	reference mean wind velocity pressure
q_x, q_y	components of the horizontal load in directions x and y
$q_{x,i}, q_{y,i}$	components of the horizontal load in directions x and y on the ith bracing element
$\bar{q}_x, \bar{q}_y$	auxiliary parameters defined by formula (5.19)
r	combination factor; modifier
r_1, r_2	critical load ratio; frequency ratio (for mode coupling); radius

r_f	reduction factor for the frequency analysis
r_{flex}	reduction factor for flexible support
r_s	reduction factor for the stability analysis
s	width of wall; distance between bays; length of arc (with closed cross-sections)
s_{flex}	reduction factor for flexible support
t	distance between the shear centre and the centre of the vertical load; wall thickness; global centroidal axis of the cross-sections of the columns; time
t^*	thickness of the equivalent wall
t_i	distance between the axis of the ith column and the global centroidal axis; wall thickness of the ith bracing element
$t_{i,k}$	wall thickness of the kth section of the ith bracing element
u	horizontal displacement of the shear centre in direction x
u_B	horizontal displacement of corner point B of the building in direction x
u_{flex}	top translation of rigid column on flexible support in direction x
u_g	horizontal displacement of the bracing element at top floor level
u_i	horizontal displacement of the shear centre of the ith bracing element in direction x
u_l	accumulative top level horizontal displacement due to the storey level displacements of the columns
u_{max}	maximum horizontal displacement in direction x
v	horizontal displacement of the shear centre in direction y
v_A	horizontal displacement of corner point A of the building in direction y
v_{flex}	top translation of rigid column on flexible support in direction y
v_i	horizontal displacement of the shear centre of the ith bracing element in direction y
v_{max}	maximum horizontal displacement in direction y
$v_{ref,0}$	basic value of the wind velocity given by means of wind maps
x	horizontal coordinate axis; horizontal coordinate
$\bar{x}$	horizontal coordinate axis; coordinate in coordinate system $\bar{x}-\bar{y}$
x_A	coordinate of corner point A of the building in direction x
x_c	coordinate of the centroid in direction x in the x-y coordinate system of the shear centre
x_i	coordinate of the shear centre of the ith bracing element in direction x
x_{max}	location of maximum translation
$\bar{x}_i, \bar{y}_i$	coordinates of the shear centre of the ith bracing element in the coordinate system $\bar{x}-\bar{y}$
$\bar{x}_o$	coordinate of the shear centre in coordinate system $\bar{x}-\bar{y}$
y	horizontal coordinate axis; horizontal coordinate
$\bar{y}$	horizontal coordinate axis; coordinate in coordinate system $\bar{x}-\bar{y}$
y_B	coordinate of corner point B of the building in direction y
y_c	coordinate of the centroid in direction y in the x-y coordinate system of the shear centre
y_i	coordinate of the shear centre of the ith bracing element in direction y
$\bar{y}_o$	coordinate of the shear centre in coordinate system $\bar{x}-\bar{y}$
z	vertical coordinate axis; vertical coordinate

z_j height of the centre of gravity of incremental area A_j
z_{max} location of maximum bending moment in the beams of the framework

GREEK LETTERS

α eigenvalue; critical load parameter for frames on fixed supports; critical load parameter for pure torsional buckling; angle between axes x and X; auxiliary parameter

α_p eigenvalue; critical load parameter for frames on pinned supports

α_S Southwell estimate for eigenvalue α; eigenvalue and critical load parameter for the sandwich model

β stiffness ratio; damping ratio; dynamic constant for seismic load; stiffness ratio for soil-structure interaction (stability; frequencies)

β_S stiffness ratio for the sandwich model

δ lateral displacement (storey drift)

ε mode coupling parameter for the monosymmetrical case

γ weight per unit volume; angular displacement

η first natural frequency parameter for pure torsional vibration; height/width ratio

η_i, η_2, η_3 ith, 2nd and 3rd natural frequency parameters for pure torsional vibration

η_q load factor (due to rotation)

η_T shear force factor (due to rotation)

$\eta_{T'}$ shear force factor (due to rotation) for the concentrated force load case

η_M bending moment factor (due to rotation)

$\eta_{M'}$ bending moment factor (due to rotation) for the concentrated force load case

κ mode coupling parameter for the 3-dimensional case

λ stiffness ratio for beams/columns of frameworks

φ rotation; angle between the diagonal and horizontal bars in cross-bracing

φ_{max} maximum rotation

μ slope of the function of the horizontal load; construction misalignment

v global critical load ratio

ρ mass density per unit volume; shape factor; air density

ρ^* mass density per unit area

σ_z compressive stress

τ_i shear stress in the ith bracing element

τ_i^b shear stress due to unsymmetrical bending

τ_i^t shear stress due to Saint-Venant torsion

τ_i^ω shear stress due to warping torsion

τ_X, τ_Y eccentricity parameters for the 3-dimensional torsional-flexural buckling

ω sectorial coordinate

ω_X, ω_Y circular frequencies for vibration in principal directions X and Y

ω_φ circular frequency for pure torsional vibration

1

Introduction

In applying physical and mathematical models which are based on the global behaviour of building structures, a unified treatment of the stress, stability and frequency analyses of bracing systems is presented for carrying out the structural analysis of buildings. In complementing the conventional 'element-based' design process, closed-form formulae and simple procedures are given for the global analysis of individual bracing elements and 3-dimensional bracing systems.

1.1 BACKGROUND

The conventional design process is normally based on the 'local' structural analysis of individual elements (columns, beams, floor slabs, walls, etc). This attitude is natural, since the structural system consists of individual elements. However, theoretical research, small-scale and large-scale tests and failures (and in some cases the lack of failures) in structural systems have indicated that complex structures cannot be considered simply as a collection of individual elements. The response of the structure is often more than the 'sum' of the responses of the individual elements since structural integrity ensures that the elements work together in a properly designed system and the structure develops some 'global' response through the complex interaction of its elements.

The 'local' or 'global' approach to structural design may affect the level of safety of the structure and can lead to considerable advantages or disadvantages as far as structural economy is concerned. A structure based on the optimum solution of the individual elements may not be economic when the system is considered as a whole. It is an interesting fact from the point of view of structural safety and economy, that when some elements in the system are supposed to fail when checked individually, they do not do so because other elements can help out. It is equally important to point out that interaction among the structural elements may also result in unfavourable phenomena.

Well-established and well-publicized methods have been made available for the analysis of individual structural elements. These methods are relatively simple and usually do not require much theoretical knowledge. Being suitable for automation, most of them have been built into computer procedures and are widely used in design offices.

The area of global structural analysis, however, is not so well developed. Many reasons have contributed to the relatively slow progress in developing methods for global design. Because of the complexity of the problem and the great number of structural elements involved in the analysis, deeper theoretical background, more sophisticated techniques and computers of great capacity are needed.

Global analysis, as structural design itself, can be carried out on two levels. An 'exact' analysis – or an analysis called exact – relies on a mathematical model as exact as possible and uses a static model which takes into account as many structural elements, material properties, geometrical and stiffness characteristics as possible. Taking everything into consideration, however, can result in problems. Even using a powerful computer, the job can be too big to handle. Because of the complexity of the results, they can be difficult to interpret. The lengthy and time-consuming procedure of handling all the data can always be a source for errors. Another disadvantage of this approach may be that the importance of the key structural elements is sometimes hidden behind the great number of input and output data.

Simplified procedures relying on carefully chosen approximations represent the other possibility for the analysis. A good approximate method relies on the most important structural characteristics and ignores those which have no real influence on the response of the structure. It is therefore simple, fast and offers a clear picture of the structural behaviour.

As both the local and the global approaches are important since they complement each other, both the exact and approximate procedures have their own significance. In the design process, the main structural characteristics are often established using an approximate method. Relying on the results of the approximate method, an exact procedure can follow, which eventually gives the final structural solution. The approximate method can fulfil another important task. As the theories behind the exact and approximate methods are often different – the exact methods are normally based on the finite element approach while the approximate methods are often based on analytical procedures – the results of the approximate analysis can be used as independent checks on the results of the exact method. When the two sets of result show the same structural behaviour, it is a strong indication that the results are correct. This endorsement is important to the structural designer, since it is sometimes not easy to detect an error with the exact analysis where thousands of data may be involved [MacLeod, 1995]. The significance of the independent verification of the results has widely been recognized and the importance of avoiding a 'Computer Aided Disaster' has been discussed at different conferences [Brohn, 1996; Knowles, 1996].

With the widespread use of more slender structural elements and lighter materials, and with the increasing demand for more economic structures, design for stability has become more and more important. In recognizing this tendency, the methods and procedures presented in this book for the stress, stability and dynamic analyses are linked together through the global critical load ratio, which is also shown to be a generic performance indicator.

With the increasing availability of more and more sophisticated and user-friendly computer procedures, the engineering society seems to be falling into two groups. Those developing the software packages manage to build on, and even develop further, their theoretical knowledge acquired at the university. On the other hand, office pressure to produce more and quickly forces some of the practising structural designers to concentrate only on pressing the right key on the keyboard. Some might say that it is not really important for the designer to understand the theory behind the analysis since the computer knows everything anyway. The tendency in the last twenty years indicates that the general knowledge of the young generation regarding basic structural behaviour is less than satisfactory [Brohn and Cowan, 1977; Brohn, 1992]. The discussion and debate on the advantages and disadvantages of the use of computers in the design office are still going on [Smart, 1997; Gardner, 1999].

1.2 GENERAL ASSUMPTIONS

The majority of the formulae and procedures are based on the application of the equivalent column concept and the summation theorems of civil engineering. The equivalent column concept is applicable to *regular* structures when the geometrical and stiffness characteristics of the bracing system do not vary over the height of the building. In addition, the following conditions also have to be fulfilled.

4 Introduction

a) The material of the structures is homogeneous, isotropic and obeys Hooke's law.
b) The floor slabs are stiff in their plane and flexible perpendicular to their plane.
c) The structures have no geometrical imperfections, they develop small deformations and the third-order effect of the axial forces is negligible.
d) The loads are applied statically and maintain their direction (conservative forces).
e) The location of the shear centre only depends on geometrical characteristics.

When applied, additional assumptions concerning the different structures and types of analysis may be given in the introduction of the corresponding chapter.

The formulae presented for the stability analysis are applicable to structures subjected to concentrated top load, distributed load (or concentrated forces) on floor levels or the two loads together. It is assumed that the frameworks and coupled shear walls are sway structures and the critical load defines the bifurcation load.

The formulae given for the dynamic analysis were derived assuming distributed mass over the floors and concentrated mass on top floor level. The horizontal load for the stress analysis is assumed to be of trapezoidal distribution.

1.3 THE STRUCTURE OF THE BOOK

The objective of the book and general assumptions are given in Chapter 1. Chapter 2 briefly describes the 3-dimensional behaviour of the load-bearing elements and introduces the equivalent column for the bracing system, which is mostly used later for the analyses. Chapter 3 covers the stability and frequency analyses of buildings. Closed-form formulae, tables and diagrams are given for the quick calculation of the basic critical loads and natural frequencies for the stability and dynamic analyses. The combination of the basic modes in both the buckling and the frequency analyses is taken into account in two ways: by a simple interaction formula and by a cubic equation. Both multistorey and single-storey buildings are covered. The effect of soil-structure interaction is taken into account approximately by simple summation formulae. An elementary approach is presented in Chapter 4 for the stress analysis of buildings braced by parallel shear walls, a system of perpendicular shear walls, or frameworks, subjected to a uniformly distributed horizontal load. Worked examples facilitate practical application. The scope of the stress analysis is extended in Chapter 5 where a comprehensive method is given for bracing systems having (Saint-Venant and warping) torsional stiffnesses as well. Closed-form solutions are derived and diagrams and tables are given for the load distribution and the stresses and deformations when the building is under horizontal load of trapezoidal distribution. Simple expressive formulae are also presented for the maximum values of the stresses and deformations. Formulae for single-storey buildings are also presented. In addition to sizing the elements of the bracing system, the technique is potentially useful both at the concept design stage and for final analysis for checking of structural adequacy, assessing the suitability of structural layouts, verifying the results of other methods, evaluating computer packages, facilitating theoretical research and developing new techniques and procedures.

Based on a series of worked examples, a comprehensive qualitative and quantitative evaluation is given in Chapter 6, which also shows how the procedures are used in practice for actual design and for finding optimum structural arrangement. The global critical load ratio, also identified as a performance indicator of the bracing system, is introduced in Chapter 7. It is demonstrated that the global critical load ratio can be used to increase the efficiency of the bracing system while ensuring adequate level of safety, leading to more economic construction. Monitoring the value of the global critical load ratio for different structural layouts offers a simple tool for increasing the global critical load and the fundamental frequencies and reducing the maximum stresses and deformations in the bracing system.

To improve the accuracy of the procedures, which are based on mathematical and physical models, alternative formulae are also derived in Chapter 8 which, instead of the theoretical values of the stiffnesses, the key elements in the structural analysis, use stiffness values based on frequencies measured on the building. These formulae are applicable to the stress and stability analyses of doubly symmetrical bracing systems.

Chapter 9 deals with the stability analysis of planar structures and presents a closed form formula for an equivalent wall. The equivalent wall can then be used in the 3-dimensional analysis. It is shown that all 'frame-like' planar structures can be characterized by four distinct deformations and the corresponding four stiffnesses. Closed-form formulae are then given for frameworks on pinned and fixed supports with and without cross bracing, for coupled shear walls and for infilled frameworks. The efficiency of planar bracing elements is investigated through two sets of representative one-bay and two-bay, four to ninety-nine storey high structures.

Chapter 10 presents the results of a series of small-scale tests and a short summary of a comprehensive accuracy analysis which show that the accuracy of the procedures is acceptable for practical structural engineering applications. A brief evaluation of the procedures is given in Chapter 11 with practical guidelines for the structural designer.

Cross-sectional characteristics for commonly used bracing elements are collected in Appendix A. Appendix B introduces the power series method and shows how a complex eigenvalue problem in stability analysis can be transformed into a simple problem of finding the smallest root of a polynomial. In providing a complete set of tables, Appendix C reduces the task of producing the critical load of combined sway-torsional buckling and the fundamental frequency of combined lateral-torsional vibration to a single-step calculation.

References cited in the book are given following the Appendices. Further reading is helped by a list of additional bibliographical items.

Name and subject indexes conclude the book.

With the exception of Chapter 4, the book does not give the detailed derivations and proofs of the formulae; these are available in separate publications cited in the text and listed in the References. Chapter 4, however, demonstrates how carefully chosen elementary static considerations can lead to the simple solution of complex problems.

2

Spatial behaviour

2.1 BASIC PRINCIPLES

The primary structural elements of buildings are the vertical and horizontal load bearing structures. These structures carry the horizontal and vertical loads of the building. The vertical (dead and live) loads are transmitted to the vertical load bearing elements (shear walls, frames and columns) by the horizontal load bearing elements (floor slabs). The horizontal loads (wind, construction misalignment, and seismic forces) are transmitted by the floor slabs to those vertical load-bearing elements that are capable of passing them on the foundation. These dedicated structural elements (shear walls, frames and cores) are called the *bracing elements* of the building, whose main task is to provide the building with adequate lateral stiffness. They represent a system, which will be referred to as the *bracing system* of the building.

Of the vertical load-bearing elements, the frameworks are basically responsible for carrying the vertical loads and the main task of the cores is to provide the necessary lateral and torsional stiffnesses. The shear walls and coupled shear walls contribute to both tasks. The floor slabs act as horizontal load bearing elements and are also responsible for transmitting the applied vertical loads to the vertical load bearing elements and for distributing the horizontal loads among the vertical load bearing elements. Compared to the shear walls and cores, the frameworks are more flexible and they are often neglected when the lateral stiffness of a building is assessed. If, however, the effect of the frameworks also has to be taken into account for some reason, then, as an approximation, they can be replaced by fictitious walls. These fictitious walls can then be included in the analysis which can be carried out in a relatively simple manner when the bracing system only comprises walls and cores. Several methods are available for the calculation of the size of the cross-section of the fictitious walls; one of them is to stipulate that the critical load of the fictitious wall and that of the framework be identical. A simple procedure is given in section 9.9 where the limitations of the use of fictitious walls are also discussed.

Subjected to external loads, a system of shear walls and cores can develop three kinds of deformation: sway in the two principal planes of inertia and rotation around the shear centre. Apart from some special symmetrical arrangements, the three modes couple resulting in a combined sway-torsional behaviour. The exact spatial analysis of these structures is rather complicated, partly because of the interaction between the horizontal and vertical load bearing systems and among the elements of the horizontal and vertical systems themselves and partly because of the great number of elements to be involved in the analysis. By applying the equivalent column approach, however, the analysis can be simplified and closed-form solutions can be produced for the stresses and deformations, the load distribution among the structural elements, the elastic critical loads and the natural frequencies.

2.2 THE EQUIVALENT COLUMN AND ITS CHARACTERISTICS

The 3-dimensional analysis is based on the analysis of the equivalent column. The equivalent column is obtained by combining the bracing shear walls and cores of the building to form a single cantilever. Its bending and torsional stiffnesses represent the whole building. As the equivalent column is situated concurrent with the shear centre (centre of stiffness) of the bracing elements, the first step is to locate this global shear centre (O in Fig. 2.1). The position of the shear centre is found by making use of the basic geometrical and stiffness characteristics of the bracing elements [Beck and Schäfer, 1969].

The calculation is carried out in the coordinate system $\bar{x} - \bar{y}$, whose origin lies in the upper left corner of the plan of the building (Fig. 2.1) and whose axes are aligned with the sides of the building:

$$\bar{x}_o = \frac{I_{xy}\left(\sum_1^n I_{y,i}\bar{y}_i - \sum_1^n I_{xy,i}\bar{x}_i\right) - I_y\left(\sum_1^n I_{xy,i}\bar{y}_i - \sum_1^n I_{x,i}\bar{x}_i\right)}{I_x I_y - I_{xy}^2}, \qquad (2.1)$$

$$\bar{y}_o = \frac{I_x\left(\sum_1^n I_{y,i}\bar{y}_i - \sum_1^n I_{xy,i}\bar{x}_i\right) - I_{xy}\left(\sum_1^n I_{xy,i}\bar{y}_i - \sum_1^n I_{x,i}\bar{x}_i\right)}{I_x I_y - I_{xy}^2}. \qquad (2.2)$$

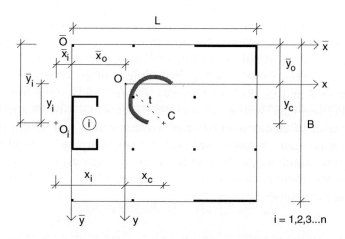

Fig. 2.1 Building layout with bracing elements and the equivalent column.

In formulae (2.1) and (2.2) $I_{x,i}$, $I_{y,i}$ and $I_{xy,i}$ represent the moments of inertia and the product of inertia of the *i*th element of the bracing system with respect to its local centroidal coordinate axes which are parallel to axes $\bar{x}$ and $\bar{y}$. Coordinates $\bar{x}_i$ and $\bar{y}_i$ stand for the location of the shear centre of the individual bracing elements in the coordinate system $\bar{x} - \bar{y}$. The sums of the moments of inertia and the product of inertia of the bracing elements are also needed in formulae (2.1) and (2.2):

$$I_x = \sum_1^n I_{x,i}, \qquad I_y = \sum_1^n I_{y,i}, \qquad I_{xy} = \sum_1^n I_{xy,i}, \qquad (2.3)$$

where $i = 1...n$, and n is the number of bracing elements. The sums of the moments of inertia of the bracing elements are important characteristics of the building in relation to its global bending. The product of inertia plays an important role in determining the orientation of the principal axes of the bracing system.

In addition to the above bending characteristics, there are two more characteristics of the equivalent column, which are associated with torsion: the Saint-Venant torsional constant and the warping (bending

10 Spatial behaviour

torsional) constant. The Saint-Venant torsional constant is the sum of the Saint-Venant torsional constants of the bracing elements and is obtained in a similar manner as with the moments of inertia:

$$J = \sum_{1}^{n} J_i , \qquad (2.4)$$

where J_i is the Saint-Venant torsional constant of the ith bracing element.

The warping constant is a weighted sum which is calculated in a coordinate system whose origin is the global shear centre [Beck, König and Reeh, 1968]. For this purpose, after making use of formulae (2.1) and (2.2), coordinate system $\bar{x} - \bar{y}$ is transferred to coordinate system x-y, whose origin coincides with the global shear centre and whose axes are parallel with axes $\bar{x}$ and $\bar{y}$ (Fig. 2.1). The warping constant of the equivalent column in this coordinate system assumes the form:

$$I_\omega = \sum_{1}^{n} (I_{\omega,i} + I_{x,i} x_i^2 + I_{y,i} y_i^2 - 2 I_{xy,i} x_i y_i) , \qquad (2.5)$$

where $I_{\omega,i}$ is the warping constant and x_i and y_i are the coordinates of the shear centre of the ith bracing element in the x-y coordinate system.

The warping (bending torsional) constant of the equivalent column represents two types of contribution: the first term in formula (2.5) represents the warping torsion of the bracing elements with regard to their own shear centre and the second, third and fourth terms stand for the bending torsion of the elements with regard to the global shear centre of the whole system. This part of contribution is realized through the bending of the bracing elements ($I_{x,i}$, $I_{y,i}$, $I_{xy,i}$) utilizing their 'torsion arm' (x_i, y_i) around the global shear centre.

Formulae (2.3), (2.4) and (2.5) reflect the assumption that the floors of the building are stiff in their own plane and flexible perpendicular to their plane. It is the out-of-plane flexibility of the floor slabs that leads to the simple sums in formulae (2.3) and (2.4) and it is their in-plane stiffness that results in the second, third and fourth terms in formula (2.5).

When the bending and torsional stiffnesses of reinforced concrete bracing elements are calculated for the establishment of the equivalent column, the effect of cracking on stiffness may have to be taken into account. The phenomenon is not discussed here as detailed information is available elsewhere; it is only mentioned that $I_e = 0.8 I_g$ is normally considered an adequate reduction in the value of the second moment of area, where I_e and I_g are the effective and the gross (uncracked) second moments of area. Detailed information and guidelines are given in [Council, 1978d].

Apart from the above constants of the equivalent column, the radius of gyration is also needed for the establishment of the equivalent column for the stability and dynamic analyses. The radius of gyration is determined by the load of the building and the area over which it is distributed. The load is represented either by the mass (for the dynamic analysis) or by the vertical load (for the stability analysis).

In the general case when the building is subjected to a load of arbitrary distribution on a layout of arbitrary shape, the radius of gyration is calculated from

$$i_p = \sqrt{\frac{\int_{(A)} q(x,y)(x^2 + y^2)dA}{\int_{(A)} q(x,y)dA}}, \qquad (2.6)$$

where $q(x,y)$ is the intensity of the load.

Concentrated forces can also be taken into consideration: when the load consists of concentrated forces formula (2.6) assumes the form

$$i_p = \sqrt{\frac{\sum_i F_i(x_i^2 + y_i^2)}{\sum_i F_i}}, \qquad (2.7)$$

where F_i is the ith concentrated force and x_i and y_i are its coordinates in the x-y coordinate system whose origin is in the shear centre.

Formula (2.6) is a general formula. In many practical cases, simpler formulae can be used for the analysis. Assuming uniformly distributed load over the plan of the building, for example, the radius of gyration is obtained from

$$i_p = \sqrt{\frac{I_o}{A}} = \sqrt{\frac{I_{xp} + I_{yp} + A(x_c^2 + y_c^2)}{A}} = \sqrt{\frac{I_{xp} + I_{yp}}{A} + t^2}, \qquad (2.8)$$

where

I_o is the polar moment of inertia of the ground plan with respect to the shear centre of the bracing system,

I_{xp}, I_{yp} are the second moments of area of the plan of the building with respect to the centroidal axes,
A is the area of the plan,
t is the distance between the shear centre (O) and the centre of vertical load (C) (Fig. 2.1), according to the formula

$$t = \sqrt{x_c^2 + y_c^2}. \tag{2.9}$$

Coordinates x_c and y_c are the coordinates of the geometrical centre of the plan of the building in coordinate system x-y whose origin is in the shear centre:

$$x_c = \frac{L}{2} - \bar{x}_o, \qquad y_c = \frac{B}{2} - \bar{y}_o. \tag{2.10}$$

When the building has a rectangular plan, the formula for the radius of gyration simplifies to

$$i_p = \sqrt{\frac{L^2 + B^2}{12} + t^2}, \tag{2.11}$$

where L and B are the plan length and breadth of the building (Fig. 2.1).

Coordinate systems $\bar{x} - \bar{y}$ and x-y have been used for convenience as they make the calculation of the basic characteristics straightforward. However, the stability and dynamic analyses can be carried out in a much simpler way in the coordinate system whose origin is placed at the global shear centre and whose horizontal axes coincide with the principal axes.

In many practical cases the product of inertia of the elements of the bracing system is zero. In such cases, axes x and y are already the principal axes and the equivalent column is established by the stiffness and geometrical characteristics defined by formulae (2.1) to (2.5) which, due to $I_{xy} = 0$, simplify considerably. [The simplified versions of formulae (2.1), (2.2) and (2.5) are given in section 5.7.] However, I_{xy} is not zero, for example, for Z and L shaped bracing cores and axes x and y are not the principal axes. A transformation of the coordinate system is needed: coordinate system x-y should be rotated around the origin (the global shear centre) in such a way that the new axes X and Y are the principal axes (Fig. 2.2). The angle of principal axis X with axis x is obtained from the formula

$$\alpha = \frac{1}{2}\arctan\frac{2I_{xy}}{I_y - I_x}. \qquad (2.12)$$

Principal axis *Y* is perpendicular to axis *X*. The change in the coordinate system only affects the moments of inertia. The moments of inertia in the new coordinate system are obtained from

$$I_X = I_x \cos^2\alpha + I_y \sin^2\alpha - I_{xy}\sin 2\alpha \qquad (2.13)$$

and

$$I_Y = I_x \sin^2\alpha + I_y \cos^2\alpha + I_{xy}\sin 2\alpha. \qquad (2.14)$$

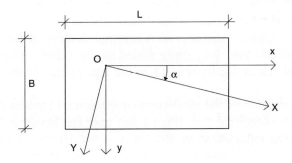

Fig. 2.2 Principal axes *X* and *Y*.

It should be noted that the product of inertia vanishes in the coordinate system whose axes are the principal axes, i.e. $I_{XY} = 0$.

Low-rise and medium-rise buildings are sometimes braced by frameworks or by a mixture of frameworks and shear walls. The equivalent column approach can still be used but the frameworks have to be replaced first by fictitious walls. When such a wall is incorporated into the equivalent column, only its in-plane bending stiffness has to be calculated and all the other stiffness characteristics are set to be zero. This procedure is considered more accurate for frameworks with cross-bracing and for infilled frameworks as they develop predominantly bending type deformations. However, when moment resistant frameworks on fixed or pinned supports are replaced by fictitious shear walls, the procedure is only approximate. The level of approximation depends on to what extent

14 Spatial behaviour

the deformation of the framework differs from the bending type deformation of the fictitious wall. Each case should be treated very carefully as an individual case. Detailed information is given in section 9.9 regarding the calculation of the size of the fictitious wall and in section 9.10 where the behaviour and efficiency of different planar structures are considered.

When the lateral stiffness of the bracing system is evaluated for practical structural engineering calculations, the contribution of the stiffnesses of the individual columns is normally neglected, being small compared to that of the shear walls and cores. This approximation simplifies the calculation and leads to conservative estimates. However, experimental evidence shows that this contribution can be significant in certain cases (e.g. when there are many columns and only few bracing walls, or when the elements of the bracing system are in a special arrangement) [Zalka and White, 1992 and 1993]. The approximate method presented below can be used in such cases. The method is based on a simple formula which converts the 'local' stiffness of a column of storey height into a 'global' stiffness of a bracing element of building height. This element can then be simply added to the other 'ordinary' bracing elements and can be incorporated into the equivalent column in the usual manner.

The accumulation of the local sway of a column of storey height h with the same cross-sectional size (Fig. 2.3/a) over the height of the building results in a total top translation of

$$u_l = 2n \frac{F(h/2)^3}{3EI_c}, \qquad (2.15)$$

where n is the number of storeys and I_c is the second moment of area of the cross-section of the column. Assuming a horizontal load of trapezoidal distribution, the top translation of a single bracing element of building height H and of a fictitious second moment of area I_{cg} (Fig. 2.3/b) is

$$u_g = \frac{q_0 H^4}{EI_{cg}} \left(\frac{1}{8} + \frac{11}{120} \mu \right), \qquad (2.16)$$

where μ is a parameter defining the slope of the function of the horizontal load, according to formula (5.2) in section 5.1 and q_0 is the intensity of the uniformly distributed part of the load (Fig. 5.1).

The equivalent column

By combining the right-hand sides of equations (2.15) and (2.16) and making use of the relationship $H=hn$, the 'local' moment of inertia of a column can be transformed into a 'global' moment of inertia for the global analysis of the building:

$$I_{cg} = 12n^2 I_c \left(\frac{1}{8} + \frac{11}{120} \mu \right), \qquad (2.17)$$

where I_{cg} is the 'global' moment of inertia of the column. The 'global' moment of inertia defined by formula (2.17) can now be used directly for the global analysis. When the stiffnesses of the equivalent column are assembled, the columns can be considered as 'ordinary' bracing elements and their 'global' moments of inertia can simply be included in the summations in formulae (2.1) to (2.5).

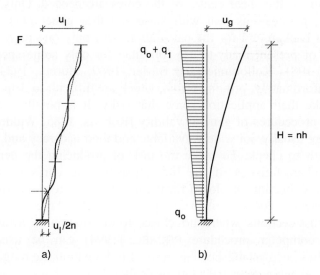

Fig. 2.3 Top translation. a) Accumulation of storey-sway over the height, b) translation of a bracing element of height H.

Formula (2.17) is considered an approximation and can only be used for the global analysis of the building. The approximation is due to the two assumptions made for the derivation of the formula, namely, it is assumed that

- the columns of storey height have fixed ends,

16 Spatial behaviour

- the global deflection of the building resulting from the double-curvature bending of the columns between the floors assumes a straight line (dashed line in Fig. 2.3/a).

These assumptions seem to represent strict restrictions but it has to be kept in mind that the effect of the columns on the global stiffnesses is of secondary nature anyway. The results of tests on small-scale, 10-storey building models given in Chapter 10 indicate that the displacements and rotations obtained by using formula (2.17) are of acceptable accuracy.

When the geometrical characteristics of the equivalent column are needed, several sources can be used for the calculation of the Saint-Venant torsional constant and the moments of inertia of the individual bracing elements [Griffel, 1966; Hrobst and Comrie, 1951; Roark and Young, 1975]. However, the situation is different when the warping constant and the location of the shear centre of the cores are needed. Only a limited number of publications deal with these and they either give a detailed theoretical background for the calculation with only one or two worked examples or present ready-to-use formulae for only some special cases [Gjelsvik, 1981; Kollbrunner and Basler, 1969; Murray, 1984; Vlasov, 1940]. Unfortunately, printing errors, which are difficult to detect in some cases, make their application somewhat risky. It is possible to develop computer procedures of general validity [Roberts, 1985; Waldron, 1986] but they are usually not widely available and their accuracy and reliability are difficult to check. To make the task of producing the bending and torsional characteristics of the individual bracing cores as simple as possible, a collection of closed-form formulae for cross-sections widely used for bracing cores is given in Appendix A. For bracing elements of special cross-sections where no closed-form solution is available, the excellent computer procedure PROSEC [1994] can be used, whose accuracy has been established and proved to be within the range required for structural engineering calculations [Zalka, 1994a].

2.3 THE SPATIAL BEHAVIOUR OF THE EQUIVALENT COLUMN

Apart from some special cases, building structures develop a combination of the three basic modes (sway in the principal planes and torsion). The nature of the behaviour (and the extent of the combination) depend on the relative position of the shear centre of the bracing system and the centre of the external load – and for lateral loads, the direction of the load.

Assuming vertical load and stability analysis, possibilities for the coupling of modes are shown in Fig. 2.4, where the whole bracing system of the building is represented by the equivalent column of open, thin-walled cross-section. The distance between the shear centre of the bracing system (*O*) and the centre of the vertical load (*C*) is marked with *t*.

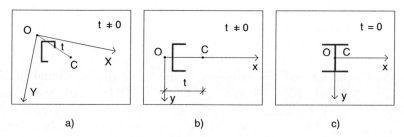

Fig. 2.4 Coupling of basic modes. a) Triple coupling, b) double coupling, c) no coupling.

When the centre of the vertical load does not lie on either principal axis (Fig. 2.4/a), sway in the two principal directions X and Y is coupled by pure torsion. The critical load which belongs to this combined sway-torsional buckling is the global critical load of the building.

When the centre of vertical load lies on one of the principal axes (x on Fig. 2.4/b), sway in that principal plane develops independently of the other two modes. Sway in the perpendicular direction combines with torsion. Both critical loads have to be calculated, i.e. the independent sway critical load in the principal plane and the combined sway-torsional critical load in the perpendicular direction, and the smaller one is the global critical load of the building. The simplest case arises when the shear centre and the centre of the vertical load coincide (Fig. 2.4/c). Sway in the principal directions and torsion about the shear centre develop independently. The global critical load is the smallest one of the independent basic critical loads.

The above principles outlined for the stability analysis can also be applied to the dynamic analysis and – assuming a horizontal load system of arbitrary direction – to the stress analysis. Sections 3.1 and 3.2 also deal with the spatial behaviour of the equivalent column when the 3-dimensional stability and vibrations of the building are investigated using the governing differential equations. In sections 5.1 to 5.3 the 3-dimensional behaviour of the equivalent column (and the building) under horizontal load is analysed in detail.

3

Stability and frequency analyses

The governing differential equations of the equivalent column – and of the bracing system it represents – have been available for some time for the stability and frequency analyses but exact solutions have only been produced for special cases and the accuracy of the approximate solutions has not been fully investigated. A number of approximate procedures are available, however, they are either too simple and only reliable in some special cases or are still too complicated for everyday use in design offices. Most investigations have treated the stability and dynamic problems separately.

Some new theoretical developments and the realization that certain similarities exist between the systems of governing differential equations of the stability and dynamic problems and between the solution procedures make it possible to develop simple, closed-form solutions for the critical loads and the natural frequencies. The differential equations and their solutions are given in the following sections. A left-handed coordinate system is used whose origin is fixed in the shear centre on the ground floor level and whose horizontal coordinate axes coincide with principal axes X-Y. Vertical coordinate axis z points upwards and the bracing system is represented by the equivalent column. The effect of soil-structure interaction is taken into account by simple summation formulae.

Priority is given to the first eigenvalues. This is justified in practical structural engineering applications. As for the stability problem, only the smallest critical load is of practical importance. Regarding dynamic behaviour, it has been shown that the response of multistorey buildings is made up predominantly of the first few modes, with the higher modes contributing only a small portion of the total, except at the top of relatively flexible buildings

The application of the procedures given in this chapter for the stability and frequency analyses is illustrated in Chapter 6 where a series of worked examples show how the procedures are used in practice.

3.1 STABILITY ANALYSIS

The problem of the combined torsional-flexural buckling of a column of thin-walled cross-section has been solved for concentrated end forces and Timoshenko presented the solution in the form of simple formulae and a cubic equation [Timoshenko and Gere, 1961]. Vlasov [1940] derived the system of three differential equations for the cantilever subjected to uniformly distributed axial load six decades ago but he did not produce a solution. Based on the power series method, an approximate solution has been made available [Zalka and Armer, 1992] which, in using tables and after some interpolation, produces the critical load. The much simpler exact solution is given in this section.

The stability of the equivalent column (representing the building) is defined by the system of fourth order, homogeneous differential equations of variable coefficients

$$EI_y u'''' + [N(z)(u' - y_c\varphi')]' = 0, \qquad (3.1)$$

$$EI_x v'''' + [N(z)(v' + x_c\varphi')]' = 0, \qquad (3.2)$$

$$EI_\omega \varphi'''' - (GJ\varphi' - N(z)i_p^2\varphi')' + [N(z)(x_c v' - y_c u')]' = 0, \qquad (3.3)$$

where the following notations are used:

- E modulus of elasticity,
- G shear modulus,
- I_X, I_Y second moments of area with respect to principal axes X and Y,
- I_ω warping constant,
- J Saint-Venant torsional constant,
- u, v translations of the shear centre in directions X and Y,
- φ rotation around the shear centre (clockwise rotation is positive),
- x_c, y_c co-ordinates of the geometrical centre of the layout,
- i_p radius of gyration.

The load of the equivalent column is obtained by distributing the floor load (of the same magnitude on each floor) over the height of the building, creating the uniformly distributed vertical load $N(z) = q(H - z)$.

The governing differential equations are accompanied by a set of boundary conditions. Lateral displacements and rotation are zero at the fixed lower end:

20 Stability and frequency analyses

$$u(0) = v(0) = \varphi(0) = 0. \tag{3.4}$$

The tangent to the column is parallel to axis z at the fixed bottom where no warping develops:

$$u'(0) = v'(0) = \varphi'(0) = 0. \tag{3.5}$$

The bending moments and warping stresses are zero at the top of the column:

$$u''(H) = v''(H) = \varphi''(H) = 0. \tag{3.6}$$

The shear forces and the torsional moments are zero at the top of the column:

$$u'''(H) = v'''(H) = EI_\omega \varphi'''(H) - GJ\varphi'(H) = 0. \tag{3.7}$$

In the boundary conditions, H is the height of the building.

The simultaneous differential equations (3.1) to (3.3) can be used to demonstrate the spatial behaviour of the equivalent column. The nature of the behaviour depends on the relative position of the shear centre of the bracing system and the centroid of the vertical load – as described in section 2.3. Rotation φ appears in all three equations, showing that the resulting deformation is composed of both sway and torsion, as a rule. Possibilities for the combination are shown in Fig. 2.4 where the whole bracing system is represented by the equivalent column of open, thin-walled cross-section.

The solution of the simultaneous differential equations (3.1) to (3.3) in the normal way would result in the eigenvalue of the problem, i.e. the critical load of the building. There is, however, a much simpler way of producing the critical load. It has been proved that the systems of differential equations (3.1) to (3.3) can be solved in two steps: the basic critical loads which belong to the basic modes have to be calculated first, then the coupling of the basic modes has to be considered [Zalka, 1994c].

3.1.1 Doubly symmetrical systems – basic critical loads

The basic critical loads are those which belong to the basic (uncoupled) critical modes: sway buckling in the principal planes and pure torsional buckling. This is the case with doubly symmetrical arrangements when the shear centre of the bracing system and the centre of the vertical load coincide and the three basic modes develop independently of one another. The governing differential equations characterizing the doubly symmetrical case (and the basic modes) are obtained from equations (3.1), (3.2) and (3.3) which, in the uncoupled case when $x_c = y_c = 0$ holds, become independent of each other and simplify to

$$EI_Y u'''' + N(z)u'' = 0, \qquad (3.8)$$

$$EI_X v'''' + N(z)v'' = 0, \qquad (3.9)$$

$$EI_\omega \varphi'''' - \left(GJ\varphi' - N(z)i_p^2 \varphi'\right)' = 0. \qquad (3.10)$$

The basic critical loads for buildings subjected to uniformly distributed load on each floor are given in the following. Based on the well-known classical formula for a cantilever subjected to uniformly distributed axial load [Timoshenko and Gere, 1961], the two sway critical loads in the principal directions are:

$$N_{cr,X} = \frac{7.84 r_s EI_Y}{H^2}, \qquad N_{cr,Y} = \frac{7.84 r_s EI_X}{H^2}, \qquad (3.11)$$

where parameter r_s is a reduction factor.

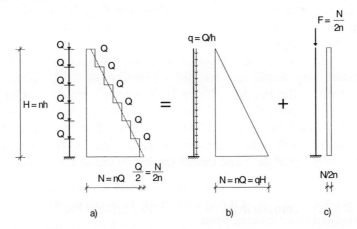

Fig. 3.1 Load diagrams. a) Load on the multistorey building, b) first part of the equivalent load: uniformly distributed load, c) second part: concentrated force on top.

Formulae (3.11) only differ from Timoshenko's formula in factor r_s. It is a reduction factor which allows for the fact that the actual load of the structure consists of concentrated forces on floor levels (Fig. 3.1/a) and is not uniformly distributed over the height (Fig. 3.1/b) as is assumed for the derivation of the original formula for buckling. The continuous load is obtained by distributing the concentrated forces *downwards* resulting in a more favourable load distribution. This unconservative manoeuvre leaves a concentrated force on top of the column (Fig. 3.1/c), which is not covered by the classical formula. The effect of this concentrated force can be accounted for by applying Dunkerley's [1894] summation theorem. (The summation theorems and their use in civil engineering are discussed in detail by Tarnai [1999].)

Dunkerley's summation theorem applied to two load systems (uniformly distributed load and concentrated top load) leads to the formula

$$\frac{1}{N_{cr}^t} = \frac{1}{F_{cr}} + \frac{1}{N_{cr}}, \qquad (3.12)$$

where F_{cr} is the critical load when the structure is subjected to a top load, N_{cr} is the critical load when the structure is subjected to a uniformly distributed load and N_{cr}^t is the total critical load when the two load systems act simultaneously. Figure 3.2 demonstrates that when the structure is under two load systems, the total critical load is always smaller than the critical load of either of the two loads.

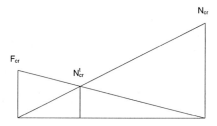

Fig. 3.2 Graphical interpretation of the Dunkerley formula.

It is advantageous to use the Dunkerley formula in a different form. When the magnitude of either of the applied loads is fixed, the magnitude of the other load can be calculated from

$$\frac{F}{F_{cr}} + \frac{N}{N_{cr}} = 1. \qquad (3.13)$$

In this case, the magnitude of the concentrated top load is $F = N/2n$, where n is the number of stories (Fig. 3.1/c). After substituting $N/2n$ for F, formula (3.13) can be rearranged as

$$N\left(\frac{1}{2nF_{cr}} + \frac{1}{N_{cr}}\right) = 1,$$

from where, in making use of the ratio $N_{cr}/F_{cr} = 3.176$ of the classical column formulae, the value of the total vertical load can be calculated as

$$N = r_s N_{cr}. \qquad (3.14)$$

In this formula which now takes into account the real load on the equivalent column according to Fig. 3.1/a,

$$r_s = \frac{n}{n + 1.588} \qquad (3.15)$$

is the reduction factor which is introduced in formulae (3.11) and whose values are given in Table 3.1. The accuracy of the (conservative) Dunkerley formula is improved in Table 3.1 by making use of the exact solution for one and two storey structures under concentrated forces – the values of r_s in Table 3.1 for $n = 1$ and $n = 2$ reflect this modification.

Table 3.1 Reduction factor r_s

n	1	2	3	4	5	6	7	8	9	10	11
r_s	0.315	0.528	0.654	0.716	0.759	0.791	0.815	0.834	0.850	0.863	0.874
n	12	13	14	15	16	18	20	25	30	50	>50
r_s	0.883	0.891	0.898	0.904	0.910	0.919	0.926	0.940	0.950	0.969	$n/(n+1.6)$

The critical load for pure torsional buckling [solution of eigenvalue problem (3.10)] is obtained from

$$N_{cr,\varphi} = \frac{\alpha r_s EI_\omega}{i_p^2 H^2} \qquad (3.16)$$

or, in the special case when the warping stiffness is zero, for example for thin-walled, closed cross-sections, from

24 Stability and frequency analyses

$$N_{cr,\varphi} = \frac{GJ}{i_p^2}. \tag{3.17}$$

In formula (3.16) the values of critical load parameter α (eigenvalue of pure torsional buckling) are given in Fig. 3.3 as a function of

$$k_s = \frac{k}{\sqrt{r_s}}, \tag{3.18}$$

where

$$k = H\sqrt{\frac{GJ}{EI_\omega}} \tag{3.19}$$

is the torsion parameter and r_s is the modifier whose values are given in Table 3.1.

The diagram in Fig. 3.3 covers the range $0 \leq k_s \leq 2$ where most practical cases fall. When the value of k_s exceeds 2.0, or greater accuracy is needed, Table 3.2 can be used. When Table 3.2 was compiled, the special solution procedure (demonstrated in Appendix B) was used which makes it possible to obtain reliable solutions of good accuracy even for ill-conditioned eigenvalue problems.

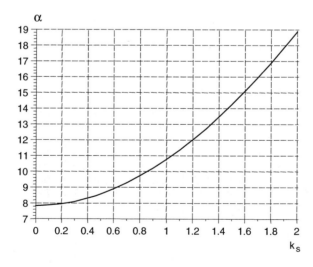

Fig. 3.3 Critical load parameter α.

Table 3.2 Critical load parameter α

k_s	α	k_s	α	k_s	α	k_s	α	k_s	α
0.00	7.837	4.5	52.48	9.2	150.9	13.9	297.4	18.6	491.6
0.01	7.838	4.6	54.10	9.3	153.5	14.0	301.0	18.7	496.2
0.05	7.845	4.7	55.73	9.4	156.1	14.1	304.7	18.8	500.9
0.10	7.867	4.8	57.39	9.5	158.8	14.2	308.4	18.9	505.6
0.20	7.957	4.9	59.06	9.6	161.5	14.3	312.1	19.0	510.3
0.30	8.107	5.0	60.75	9.7	164.2	14.4	315.8	19.1	515.0
0.40	8.316	5.1	62.47	9.8	166.9	14.5	319.5	19.2	519.8
0.50	8.583	5.2	64.20	9.9	169.7	14.6	323.3	19.3	524.6
0.60	8.909	5.3	65.96	10.0	172.4	14.7	327.1	19.4	529.4
0.70	9.291	5.4	67.73	10.1	175.2	14.8	330.9	19.5	534.2
0.80	9.730	5.5	69.52	10.2	178.0	14.9	334.7	19.6	539.0
0.90	10.22	5.6	71.34	10.3	180.9	15.0	338.6	19.7	543.9
1.0	10.77	5.7	73.17	10.4	183.7	15.1	342.5	19.8	548.8
1.1	11.37	5.8	75.03	10.5	186.6	15.2	346.4	19.9	553.7
1.2	12.02	5.9	76.90	10.6	189.5	15.3	350.3	20	558.6
1.3	12.72	6.0	78.80	10.7	192.4	15.4	354.2	21	609.0
1.4	13.47	6.1	80.72	10.8	195.4	15.5	358.2	22	661.5
1.5	14.27	6.2	82.66	10.9	198.3	15.6	362.2	23	716.2
1.6	15.11	6.3	84.62	11.0	201.3	15.7	366.2	24	772.9
1.7	15.99	6.4	86.60	11.1	204.3	15.8	370.2	25	831.8
1.8	16.91	6.5	88.60	11.2	207.3	15.9	374.2	26	892.8
1.9	17.87	6.6	90.63	11.3	210.4	16.0	378.3	27	955.9
2.0	18.87	6.7	92.68	11.4	213.5	16.1	382.4	28	1021.1
2.1	19.91	6.8	94.74	11.5	216.6	16.2	386.5	29	1088.4
2.2	20.98	6.9	96.83	11.6	219.7	16.3	390.6	30	1157.8
2.3	22.08	7.0	98.94	11.7	222.8	16.4	394.8	35	1536.3
2.4	23.21	7.1	101.1	11.8	226.0	16.5	399.0	40	1967.1
2.5	24.38	7.2	103.2	11.9	229.2	16.6	403.2	50	2984.7
2.6	25.57	7.3	105.4	12.0	232.4	16.7	407.4	60	4209.3
2.7	26.79	7.4	107.6	12.1	235.6	16.8	411.6	70	5640.9
2.8	28.03	7.5	109.8	12.2	238.8	16.9	415.9	80	7278.1
2.9	29.30	7.6	112.1	12.3	242.1	17.0	420.2	90	9120.7
3.0	30.59	7.7	114.3	12.4	245.4	17.1	424.5	100	11168
3.1	31.91	7.8	116.6	12.5	248.7	17.2	428.8	150	24471
3.2	33.25	7.9	118.9	12.6	252.1	17.3	433.1	200	42864
3.3	34.61	8.0	121.2	12.7	255.4	17.4	437.5	300	94863
3.4	35.99	8.1	123.6	12.8	258.8	17.5	441.9	400	167093
3.5	37.39	8.2	126.0	12.9	262.2	17.6	446.3	500	259498
3.6	38.81	8.3	128.4	13.0	265.6	17.7	450.7	700	504824
3.7	40.25	8.4	130.8	13.1	269.1	17.8	455.2	1000	1023750
3.8	41.71	8.5	133.2	13.2	272.5	17.9	459.7	1500	2290629
3.9	43.19	8.6	135.7	13.3	276.0	18.0	464.2	2000	4059499
4.0	44.69	8.7	138.2	13.4	279.5	18.1	468.7	2500	6330008
4.1	46.21	8.8	140.7	13.5	283.0	18.2	473.2	3000	9101926
4.2	47.75	8.9	143.2	13.6	286.6	18.3	477.8	3500	12375092
4.3	49.31	9.0	145.7	13.7	290.2	18.4	482.4	4000	16149383
4.4	50.89	9.1	148.3	13.8	293.8	18.5	487.0	>4000	$k_s^2 + 8$

26 Stability and frequency analyses

The evaluation of the formulae of the basic critical loads (3.11) and (3.16) shows that the most important characteristics that influence the values of the basic critical loads are:

- the height of the building,
- the bending stiffnesses of the bracing system,
- the warping stiffness of the bracing system,
- the radius of gyration.

The sway critical loads are in direct proportion to the bending stiffnesses of the bracing system and in inverse proportion to the square of the height of the building. In a similar manner, the pure torsional critical load is in direct proportion to the warping stiffness of the bracing system and in inverse proportion to the square of the height. The Saint-Venant torsional stiffness affects the value of the critical load through the critical load parameter $\alpha(k_s)$ but its effect is normally small as in most practical cases $k_s < 1$ holds. There is, however, a significant difference between the sway- and pure torsional critical loads. The value of the pure torsional critical load also depends on the radius of gyration. The effect of the radius of gyration is best shown by formula (2.11). According to the formula which assumes uniformly distributed floor load, the greater the size of the building (and the distance between the shear centre and the centre of vertical load), the greater the radius of gyration and consequently the smaller the pure torsional critical load. This is in sharp contrast to sway buckling where the geometrical characteristics of the layout of the building do not influence the critical load.

Formula (3.17) shows another interesting fact. When a structure is braced by a bracing element of zero warping stiffness (e.g. a core of thin-walled, closed cross-section), the value of the critical load for pure torsional buckling does not depend on the height of the building nor on the distribution of the load.

3.1.2 Coupling of the basic modes; combined sway-torsional buckling

As the simultaneous differential equations (3.1) to (3.3) show, the basic modes combine in the general case. The coupling of the basic modes can be taken into account in two ways: approximately or exactly. If the main aim of the investigation is to show whether or not the building is in a stable state, then the approximate method can be used (first). It is very quick and simple, albeit conservative. If it indicates an unstable building or a building with an insufficient safety margin, then, as a second step, the exact method may still prove that the building is in a stable state. (Chapter 7 deals with the necessary level of safety against buckling.)

(a) An approximate method

The combination of the basic critical loads may be taken into account by using the Föppl–Papkovich theorem. According to the theorem, if a structure is characterized by n stiffness parameters, the critical load can be approximated by combining the n corresponding part critical loads [Tarnai, 1999]. Each part critical load belongs to a case where all but one stiffness parameters are assumed to be infinitely great. In applying the theorem to 3-dimensional buckling, the combined critical load can be obtained from

$$\frac{1}{N_{cr}} = \frac{1}{N_{cr,X}} + \frac{1}{N_{cr,Y}} + \frac{1}{N_{cr,\varphi}}, \qquad (3.20)$$

where $N_{cr,X}$, $N_{cr,Y}$ and $N_{cr,\varphi}$ are the basic critical loads defined by formulae (3.11) and (3.16). The advantage of formula (3.20) is that it is easy to use and it is always conservative. However, its use in certain cases can lead to considerably uneconomical structural solutions, as the error of the formula can be as great as 67%. The more sophisticated and only slightly more complicated exact solution is given in the next section.

(b) The exact method

The exact solution of the simultaneous differential equations (3.1) to (3.3) leads to the determinant

$$\begin{vmatrix} N - N_{cr,X} & 0 & -Ny_c \\ 0 & N - N_{cr,Y} & Nx_c \\ -Ny_c & Nx_c & i_p^2(N - N_{cr,\varphi}) \end{vmatrix} = 0 \qquad (3.21)$$

which defines the coupling of the basic modes and which can be used for the calculation of the critical load of the combined sway-torsional buckling if the basic critical loads ($N_{cr,X}$, $N_{cr,Y}$ and $N_{cr,\varphi}$) are known. The expansion of the determinant results in a cubic equation in the form

$$N^3 + a_2 N^2 + a_1 N - a_0 = 0, \qquad (3.22)$$

where the coefficients are

$$a_2 = \frac{N_{cr,X}\tau_X^2 + N_{cr,Y}\tau_Y^2 - N_{cr,\varphi} - N_{cr,X} - N_{cr,Y}}{1-\tau_X^2-\tau_Y^2}, \quad (3.23)$$

$$a_1 = \frac{N_{cr,X}N_{cr,Y} + N_{cr,\varphi}N_{cr,X} + N_{cr,\varphi}N_{cr,Y}}{1-\tau_X^2-\tau_Y^2}, \qquad a_0 = \frac{N_{cr,\varphi}N_{cr,X}N_{cr,Y}}{1-\tau_X^2-\tau_Y^2}. \quad (3.24)$$

In formulae (3.23) and (3.24) parameters

$$\tau_X = \frac{x_c}{i_p} \quad \text{and} \quad \tau_Y = \frac{y_c}{i_p} \quad (3.25)$$

characterize the eccentricity of the load. Equation (3.22) can be used to take into account the coupling of the basic critical loads. Of the three roots of the cubic equation, the smallest one is the most important one in practical application, being the smallest critical load of the building.

To speed up the calculation, cubic equation (3.22) has been solved for a great number of basic critical load ratios

$$r_1 = \frac{N_{cr,\varphi}}{N_{cr,X}} \quad \text{and} \quad r_2 = \frac{N_{cr,\varphi}}{N_{cr,Y}} \quad (3.26)$$

and eccentricity parameters τ_X and τ_Y, and the resulting mode coupling parameters κ are given in Tables C1 to C11 in Appendix C as a function of r_1, r_2, τ_X and τ_Y. When the mode coupling parameter is known, the critical load of combined sway-torsional buckling is obtained from

$$N_{cr} = \kappa N_{cr,\varphi} \quad (3.27)$$

where $N_{cr,\varphi}$ is the basic critical load for pure torsional buckling.

The evaluation of cubic equation (3.22) leads to interesting observations. Cubic equation (3.22) gives three values of critical load for each mode. It can be easily demonstrated by the evaluation of the equation [Timoshenko and Gere, 1961] that one critical load is always lower than any of the uncoupled critical loads, in other words the coupling of modes always results in a critical load which is smaller than the smallest uncoupled critical load. One critical load is always higher than the three uncoupled critical loads and the third critical load is intermediate between the uncoupled critical loads $N_{cr,X}$ and $N_{cr,Y}$. Each of the three critical loads for a given mode corresponds to a particular pattern of buckling. Figure 3.4 shows the left-hand side of equation (3.22) of N with the uncoupled critical loads $N_{cr,X}$, $N_{cr,Y}$ and $N_{cr,\varphi}$, and the three coupled critical loads $N_{cr,1}^{(1)}$, $N_{cr,1}^{(2)}$ and $N_{cr,1}^{(3)}$ which belong to the three patterns of the first mode.

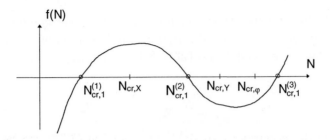

Fig. 3.4 Relationship between the coupled and uncoupled critical loads.

It is only mentioned here that cubic equation (3.22) can also be used for the stability analysis of bracing systems developing shear deformation [Hegedűs and Kollár, 1987].

The coupling of the basic modes clearly reduces the value of the critical loads. The magnitude of the reduction mainly depends on two factors: the relative value of the uncoupled critical loads compared to one another and the eccentricity of the bracing system in relation to the centroid of the load of the building. The nearer the values of the three uncoupled critical loads, the bigger the reduction compared to the smallest of the three values. Similarly, the greater the distance between the shear centre and the centroid of the load, the smaller the coupled critical load.

The importance of taking into account the effect of mode coupling cannot be overstated: the maximum error made by neglecting the effect of coupling is 141% overestimation in the value of the smallest critical load.

(c) Special case: monosymmetric arrangements

Cubic equation (3.22) can always be used for taking into account the effect of interaction. However, the solution of cubic equation (3.22) simplifies in the monosymmetric case when the centre of the vertical load lies on one of the principal axes (Fig. 2.4/b) and a very simple formula can be produced for dealing with the interaction. In such cases sway buckling in the plane of symmetry develops independently and only sway buckling perpendicular to the axis of symmetry and pure torsional buckling combine. Assuming uniformly distributed floor load and that axis X is an axis of symmetry, i.e. that $y_c = 0$ holds, the combined critical load can be expressed by the closed form formula

$$N_{combined} = \varepsilon N_{cr,Y}, \qquad (3.28)$$

where the values for the mode coupling parameter ε are given in Fig. 3.5 and, if the range covered by the figure is not wide enough or if more accurate values are needed, in Table 3.3, as a function of τ_X [formula (3.25)] and

$$r_2 = \frac{N_{cr,\varphi}}{N_{cr,Y}}. \qquad (3.29)$$

The basic critical loads $N_{cr,Y}$ and $N_{cr,\varphi}$ are defined by formulae (3.11) and (3.16). Table 3.3 covers the range $0 \le r_2 \le 100$. When $r_2 > 100$, $r_2 = 100$ can be used. The error made by this approximation is always smaller than 1%. The global critical load for monosymmetrical bracing systems is the smaller one of the combined critical load and the independent sway critical load in the plane of symmetry.

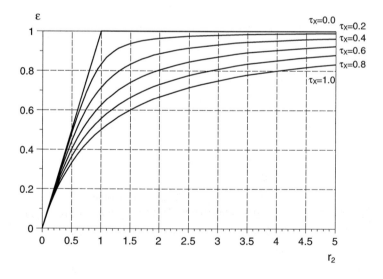

Fig. 3.5 Mode coupling parameter ε for the monosymmetrical case.

Table 3.3 Mode coupling parameter ε for the monosymmetrical case

r_2	0.0	0.1	0.2	0.3	0.4	0.5	0.6	0.7
$\tau_X=0.0$	0.0000	0.1000	0.2000	0.3000	0.4000	0.5000	0.6000	0.7000
$\tau_X=0.1$	0.0000	0.0999	0.1995	0.2987	0.3974	0.4951	0.5914	0.6851
$\tau_X=0.2$	0.0000	0.0996	0.1980	0.2951	0.3900	0.4821	0.5698	0.6513
$\tau_X=0.3$	0.0000	0.0990	0.1957	0.2894	0.3792	0.4639	0.5422	0.6127
$\tau_X=0.4$	0.0000	0.0983	0.1926	0.2822	0.3662	0.4435	0.5134	0.5753
$\tau_X=0.5$	0.0000	0.0974	0.1890	0.2741	0.3521	0.4226	0.4855	0.5408
$\tau_X=0.6$	0.0000	0.0963	0.1849	0.2655	0.3379	0.4024	0.4594	0.5095
$\tau_X=0.7$	0.0000	0.0951	0.1805	0.2566	0.3239	0.3833	0.4354	0.4812
$\tau_X=0.8$	0.0000	0.0938	0.1760	0.2478	0.3105	0.3654	0.4135	0.4557
$\tau_X=0.9$	0.0000	0.0924	0.1713	0.2391	0.2977	0.3487	0.3934	0.4327
$\tau_X=1.0$	0.0000	0.0909	0.1667	0.2308	0.2857	0.3333	0.3750	0.4118
r_2	0.8	0.9	1.0	1.2	1.4	1.6	1.8	2.0
$\tau_X=0.0$	0.8000	0.9000	1.0000	1.0000	1.0000	1.0000	1.0000	1.0000
$\tau_X=0.1$	0.7736	0.8513	0.9091	0.9613	0.9774	0.9843	0.9880	0.9903
$\tau_X=0.2$	0.7240	0.7852	0.8333	0.8950	0.9273	0.9454	0.9566	0.9641
$\tau_X=0.3$	0.6743	0.7264	0.7692	0.8313	0.8710	0.8970	0.9149	0.9278
$\tau_X=0.4$	0.6292	0.6753	0.7143	0.7745	0.8169	0.8474	0.8698	0.8869
$\tau_X=0.5$	0.5890	0.6307	0.6667	0.7243	0.7673	0.8000	0.8253	0.8453
$\tau_X=0.6$	0.5533	0.5916	0.6250	0.6800	0.7226	0.7561	0.7830	0.8049
$\tau_X=0.7$	0.5215	0.5569	0.5882	0.6406	0.6822	0.7159	0.7436	0.7666
$\tau_X=0.8$	0.4931	0.5261	0.5556	0.6054	0.6459	0.6793	0.7072	0.7307
$\tau_X=0.9$	0.4675	0.4985	0.5263	0.5739	0.6131	0.6459	0.6737	0.6975
$\tau_X=1.0$	0.4444	0.4737	0.5000	0.5455	0.5833	0.6154	0.6429	0.6667
r_2	3.0	4.0	5.0	10.0	20.0	30.0	50.0	100.0
$\tau_X=0.0$	1.0000	1.0000	1.0000	1.0000	1.0000	1.0000	1.0000	1.0000
$\tau_X=0.1$	0.9951	0.9967	0.9975	0.9989	0.9995	0.9997	0.9998	0.9999
$\tau_X=0.2$	0.9809	0.9871	0.9902	0.9956	0.9979	0.9986	0.9992	0.9996
$\tau_X=0.3$	0.9594	0.9719	0.9786	0.9902	0.9953	0.9969	0.9982	0.9991
$\tau_X=0.4$	0.9327	0.9524	0.9632	0.9829	0.9917	0.9945	0.9968	0.9984
$\tau_X=0.5$	0.9028	0.9296	0.9449	0.9737	0.9872	0.9915	0.9949	0.9975
$\tau_X=0.6$	0.8715	0.9048	0.9245	0.9631	0.9818	0.9879	0.9928	0.9964
$\tau_X=0.7$	0.8400	0.8788	0.9026	0.9510	0.9755	0.9837	0.9902	0.9951
$\tau_X=0.8$	0.8089	0.8523	0.8798	0.9379	0.9685	0.9789	0.9873	0.9936
$\tau_X=0.9$	0.7788	0.8259	0.8566	0.9238	0.9607	0.9736	0.9840	0.9920
$\tau_X=1.0$	0.7500	0.8000	0.8333	0.9091	0.9524	0.9677	0.9804	0.9901

32 Stability and frequency analyses

The special case of doubly symmetrical bracing systems (Fig. 2.4/c) is automatically covered in Table 3.3 by the rows defined by $\tau_X = 0.0$. Values in Table 3.3 can also be used to show the effect of neglecting the coupling of the basic modes in the monosymmetrical case. According to the values, it is always unconservative to neglect coupling. The maximum error made by neglecting coupling is 100%; when the two basic critical loads $N_{cr,Y}$ and $N_{cr,\varphi}$ are equal and eccentricity is maximum ($\tau_X = 1.0$), coupling reduces the combined critical load to half of the basic critical load.

3.1.3 Concentrated top load; single-storey buildings

The formulae presented for the global critical loads were derived on the assumption that the load of the structures was uniformly distributed over the floors, being the practical case with multistorey buildings. In certain cases, however, concentrated load on top of the building may need be considered. A panorama restaurant, a swimming pool or water tanks may represent some extra load on the top of the building which may not be covered by the uniformly distributed floor load, considered to be the same at each floor level. Even a relatively small amount of extra concentrated load on top of the building should be taken into account as it represents a more dangerous load case than the UDL case.

The concentrated top load case is also of practical importance when single-storey buildings are investigated as the majority of the vertical load is on the (top) floor, which is represented by a concentrated force on top of the structure.

When the structure is under concentrated top load and the system develops predominantly bending type deformation, the basic critical loads are as follows. The sway critical loads in the principal directions are:

$$F_{cr,X} = \frac{\pi^2 EI_Y}{4H^2}, \qquad F_{cr,Y} = \frac{\pi^2 EI_X}{4H^2} \qquad (3.30)$$

and the critical load for pure torsional buckling is given by

$$F_{cr,\varphi} = \frac{EI_\omega}{i_p^2 H^2}\left(\frac{\pi^2}{4} + k^2\right), \qquad (3.31)$$

where H is the height of the building.

When the warping stiffness of the bracing system is zero, the critical load of pure torsional buckling is obtained from

$$F_{cr,\varphi} = \frac{GJ}{i_p^2}. \qquad (3.31a)$$

The basic critical loads may interact. Once again, the interaction of the basic critical loads can be taken into account by using the approximate formula (3.20) or the exact cubic equation (3.22), where $F_{cr,X}$, $F_{cr,Y}$ and $F_{cr,\varphi}$ are to be substituted for $N_{cr,X}$, $N_{cr,Y}$ and $N_{cr,\varphi}$. The smallest root of the equation is the global critical load F_{cr}.

When the concentrated top load and the UDL on the floors act simultaneously, their effect should be combined. Dunkerley's summation theorem (cf. section 3.1.1) offers a simple formula for the combination of the two load cases:

$$N_{cr}^D = \left(\frac{1}{2nF_{cr}} + \frac{1}{N_{cr}} \right)^{-1}, \qquad (3.32)$$

where

- F_{cr} is the critical concentrated load,
- N_{cr} is the critical uniformly distributed load,
- n is the number of storeys,
- N_{cr}^D is the combined global critical load of the bracing system.

3.1.4 Shear mode situations

Closed-form formulae for the critical load of building structures under uniformly distributed floor load, developing predominantly bending deformations were given in the previous sections. This section shows how the theory can be extended and the formulae given can be used with or without modification for different special cases.

Bracing systems which (also) develop shear deformation are not discussed in detail as their practical importance is relatively small. However, simple approximate solutions are given below for some special cases.

The two typical cases when shear modes have to be considered are as follows. Shear type deformations are

a) concentrated on one storey level – 'local' shear (Fig. 3.6/a),
b) 'distributed' over the height of the building – 'global' shear (Fig. 3.6/b).

(a) Concentrated shear on one storey level – 'local' shear

Shear walls in multistorey buildings are sometimes omitted on the ground floor level (or somewhere else) and also the height of the first storey (or some other storey) may be bigger to provide more space. Such arrangements result in reduced first storey stiffness and the building may develop 'local' shear mode deformations between the ground floor and the first floor levels (Fig. 3.6/a). It is difficult to produce the exact critical loads for such irregular situations, but the formulae

$$F_{cr,X} = \frac{\pi^2 EI_Y}{h^2}, \qquad F_{cr,Y} = \frac{\pi^2 EI_X}{h^2}, \qquad (3.33)$$

$$F_{cr,\varphi} = \frac{1}{i_p^2}\left(\frac{\pi^2 EI_\omega}{h^2} + GJ\right) \qquad (3.34)$$

offer good approximations for the basic critical loads, where h is the storey height at ground floor level (or at any other level where local shear is considered). As the 'local' shear deformation is in fact bending type deformation as far as the bracing elements on the storey level are concerned, the location of the shear centre can be determined by using formulae (2.1) and (2.2).

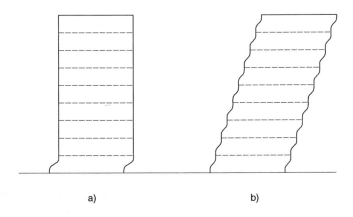

a)　　　　　　　　　　b)

Fig. 3.6 Shear mode situations. a) Local shear, b) global shear.

With the basic critical loads (3.33) and (3.34), the analysis is carried out exactly as described in the previous sections. It has been proved that the interaction of the basic critical loads can be taken into account by using the cubic equation (3.22) [or the approximate formula (3.20)], where $F_{cr,X}$,

$F_{cr,Y}$ and $F_{cr,\varphi}$ are to be substituted for $N_{cr,X}$, $N_{cr,Y}$ and $N_{cr,\varphi}$. The smallest root of the equation is the global critical load. This procedure can be applied to any other storey where the stiffness suffers a sudden reduction. If there are more than one storey with sudden changes in stiffness, each storey has to be considered and the investigation may have to be carried out several times.

(b) 'Distributed' shear over the height – 'global' shear
When buildings are braced by moment resistant frameworks, all the storeys may be involved in a global type shear deformation (Fig. 3.6/b). The analysis is best carried out by replacing the frameworks by fictitious walls based on identical critical loads. The torsional critical load is then obtained from formula (3.16), bearing in mind that the fictitious walls do not have Saint-Venant torsional stiffness ($k = 0$ and therefore $\alpha = 7.84$). Neither do the fictitious walls have the product of inertia and their own warping stiffness, i.e. $I_{xy,i}$ and $I_{\omega,i}$ are zero in formulae (2.3) and (2.5). It has been shown that the cubic equation (3.22) is also applicable to shear mode situations [Hegedűs and Kollár, 1987] so, having established the basic critical loads, equation (3.22) is used for the calculation of the combined critical load.

The sway critical loads of the individual frameworks can be easily determined using the simple closed-form solutions given in Chapter 9.

(c) Combined shear and bending mode situations
Frameworks are usually much more flexible compared to shear walls and cores and their contribution to the lateral stiffness of the bracing system is usually small and is often neglected. In certain cases, however, moment resistant frameworks and coupled shear walls can alter the predominantly bending type behaviour of the shear walls and cores, making the overall deformation a mixture of shear and bending. Such situations may emerge when the number of frameworks and coupled shear walls is relatively great compared to the number of shear walls and cores or when their position is such that they play a key role in resisting the applied loads.

The procedure presented in sections 3.1.1 and 3.1.2 can be applied to such cases. Making use of the sway critical load of a framework or coupled shear walls system, each framework or coupled shear wall system is replaced by a fictitious wall of equivalent thickness – see Chapter 9 for details. The fictitious walls are then treated like any other ordinary solid shear walls of the bracing system for the establishment of the equivalent column and the original procedure can proceed.

36 Stability and frequency analyses

The main approximation in this procedure regards the location of the shear centre which is different for systems developing bending and systems with shear type deformations, even if the calculation of the corresponding critical loads takes account of the different nature of the deformations.

Kollár and Póth [1994] presented a simple method for the stress analysis and they also gave guidelines for the simple, albeit approximate, determination of the location of the shear centre.

A closed-form solution has been derived for the stability analysis of buildings under concentrated top load and developing both bending and shear deformation [Hegedűs and Kollár, 1987] but further investigation of the treatment of combined shear and bending mode systems is still needed.

3.1.5 Soil-structure interaction

It was assumed for the derivation of the formulae given for the critical load in the previous sections that the elements of the bracing system were founded on a rigid base where no rotations occur. However, buildings are sometimes constructed on loose soil and compliant soil adds flexibility to the system at the base and the soil-structure interaction may have a significant effect on the overall behaviour of the bracing system and should be considered in the analysis.

Support restraint conditions at the bottom end of the structure may be set as intermediate between zero restraint and full restraint. If such restraints against translations or rotations are linearly elastic, they may be idealized as springs [Key, 1988; Wolf 1985].

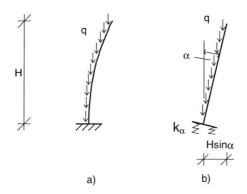

Fig. 3.7 Soil-structure interaction. a) Flexible column on fixed support, b) stiff column on flexible support.

The application of the Föppl–Papkovich theorem to the equivalent column leads to a simple conservative formula to handle soil-structure interaction. According to the theorem, the critical load which allows for the effect of soil-structure interaction can be obtained using two models with the associated part critical loads. It is assumed for the first model that the flexible equivalent column is based on rigid foundation – the formulae given earlier in this chapter are in line with this model. Second, it is assumed that the superstructure is infinitely stiff and is based on flexible foundation. Figure 3.7 shows the two models, assuming sway buckling.

When the two part critical loads (N_{cr} and N_{cr}^{flex}) are known, the Föppl–Papkovich theorem can be applied and the critical load which takes into account the effect of elastic foundation (N_{cr}^{int}) is calculated from

$$\frac{1}{N_{cr}^{int}} = \frac{1}{N_{cr}} + \frac{1}{N_{cr}^{flex}}. \tag{3.35}$$

The critical load which is associated with the simple model shown in Fig. 3.7/b is obtained using static considerations. Moment equilibrium of the column developing sway buckling is expressed by

$$qH\frac{H\sin\alpha}{2} - k_\alpha \alpha = 0,$$

where the effect of the flexible support is represented by spring constant k_α. Barkan [1962] investigated the elastic properties of soil and made recommendations for spring constants for different motions (Table 3.4).

According to his recommendation, the spring constant for the model shown in Fig. 3.7/b is calculated from

$$k_\alpha = k'_\alpha I_f, \tag{3.36}$$

where

k'_α is the spring coefficient for rocking motion,
I_f is the second moment of the contact area between foundation and soil in the relevant direction.

In Table 3.4

A_f is the contact area between foundation and soil,

38 Stability and frequency analyses

I_{fo} is the polar moment of inertia of the contact area about between foundation and soil.

Table 3.4 Spring constants and coefficients for different motions

Type of motion		Spring constant	Spring coefficient
Vertical		$k_z = k_s A_f$	–
Horizontal		$k_x = k'_x A_f$	$k'_x = 0.5 k_s$
Rocking		$k_\alpha = k'_\alpha I_f$	$k'_\alpha = 2 k_s$
Torsional		$k_\varphi = k'_\varphi I_{fo}$	$k'_\varphi = 1.5 k_s$

Values for spring coefficient k'_α for rocking motion – and for other spring coefficients for other types of motion – should be determined from soil data for accurate design. If no reliable soil data is available, coefficients k'_α, k'_x and k'_φ can be estimated using the third column in Table 3.4, where k_s is the modulus of subgrade reaction (also called 'dynamic subgrade reaction' or 'coefficient of elastic uniform compression'). Recommended values of the modulus of subgrade reaction are given in Table 3.5, using Barkan's [1962] data.

If reliable soil information is available, the procedures outlined in [Richart, Woods and Hall, 1970] can be used for determining more accurate spring constants and coefficients for the analysis of vertical and rocking motions.

After substituting for the spring constant in the moment equilibrium equation and taking into consideration that $\sin\alpha/\alpha = 1$ holds at $\alpha = 0$, the sway critical load of the stiff column on flexible support is obtained as

$$N_{cr}^{flex} = \frac{2 k'_\alpha I_f}{H}. \tag{3.37}$$

Table 3.5 Recommended values of the modulus of subgrade reaction k_s

Soil group category	Soil group	Modulus of subgrade reaction k_s [kN/m³]
I	Weak soils (clay and silty clays with sand, in a plastic state; clayey and silty sands; also soils of categories II and III with laminae of organic silt and of peat)	up to 30000
II	Soils of medium strength (clays and silty clays with sand, close to the plastic limit; sand)	30000–50000
III	Strong soils (clay and silty clays with sand, of hard consistency; gravels and gravelly sands; loess and loessial soils)	50000–100000
IV	Rocks	over 100000

In combining this formula and, for example, the formula for sway buckling in direction X using the Föppl–Papkovich formula (3.35) above, the formula for the critical load which takes into account the effect of soil-structure interaction in direction X is obtained as

$$N_{cr,X}^{int} = s_{flex} N_{cr,X}, \qquad (3.38)$$

where $N_{cr,X}$ is the sway critical load in direction X [formula (3.11)] and

$$s_{flex} = \frac{1}{1+3.92\beta} \qquad (3.39)$$

is the reduction factor which is responsible for the effect of the flexible foundation.

Table 3.6 Reduction factor s_{flex}

β	0	0.1	0.2	0.3	0.4	0.5	0.6	0.7	0.8	0.9	1.0
s_{flex}	1.000	0.718	0.561	0.460	0.389	0.338	0.298	0.267	0.242	0.221	0.203
β	2	3	4	5	6	7	8	9	10	50	100
s_{flex}	0.113	0.078	0.060	0.049	0.041	0.035	0.031	0.028	0.025	0.005	0.003

40 Stability and frequency analyses

Values for factor s_{flex} are given in Table 3.6 as a function of stiffness ratio β:

$$\beta = \frac{r_s EI_Y}{Hk'_\alpha I_{f,Y}}, \qquad (3.40)$$

where

- I_Y is the second moment of area of the bracing system with respect to axis Y,
- $I_{f,Y}$ is the second moment of area of the foundation with respect to axis Y,
- r_s reduction factor according to Table 3.1.

Graphical interpretation of the Föppl–Papkovich formulae is to be seen in Fig. 3.8. The diagrams clearly show that the interaction always reduces the value of the critical load. The maximum reduction is 50%; it occurs when the two part critical loads are of the same value (Fig. 3.8/a). When one of the part critical loads is much smaller than the other one, the value of the critical load is controlled by the smaller value (Fig. 3.8/b/c).

The situation with torsion is somewhat different. The nature of the interaction between soil and structure is much more complicated and it would depend on several factors, e.g. the structure and geometry of the foundation (separate footings, flat-slab foundation, rigid-box foundation, raft foundation, piled foundation). Each type should be handled individually. The treatment of torsional interaction is outside the scope of this book; it is only mentioned here that equivalent spring constants for horizontal and torsional deformations are given by Key [1988] for circular and rectangular bases. Wolf's [1985] monograph offers a detailed analysis of soil-structure interaction.

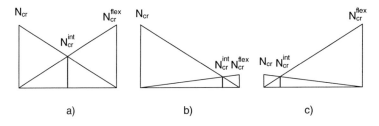

Fig. 3.8 Graphical interpretation of the Föppl–Papkovich formula. a) Part critical loads of the same magnitude, b) $N_{cr}^{flex} \ll N_{cr}$ and $N_{cr}^{int} \approx N_{cr}^{flex}$, c) $N_{cr} \ll N_{cr}^{flex}$ and $N_{cr}^{int} \approx N_{cr}$.

Stability analysis

It should be pointed out here that it is very difficult to estimate appropriate soil parameters. The accuracy of data related to soil is of one order of magnitude less reliable than that of data on the geometrical and stiffness characteristics of the superstructure.

3.1.6 Individual beam-columns

(a) Stability of columns of thin-walled cross-section

The formulae given in sections 3.1.1 and 3.1.2 for multistorey buildings can also be used with some modifications for the stability analysis of beam-columns of thin-walled cross section. The sway critical load of columns subjected to uniformly distributed axially load may be calculated from

$$N_{cr,X} = \frac{cEI_Y}{h^2}, \qquad N_{cr,Y} = \frac{cEI_X}{h^2}, \qquad (3.41)$$

where h is the height of the column. Values for parameter c are given in Fig. 3.9 for different end conditions.

The value of pure torsional buckling can be approximated using Southwell's summation formula from

$$N_{cr,\varphi} = \frac{1}{i_p^2}\left(GJ + \frac{cEI_\omega}{h^2}\right), \qquad (3.41a)$$

where i_p is the radius of gyration of the cross-section of the column and values of parameter c are again given in Fig. 3.9. Formula (3.41a) is always conservative.

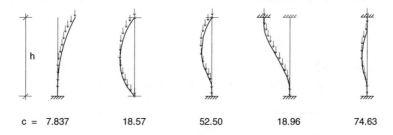

c = 7.837   18.57   52.50   18.96   74.63

Fig. 3.9 Parameter c for different end conditions.

42   Stability and frequency analyses

Cubic equation (3.22) can be used for taking into account the coupling of the basic modes. It is interesting to note that cubic equation (3.22) can be considered as a generalization of equation (5.5) in Eurocode 3, Part 1.3 [1992] for the flexural-torsional buckling analysis of thin-walled columns subjected to concentrated end forces. The generalization is twofold. First, while the corresponding equation in Eurocode 3 is only valid for monosymmetrical cross-sections, cubic equation (3.22) can be used for cross-sections of any shape with no restriction at all, as far as symmetry is concerned. Second, apart from the concentrated load case, cubic equation (3.22) is also applicable to columns subjected to uniformly distributed axial load.

(b)   Stability of storey-height columns in multistorey buildings

The storey-height columns of a multistorey building can lose stability in two ways: they can develop sway buckling (when the joints develop relative translation) or non-sway buckling. Of the critical loads of the two types, the smaller one is of practical importance. Clearly, the column must have the necessary stiffness against non-sway buckling. As for sway buckling, engineering common sense suggests that if the bracing system is adequate, i.e. if it has the necessary stiffness against the full-height buckling of the building (cf. Chapter 7), then it is also ensured that the columns do not develop sway buckling or, to be more precise, the sway buckling critical load is always greater than that of the non-sway buckling. Detailed discussion of this phenomenon is available elsewhere [Kollár and Zalka, 1999] and only the findings are summarized here.

The assumption that the storey-height columns of a building with adequate bracing system develop non-sway buckling is

- conservative when the columns have the same type of joints at both ends, i.e. fixed joints at both ends, or pinned joints at both ends and
- unconservative when the columns have a fixed lower end and a pinned upper end.

The unconservative error can be eliminated by assuming that in the mixed case (fixed lower end and pinned upper end) the columns are pinned at both ends.

3.2 FREQUENCY ANALYSIS

The dynamic problem of beam-columns of thin-walled cross-section has been investigated by several scientists. Garland [1940] applied the Rayleigh–Ritz method to cantilever beams and derived a simple approximate solution but assumed infinitely great stiffness in one of the principal directions. Gere and Lin [1958] set up the complete system of governing differential equations but only produced a solution for beam-columns on pinned supports. Their difficulty in providing a solution for other support systems might have lain with handling pure torsional vibration. Gere [1954] had earlier published the differential equation of pure torsional vibration but only presented a solution for the simply supported case. Several approximate solutions have been recommended for the pure torsional vibration of cantilevers [Southwell, 1922; Gere, 1954; Kollár, 1979; Goschy, 1981] but the exact solution has not been produced and the accuracy of the approximate procedures has not been investigated either. The exact solution to pure torsional vibrations is presented in this section while the results of a comprehensive accuracy analysis regarding the approximate solutions mentioned above are available elsewhere. For lack of the exact solution to pure torsional vibration, solutions to coupled vibrations [Kollár, 1979; Rosman, 1980 and 1981; Vértes, 1985; Goschy, 1981] could only be approximate.

The procedure presented in this section can also be used for a simplified dynamic analysis of buildings in seismic zones, where one of the most important input data is the fundamental frequency of the building [Eurocode 8, 1996; Zalka, 1988].

Assuming uniformly distributed mass, the vibrations of the building are defined by the simultaneous partial differential equations [Gere and Lin, 1958]:

$$EI_Y u'''' + \rho A(\ddot{u} - y_c \ddot{\varphi}) = 0, \tag{3.42}$$

$$EI_X v'''' + \rho A(\ddot{v} + x_c \ddot{\varphi}) = 0, \tag{3.43}$$

$$EI_\omega \varphi'''' - GJ\varphi'' + \rho A i_p^2 \ddot{\varphi} + \rho A(x_c \ddot{v} - y_c \ddot{u}) = 0. \tag{3.44}$$

In the above differential equations primes and dots mark differentiation by z and t, respectively, and

A is the area of the plan of the building,
u, v, φ are the lateral and torsional motions,

ρ is the mass density per unit volume of the material of the building, defined by

$$\rho = \frac{\gamma}{g}, \qquad (3.45)$$

where

γ is the weight per unit volume of the building [kN/m³],
g is the gravity acceleration [$g = 9.81$ m/s²].

The similarity between the governing differential equations of stability [equations (3.1) to (3.3)] and those of vibrations [equations (3.42) to (3.44)] is remarkable. In view of this similarity, it is not surprising that the boundary conditions given for the stability analysis [equations (3.4) to (3.7)] also apply to the frequency analysis.

The frequency analysis can be carried out in a way similar to the stability analysis. The first step is to produce the basic natural frequencies which belong to the basic (independent) modes: lateral vibrations in the principal planes and pure torsional vibrations. Second, the coupling of the basic modes has to be taken into account.

3.2.1 Doubly symmetrical systems – basic natural frequencies

The differential equations characterizing the basic modes are obtained from the general equations (3.42), (3.43) and (3.45) by using the substitution $x_c = y_c = 0$. After separating the variables and eliminating the time dependent functions – which are not needed for the determination of the natural frequencies – the general differential equations simplify to

$$EI_Y u_1'''' - \rho A \omega_X^2 u_1 = 0, \qquad (3.46)$$

$$EI_X v_1'''' - \rho A \omega_Y^2 v_1 = 0, \qquad (3.47)$$

$$EI_\omega \varphi_1'''' - GJ \varphi_1'' - \rho A \omega_\varphi^2 i_p^2 \varphi_1 = 0, \qquad (3.48)$$

where u_1, v_1 and φ_1 characterize the lateral and torsional motions and ω_X, ω_Y and ω_φ are the circular frequencies.

Assuming uniformly distributed weight on the floors, the basic natural frequencies are given as follows. In modifying the classical solution for the lateral vibrations of a cantilever under its uniformly distributed weight over the height [Timoshenko and Young, 1955], the circular frequencies are obtained as

$$\omega_{X,i} = \frac{l_i r_f}{H^2}\sqrt{\frac{EI_Y}{\rho A}}, \qquad \omega_{Y,i} = \frac{l_i r_f}{H^2}\sqrt{\frac{EI_X}{\rho A}}, \qquad (3.49)$$

where values for the first three frequencies are given in Table 3.7.

Making use of the relationship

$$f = \frac{\omega}{2\pi}, \qquad (3.50)$$

the formulae for the natural frequencies are:

$$f_{X,i} = \frac{l_i r_f}{2\pi H^2}\sqrt{\frac{EI_Y}{\rho A}}, \qquad f_{Y,i} = \frac{l_i r_f}{2\pi H^2}\sqrt{\frac{EI_X}{\rho A}}. \qquad (3.51)$$

Table 3.7 Factors for the first three lateral frequencies

i	1	2	3
l_i	3.5160	22.0345	61.6972
$l_i/2\pi$	0.5596	3.5069	9.8194
$2\pi/l_i$	1.7870	0.2852	0.1018

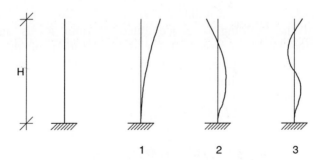

Fig. 3.10 Mode shapes for lateral vibrations.

According to Table 3.7, the first natural frequencies of the equivalent column in the principal planes are obtained from

$$f_X = \frac{0.56 r_f}{H^2} \sqrt{\frac{EI_Y}{\rho A}}, \qquad f_Y = \frac{0.56 r_f}{H^2} \sqrt{\frac{EI_X}{\rho A}} \qquad (3.52)$$

and the first natural periods are

$$T_X = \frac{1.787 H^2}{r_f} \sqrt{\frac{\rho A}{EI_Y}}, \qquad T_Y = \frac{1.787 H^2}{r_f} \sqrt{\frac{\rho A}{EI_X}}. \qquad (3.53)$$

The mode shapes for the first three modes are shown in Fig. 3.10. Studies of the elastic response of multistorey buildings indicate that for most buildings the fundamental mode contributes about 80%, while the second and third modes about 15% of the total response [Fintel, 1974].

Table 3.8 Reduction factor r_f

n	1	2	3	4	5	6	7	8	9	10	11
r_f	0.493	0.653	0.770	0.812	0.842	0.863	0.879	0.892	0.902	0.911	0.918
n	12	13	14	15	16	18	20	25	30	50	>50
r_f	0.924	0.929	0.934	0.938	0.941	0.947	0.952	0.961	0.967	0.980	$\sqrt{n/(n+2.06)}$

Formulae (3.51) show that the value of the lateral frequencies depends on three factors: the height, the lateral stiffness and the weight of the building. Of the three factors, the height of the building is far the most important factor. The value of the frequencies is in inverse proportion to the square of the height and to the square root of the mass density and in direct proportion to the square root of the lateral stiffness.

Reduction factor r_f in formulae (3.49) to (3.53) allows for the fact that the mass of the building is concentrated at floor levels and is not uniformly distributed as assumed for the derivation of the classical formula of a cantilever subjected to its weight. The phenomenon is similar to the one discussed earlier in section 3.1.1 in detail. The application of the Dunkerley summation theorem now leads to

$$r_f = \sqrt{\frac{n}{n+2.06}}, \qquad (3.53a)$$

where n is the number of storeys. Values for r_f are given in Table 3.8, where the values for $n = 1$ and $n = 2$ have been modified, based on the exact solution, to compensate for the conservative nature of the Dunkerley formula.

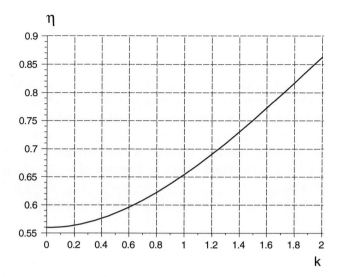

Fig. 3.11 Frequency parameter η.

The solution of equation (3.48) and its modification by factor r_f [Zalka, 1994b] lead to the natural frequencies of pure torsional vibrations:

$$f_{\varphi,i} = \frac{\eta_i r_f}{i_p H^2}\sqrt{\frac{EI_\omega}{\rho A}}, \qquad (3.54)$$

where η_i is the frequency parameter.

Values for the first frequency parameter η (the first eigenvalue of the problem) are given in Fig. 3.11 as a function of torsion parameter k. Values in the range of $0 \leq k \leq 2$ are given offering a solution for most practical cases. Table 3.9 covers the much wider range of $0 \leq k \leq 1000$ making it possible to handle even extreme cases.

In making use of Fig. 3.11 or Table 3.9, the fundamental frequency is obtained from

$$f_\varphi = \frac{\eta r_f}{i_p H^2}\sqrt{\frac{EI_\omega}{\rho A}}. \qquad (3.55)$$

Frequency parameters for the second and third natural frequencies for pure torsional vibrations are given in Tables 3.10 and 3.11.

The tables are limited to $k_{max} = 1000$. When the value of parameter k exceeds 1000, the warping stiffness is negligible and formula

48 Stability and frequency analyses

$$f_{\varphi,i} = \frac{(i-0.5)}{2Hi_p}\sqrt{\frac{GJ}{\rho A}} \qquad (3.56)$$

can be used instead of formula (3.54), where $i = 1, 2$ and 3.

The fundamental mode shape is given in Fig. 3.12 as a function of parameter k. The mode shape is a combination of the two functions: one is associated with the Saint-Venant torsional stiffness and the other one is associated with the warping stiffness. With great k values, e.g. for $k \gg 100$, the warping stiffness dominates the response and when k is small, e.g. for $k \ll 1$, the mode shape is determined by the Saint-Venant torsional stiffness.

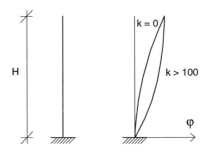

Fig. 3.12 Fundamental mode shape for pure torsional vibrations.

Formula (3.54) shows that the value of the pure torsional frequencies depends on five factors: the height, the warping stiffnesses, the mass density of the building, the radius of gyration of the ground plan of the building and the frequency parameter. As with the lateral frequencies, the height of the building is the most important factor as its increase reduces the value of the frequencies quadratically. The value of the frequencies is in proportion to the square root of the warping stiffness and in inverse proportion to the mass density of the building. The effect of the Saint-Venant stiffness enters the equation indirectly, through the frequency parameter. It should be mentioned here that, according to practical examples, the Saint-Venant torsional stiffness is much less significant compared to the warping stiffness.

Table 3.9 Frequency parameter η

k	η	k	η	k	η	k	η	k	η
0.00	0.5596	4.5	1.465	9.2	2.607	13.9	3.764	18.6	4.930
0.01	0.5596	4.6	1.489	9.3	2.631	14.0	3.789	18.7	4.955
0.05	0.5599	4.7	1.513	9.4	2.656	14.1	3.814	18.8	4.979
0.10	0.5606	4.8	1.537	9.5	2.680	14.2	3.838	18.9	5.004
0.20	0.5638	4.9	1.561	9.6	2.705	14.3	3.863	19.0	5.029
0.30	0.5690	5.0	1.586	9.7	2.729	14.4	3.888	19.1	5.054
0.40	0.5761	5.1	1.610	9.8	2.754	14.5	3.913	19.2	5.079
0.50	0.5851	5.2	1.634	9.9	2.778	14.6	3.937	19.3	5.104
0.60	0.5959	5.3	1.658	10.0	2.803	14.7	3.962	19.4	5.128
0.70	0.6084	5.4	1.682	10.1	2.827	14.8	3.987	19.5	5.153
0.80	0.6223	5.5	1.706	10.2	2.852	14.9	4.012	19.6	5.178
0.90	0.6376	5.6	1.731	10.3	2.876	15.0	4.036	19.7	5.203
1.0	0.6542	5.7	1.755	10.4	2.901	15.1	4.061	19.8	5.228
1.1	0.6718	5.8	1.779	10.5	2.926	15.2	4.086	19.9	5.253
1.2	0.6905	5.9	1.803	10.6	2.950	15.3	4.111	20	5.278
1.3	0.7100	6.0	1.827	10.7	2.975	15.4	4.136	21	5.526
1.4	0.7302	6.1	1.852	10.8	2.999	15.5	4.160	22	5.775
1.5	0.7511	6.2	1.876	10.9	3.024	15.6	4.185	23	6.024
1.6	0.7726	6.3	1.900	11.0	3.049	15.7	4.210	24	6.273
1.7	0.7946	6.4	1.924	11.1	3.073	15.8	4.235	25	6.522
1.8	0.8170	6.5	1.949	11.2	3.098	15.9	4.259	26	6.771
1.9	0.8397	6.6	1.973	11.3	3.122	16.0	4.284	27	7.021
2.0	0.8628	6.7	1.997	11.4	3.147	16.1	4.309	28	7.270
2.1	0.8860	6.8	2.021	11.5	3.172	16.2	4.334	29	7.519
2.2	0.9095	6.9	2.046	11.6	3.196	16.3	4.359	30	7.769
2.3	0.9332	7.0	2.070	11.7	3.221	16.4	4.383	31	8.018
2.4	0.9570	7.1	2.094	11.8	3.246	16.5	4.408	32	8.267
2.5	0.9809	7.2	2.119	11.9	3.270	16.6	4.433	33	8.517
2.6	1.0049	7.3	2.143	12.0	3.295	16.7	4.458	34	8.766
2.7	1.0290	7.4	2.167	12.1	3.320	16.8	4.483	35	9.016
2.8	1.0531	7.5	2.192	12.2	3.344	16.9	4.507	36	9.265
2.9	1.0772	7.6	2.216	12.3	3.369	17.0	4.532	37	9.515
3.0	1.1014	7.7	2.240	12.4	3.394	17.1	4.557	38	9.765
3.1	1.1257	7.8	2.265	12.5	3.418	17.2	4.582	39	10.01
3.2	1.1499	7.9	2.289	12.6	3.443	17.3	4.607	40	10.26
3.3	1.1741	8.0	2.313	12.7	3.468	17.4	4.632	50	12.76
3.4	1.1984	8.1	2.338	12.8	3.492	17.5	4.656	60	15.26
3.5	1.2226	8.2	2.362	12.9	3.517	17.6	4.681	70	17.76
3.6	1.2468	8.3	2.387	13.0	3.542	17.7	4.706	80	20.26
3.7	1.2711	8.4	2.411	13.1	3.566	17.8	4.731	90	22.76
3.8	1.2953	8.5	2.435	13.2	3.591	17.9	4.756	100	25.26
3.9	1.3195	8.6	2.460	13.3	3.616	18.0	4.781	200	50.25
4.0	1.3437	8.7	2.484	13.4	3.641	18.1	4.805	300	75.25
4.1	1.3679	8.8	2.509	13.5	3.665	18.2	4.830	400	100.25
4.2	1.3921	8.9	2.533	13.6	3.690	18.3	4.855	500	125.25
4.3	1.4163	9.0	2.558	13.7	3.715	18.4	4.880	1000	250.25
4.4	1.4405	9.1	2.582	13.8	3.739	18.5	4.905	>1000	k/4

50 Stability and frequency analyses

Table 3.10 Second natural frequency parameter η_2

k	η_2	k	η_2	k	η_2	k	η_2	k	η_2
0.00	3.507	4.5	5.290	9.2	8.435	13.9	11.76	18.6	15.15
0.01	3.507	4.6	5.352	9.3	8.504	14.0	11.83	18.7	15.23
0.05	3.507	4.7	5.415	9.4	8.574	14.1	11.90	18.8	15.30
0.10	3.508	4.8	5.478	9.5	8.643	14.2	11.97	18.9	15.37
0.20	3.512	4.9	5.542	9.6	8.713	14.3	12.04	19.0	15.45
0.30	3.517	5.0	5.606	9.7	8.783	14.4	12.11	19.1	15.52
0.40	3.526	5.1	5.670	9.8	8.853	14.5	12.19	19.2	15.59
0.50	3.536	5.2	5.734	9.9	8.922	14.6	12.26	19.3	15.66
0.60	3.549	5.3	5.799	10.0	8.992	14.7	12.33	19.4	15.74
0.70	3.564	5.4	5.864	10.1	9.062	14.8	12.40	19.5	15.81
0.80	3.581	5.5	5.929	10.2	9.132	14.9	12.47	19.6	15.88
0.90	3.600	5.6	5.994	10.3	9.202	15.0	12.55	19.7	15.96
1.0	3.622	5.7	6.060	10.4	9.272	15.1	12.62	19.8	16.03
1.1	3.645	5.8	6.125	10.5	9.342	15.2	12.69	19.9	16.10
1.2	3.671	5.9	6.191	10.6	9.413	15.3	12.76	20	16.18
1.3	3.699	6.0	6.257	10.7	9.483	15.4	12.83	21	16.91
1.4	3.728	6.1	6.323	10.8	9.553	15.5	12.91	22	17.64
1.5	3.760	6.2	6.390	10.9	9.624	15.6	12.98	23	18.38
1.6	3.793	6.3	6.457	11.0	9.694	15.7	13.05	24	19.11
1.7	3.828	6.4	6.523	11.1	9.764	15.8	13.12	25	19.85
1.8	3.865	6.5	6.590	11.2	9.835	15.9	13.19	26	20.59
1.9	3.903	6.6	6.657	11.3	9.905	16.0	13.27	27	21.32
2.0	3.943	6.7	6.724	11.4	9.976	16.1	13.34	28	22.06
2.1	3.985	6.8	6.792	11.5	10.05	16.2	13.41	29	22.80
2.2	4.028	6.9	6.859	11.6	10.12	16.3	13.48	30	23.54
2.3	4.072	7.0	6.926	11.7	10.19	16.4	13.56	31	24.28
2.4	4.118	7.1	6.994	11.8	10.26	16.5	13.63	32	25.03
2.5	4.165	7.2	7.062	11.9	10.33	16.6	13.70	33	25.77
2.6	4.213	7.3	7.130	12.0	10.40	16.7	13.77	34	26.51
2.7	4.262	7.4	7.198	12.1	10.47	16.8	13.85	35	27.25
2.8	4.312	7.5	7.266	12.2	10.54	16.9	13.92	36	28.00
2.9	4.364	7.6	7.334	12.3	10.61	17.0	13.99	37	28.74
3.0	4.416	7.7	7.402	12.4	10.69	17.1	14.06	38	29.48
3.1	4.470	7.8	7.470	12.5	10.76	17.2	14.14	39	30.23
3.2	4.524	7.9	7.539	12.6	10.83	17.3	14.21	40	30.97
3.3	4.579	8.0	7.607	12.7	10.90	17.4	14.28	50	38.43
3.4	4.635	8.1	7.676	12.8	10.97	17.5	14.35	60	45.90
3.5	4.691	8.2	7.744	12.9	11.04	17.6	14.43	70	53.38
3.6	4.749	8.3	7.813	13.0	11.11	17.7	14.50	80	60.86
3.7	4.807	8.4	7.882	13.1	11.18	17.8	14.57	90	68.35
3.8	4.865	8.5	7.951	13.2	11.25	17.9	14.64	100	75.84
3.9	4.924	8.6	8.020	13.3	11.33	18.0	14.72	200	150.80
4.0	4.984	8.7	8.089	13.4	11.40	18.1	14.79	300	225.78
4.1	5.044	8.8	8.158	13.5	11.47	18.2	14.86	400	300.78
4.2	5.105	8.9	8.227	13.6	11.54	18.3	14.94	500	375.77
4.3	5.166	9.0	8.296	13.7	11.61	18.4	15.01	1000	750.76
4.4	5.228	9.1	8.366	13.8	11.68	18.5	15.08	>1000	$3k/4$

Frequency analysis

Table 3.11 Third natural frequency parameter η_3

k	η_3	k	η_3	k	η_3	k	η_3	k	η_3
0.00	9.819	4.5	11.65	9.2	15.96	13.9	21.04	18.6	26.42
0.01	9.819	4.6	11.73	9.3	16.06	14.0	21.15	18.7	26.54
0.05	9.820	4.7	11.80	9.4	16.17	14.1	21.27	18.8	26.65
0.10	9.820	4.8	11.88	9.5	16.27	14.2	21.38	18.9	26.77
0.20	9.823	4.9	11.96	9.6	16.37	14.3	21.49	19.0	26.89
0.30	9.828	5.0	12.04	9.7	16.48	14.4	21.60	19.1	27.00
0.40	9.835	5.1	12.12	9.8	16.58	14.5	21.72	19.2	27.12
0.50	9.844	5.2	12.20	9.9	16.69	14.6	21.83	19.3	27.24
0.60	9.855	5.3	12.28	10.0	16.79	14.7	21.94	19.4	27.36
0.70	9.868	5.4	12.36	10.1	16.90	14.8	22.06	19.5	27.47
0.80	9.883	5.5	12.44	10.2	17.00	14.9	22.17	19.6	27.59
0.90	9.900	5.6	12.53	10.3	17.11	15.0	22.28	19.7	27.71
1.0	9.919	5.7	12.61	10.4	17.21	15.1	22.39	19.8	27.82
1.1	9.939	5.8	12.70	10.5	17.32	15.2	22.51	19.9	27.94
1.2	9.962	5.9	12.79	10.6	17.43	15.3	22.62	20	28.06
1.3	9.987	6.0	12.87	10.7	17.53	15.4	22.74	21	29.24
1.4	10.01	6.1	12.96	10.8	17.64	15.5	22.85	22	30.42
1.5	10.04	6.2	13.05	10.9	17.75	15.6	22.96	23	31.60
1.6	10.07	6.3	13.14	11.0	17.85	15.7	23.08	24	32.79
1.7	10.10	6.4	13.23	11.1	17.96	15.8	23.19	25	33.99
1.8	10.14	6.5	13.32	11.2	18.07	15.9	23.31	26	35.19
1.9	10.17	6.6	13.41	11.3	18.18	16.0	23.42	27	36.39
2.0	10.21	6.7	13.50	11.4	18.29	16.1	23.53	28	37.60
2.1	10.25	6.8	13.60	11.5	18.39	16.2	23.65	29	38.80
2.2	10.29	6.9	13.69	11.6	18.50	16.3	23.76	30	40.01
2.3	10.33	7.0	13.78	11.7	18.61	16.4	23.88	31	41.23
2.4	10.38	7.1	13.88	11.8	18.72	16.5	23.99	32	42.44
2.5	10.42	7.2	13.97	11.9	18.83	16.6	24.11	33	43.66
2.6	10.47	7.3	14.07	12.0	18.94	16.7	24.22	34	44.87
2.7	10.52	7.4	14.16	12.1	19.05	16.8	24.34	35	46.09
2.8	10.57	7.5	14.26	12.2	19.16	16.9	24.45	36	47.32
2.9	10.62	7.6	14.36	12.3	19.27	17.0	24.57	37	48.54
3.0	10.68	7.7	14.45	12.4	19.38	17.1	24.68	38	49.76
3.1	10.73	7.8	14.55	12.5	19.49	17.2	24.80	39	50.99
3.2	10.79	7.9	14.65	12.6	19.60	17.3	24.91	40	52.21
3.3	10.85	8.0	14.75	12.7	19.71	17.4	25.03	50	64.53
3.4	10.91	8.1	14.85	12.8	19.82	17.5	25.15	60	76.90
3.5	10.97	8.2	14.95	12.9	19.93	17.6	25.26	70	89.31
3.6	11.03	8.3	15.05	13.0	20.04	17.7	25.38	80	101.74
3.7	11.10	8.4	15.15	13.1	20.15	17.8	25.49	90	114.19
3.8	11.16	8.5	15.25	13.2	20.26	17.9	25.61	100	126.65
3.9	11.23	8.6	15.35	13.3	20.37	18.0	25.72	200	251.45
4.0	11.30	8.7	15.45	13.4	20.48	18.1	25.84	300	376.38
4.1	11.36	8.8	15.55	13.5	20.59	18.2	25.96	400	501.35
4.2	11.44	8.9	15.65	13.6	20.71	18.3	26.07	500	626.33
4.3	11.51	9.0	15.75	13.7	20.82	18.4	26.19	1000	1251.3
4.4	11.58	9.1	15.86	13.8	20.93	18.5	26.30	>1000	$5k/4$

52 Stability and frequency analyses

Finally, the value of the frequencies is in inverse proportion to the radius of gyration of the ground plan. Formulae of the warping constant (2.5) and the radius of gyration (2.6) to (2.11) show that the arrangement of the bracing elements has a great effect on the torsional response of the building. The torsional performance can be significantly improved in two ways. Reducing the distance between the shear centre of the bracing system and the centroid of the mass of the building leads to a bracing system arrangement which is doubly symmetrical or nearly doubly symmetrical. Investigations show that multistorey buildings are very sensitive to eccentricity [Jeary, 1981]. Even 10 per cent eccentricity can ensure that the structure becomes susceptible to torsional vibrations [Zhang, Xu and Kwok, 1993].

Perhaps equally important is to place the bracing shear walls in such a way that their 'torsion arm' (the perpendicular distance between their planes and the shear centre of the system) is maximum. Incidentally, this is exactly the case when efficiency, as far as stability is concerned, is considered. This principle is shown in Fig. 4.13 where a bracing system comprising four shear walls are arranged in two ways. Arrangement 'a' is not efficient, as it represents practically zero warping stiffness. On the other hand, arrangement 'b' is highly efficient as the system has significant warping stiffness.

3.2.2 Coupling of the basic modes; combined lateral-torsional vibrations

The similarity between the governing differential equations of stability and vibrations and between the structures of the formulae of the basic critical loads and natural frequencies indicate that the handling of the coupling of basic modes may also be similar. Indeed, the similarity still exists and simple procedures, very similar to those presented in section 3.1.2 for taking into account the coupling of the basic critical modes for the stability analysis, can be used for taking into account the coupling of basic modes for the frequency analysis.

(a) An approximate method
The combination of the basic frequencies may be taken into account by the Föppl–Papkovich theorem and the combined lateral-torsional frequencies can be obtained from

$$\frac{1}{f^2} = \frac{1}{f_X^2} + \frac{1}{f_Y^2} + \frac{1}{f_\varphi^2}, \tag{3.57}$$

where f_X, f_Y and f_φ are the basic (uncoupled) frequencies defined by formulae (3.51) and (3.54). The advantage of formula (3.57) is that it is easy to use and it is always conservative. However, its use in certain cases can lead to considerably uneconomical structural solutions as the error of the formula can be as great as 42%. A more accurate and only slightly more complicated solution is given in the next section.

(b) A more accurate method

The solution of the simultaneous differential equations (3.42) to (3.44) leads to the determinant

$$\begin{vmatrix} \omega^2 - \omega_X^2 & 0 & -\omega^2 y_c \\ 0 & \omega - \omega_Y^2 & \omega^2 x_c \\ -\omega^2 y_c & \omega^2 x_c & i_p^2(\omega - \omega_\varphi^2) \end{vmatrix} = 0, \tag{3.58}$$

where ω_X, ω_Y and ω_φ are the basic lateral and torsional circular frequencies. After making use of the relationship $f = \omega/2\pi$, the expansion of the determinant results in the cubic equation

$$(f^2)^3 + a_2(f^2)^2 + a_1 f^2 - a_0 = 0, \tag{3.59}$$

where the coefficients are

$$a_2 = \frac{f_X^2 \tau_X^2 + f_Y^2 \tau_Y^2 - f_\varphi^2 - f_X^2 - f_Y^2}{1 - \tau_X^2 - \tau_Y^2}, \tag{3.60}$$

$$a_1 = \frac{f_X^2 f_Y^2 + f_\varphi^2 f_X^2 + f_\varphi^2 f_Y^2}{1 - \tau_X^2 - \tau_Y^2}, \qquad a_0 = \frac{f_X^2 f_Y^2 f_\varphi^2}{1 - \tau_X^2 - \tau_Y^2}. \tag{3.61}$$

In formulae (3.60) and (3.61) parameters

$$\tau_X = \frac{x_c}{i_p} \quad \text{and} \quad \tau_Y = \frac{y_c}{i_p} \tag{3.62}$$

characterize the eccentricity of the mass and f_X, f_Y and f_ω are the basic frequencies. Of the three roots of the cubic equation, the smallest one is

54 Stability and frequency analyses

the most important in practical application being the fundamental frequency of the building – cf. notes on the cubic equation for stability in section 3.1.2.

The only difference between cubic equations (3.22) and (3.59) is that equation (3.22) is exact in taking into account the interaction among the basic critical loads for the stability analysis and equation (3.59) is slightly approximate for the dynamic analysis. The level of approximation in the dynamic analysis is very good and is well within practical structural engineering requirements: the error is below 1% with double coupling [Gere and Lin, 1958] and does not exceed 2% in the general case when all three basic frequencies are coupled.

The similarity between the buckling and dynamic behaviour can once more be exploited and the calculation of the fundamental (combined) frequency can be speeded up by using Tables C1 to C11 in Appendix C. The comparison of formulae (3.21) to (3.24) and (3.58) to (3.61) demonstrates that any mode coupling parameter in the tables is in fact a mode coupling parameter for the frequency analysis as well, if it is obtained as a function of

$$r_1 = \frac{f_\varphi^2}{f_X^2} \quad \text{and} \quad r_2 = \frac{f_\varphi^2}{f_Y^2} \qquad (3.63)$$

and eccentricity parameters τ_X and τ_Y defined by formulae (3.62).

Knowing the mode coupling parameter κ, the combined (lateral-torsional) fundamental frequency is obtained from

$$f = \sqrt{\kappa} f_\varphi, \qquad (3.64)$$

where f_φ is the basic frequency for pure torsional vibration.

The evaluation of cubic equation (3.59) leads to observations similar to those presented in section 3.1.2.b.

As with the case with the coupling of basic critical loads, the coupling of the basic frequencies clearly reduces the value of the natural frequencies. The magnitude of the reduction mainly depends on two factors: the relative value of the uncoupled frequencies compared to one another and the eccentricity of the bracing system in relation to the centroid of the load (mass of the building). The nearer the values of the three uncoupled frequencies, the bigger the reduction compared to the smallest of the three values. Similarly, the greater the distance between the shear centre and the centroid of the load, the smaller the coupled frequency. It is very important to take the coupling of the basic modes into consideration: the maximum error made by failing to do so may result in an unconservative error of 55% in the value of the fundamental frequency.

(c) Special case: monosymmetric arrangements

The calculation can be simplified in monosymmetrical cases (Fig. 2.4/b) and the mode coupling parameters in Fig. 3.5 and in Table 3.3 in section 3.1.2.c can be used for the prompt calculation of the combined frequency.

In the monosymmetrical case lateral vibration in the plane of symmetry develops independently and only the lateral vibration perpendicular to the axis of symmetry and pure torsional vibration combine. Assuming uniformly distributed mass and that axis X is an axis of symmetry, i.e. that $y_c = 0$ holds, the combined frequency can be expressed by the closed form formula

$$f_{combined} = \sqrt{\varepsilon} f_Y \quad (3.65)$$

where the values for the mode coupling parameter ε are given in Fig. 3.5 and in Table 3.3, as a function of τ_X [formula (3.62)] and

$$r_2 = \frac{f_\varphi^2}{f_Y^2}. \quad (3.66)$$

The basic frequencies f_Y and f_φ are defined by formulae (3.51) and (3.54). Table 3.3 covers the range $0 \le r_2 \le 100$. When $r_2 > 100$, $r_2 = 100$ can be used. The error made by this approximation is always smaller than 1%. The fundamental frequency for monosymmetrical bracing systems is the smaller one of the combined frequency and the independent lateral frequency in the plane of symmetry.

3.2.3 Concentrated mass at top level; single-storey buildings

Single-storey buildings represent a special case as most of the mass of the building is concentrated in the floor slab. The governing differential equations (3.42) to (3.44) (and also their solution) simplify and the basic natural frequencies of the equivalent column are readily available using classical formulae presented for cantilevers [Timoshenko and Young, 1955]. The fundamental frequencies for lateral vibrations are

56 Stability and frequency analyses

$$f_X = \frac{\sqrt{3}}{2\pi H}\sqrt{\frac{EI_Y}{AH\rho^*}}, \qquad f_Y = \frac{\sqrt{3}}{2\pi H}\sqrt{\frac{EI_X}{AH\rho^*}}, \qquad (3.67)$$

where ρ^* is the mass density per unit area of the top floor, defined by

$$\rho^* = \frac{Q}{g}, \qquad (3.68)$$

where

Q is the weight per unit area of the top floor [kN/m²],
g is the gravity acceleration [$g = 9.81$ m/s²].

The situation is slightly more complicated for pure torsional vibrations as both the Saint-Venant and the warping torsional stiffnesses have a contribution. When the warping stiffness is considered the fundamental torsional frequency is

$$f_\varphi^\omega = \frac{\sqrt{3}}{2\pi i_p H}\sqrt{\frac{EI_\omega}{AH\rho^*}} \qquad (3.69)$$

and when the Saint-Venant stiffness is taken into consideration the fundamental torsional frequency is

$$f_\varphi^t = \frac{1}{2\pi i_p}\sqrt{\frac{GJ}{AH\rho^*}}. \qquad (3.70)$$

The equivalent column (i.e. the bracing system) normally has both types of stiffness. Their simultaneous effects can be taken into account approximately by using the Southwell theorem. Accordingly, a conservative estimate of the fundamental torsional frequency is obtained from

$$f_\varphi^2 = (f_\varphi^\omega)^2 + (f_\varphi^t)^2 = \frac{3}{4\pi^2 i_p^2 H^2}\frac{EI_\omega}{AH\rho^*} + \frac{1}{4\pi^2 i_p^2}\frac{GJ}{AH\rho^*} \qquad (3.71)$$

as:

$$f_\varphi = \frac{\sqrt{3+k}}{2\pi i_p H} \sqrt{\frac{EI_\omega}{AH\rho^*}}.$$ (3.72)

3.2.4 Soil-structure interaction

The effect of the foundation medium in modifying the response of a structure, in relation to its behaviour when founded on an essentially rigid base on firm or moderately firm ground is relatively small [Parmelee, Perelman and Lee, 1969]. This is particularly true for multistorey buildings which are relatively flexible compared to the supporting medium. For such cases, the assumption that the structure is rigidly built-in at the bottom is justifiable.

For stiff structures, particularly those resting on relatively soft ground, the effects of soil-structure interaction can be significant. It cannot be predicted, however, whether the interaction is beneficial or detrimental to the structural system as several aspects are involved, including the properties of the structure, the supporting soil, damping, coupling between substructure and its supporting soil, etc. It is especially difficult to establish reliable properties of soil. Theoretical research and the results of full-scale tests indicate that foundation flexibility has a greater effect on the first frequency than on the second one [Maciag and Kuzniar, 1993].

Because of its reliability, the following simple formula based on the Föppl–Papkovich theorem can be used to obtain a conservative estimate of the effect of the flexible supports on the lateral frequencies of the building. According to the Föppl–Papkovich theorem, the fundamental frequency of the building can be calculated from two parts. First, it is assumed that the flexible superstructure (modelled by the equivalent column) is on a fixed support (Fig. 3.13/a). The frequency which belongs to this case is

$$f = \frac{0.56 r_f}{H^2} \sqrt{\frac{EI}{\rho A}}.$$ (3.73)

Second, it is assumed that the superstructure is totally stiff and is on a flexible support (Fig. 3.13/b). The fundamental frequency which belongs to this case [Kollár, 1979] assumes the form

58 Stability and frequency analyses

$$f_{flex} = \frac{\sqrt{3}}{2\pi H} \sqrt{\frac{k'_\alpha I_f}{\rho A H}},\qquad(3.74)$$

where

- k'_α is the spring coefficient for rocking motion (cf. section 3.1.5),
- I_f is the second moment of contact area between the foundation and soil in the relevant direction.

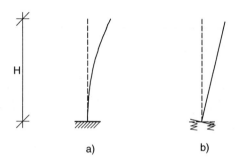

Fig. 3.13 Soil-structure interaction. a) Flexible superstructure on fixed support, b) stiff superstructure on flexible support.

Table 3.4 and Table 3.5 in section 3.1.5 offer suitable values for the spring coefficient in formula (3.74).

The Föppl–Papkovich formula can now be applied to the 'part' frequencies:

$$\frac{1}{f_{int}^2} = \frac{1}{f^2} + \frac{1}{f_{flex}^2}.\qquad(3.75)$$

After substituting for f and f_{flex} in formula (3.75) and some rearrangements, the formula for the fundamental frequency can be obtained as

$$f_{int} = r_{flex} f,\qquad(3.76)$$

where

$$r_{flex} = \frac{1}{\sqrt{1+4.13\beta}} \tag{3.77}$$

is the reduction factor which takes into account the effect of the flexible foundation. Values for factor r_{flex} are given in Table 3.12 as a function of β:

$$\beta = \frac{r_f^2 EI}{Hk'_\alpha I_f}, \tag{3.78}$$

where values for r_f are given in Table 3.8.

Table 3.12 Reduction factor r_{flex}

β	0	0.1	0.2	0.3	0.4	0.5	0.6	0.7	0.8	0.9	1.0
r_{flex}	1.000	0.841	0.740	0.668	0.614	0.571	0.536	0.507	0.482	0.460	0.442
β	2	3	4	5	6	7	8	9	10	50	100
r_{flex}	0.329	0.273	0.239	0.215	0.197	0.183	0.171	0.162	0.154	0.069	0.049

Once again, the effect of the rotational interaction between soil and structure also has to be considered. No simple solution is available for torsional interaction and each case should be carefully examined after taking into account the type of foundation which basically affects the nature of the phenomenon.

Ellis [1986] recommends a simple summation formula for taking into account both the translational and torsional interaction; he also investigates the significance of soil-structure interaction using measurements on four buildings and proposes an approximate method of quantifying soil-structure interaction.

Although other simple approximate methods [Ellis, 1984] and sophisticated computer procedures [Coull and Mukherjee, 1978] are available for the analysis of structures on flexible supports, it should be pointed out that it is very difficult to estimate appropriate soil parameters. Accuracy of data related to soil is of one order of magnitude less reliable than data on geometrical and stiffness characteristics of the superstructure.

3.2.5 Supplementary remarks

The procedures presented in the previous sections for the frequency analysis concentrate on the dominant characteristics of the structure while items whose influence on the dynamic response is normally small are neglected. Under certain circumstances, however, some phenomena and characteristics of usually secondary importance may increase their effect and therefore they may be no longer negligible. Such cases are discussed briefly in this section.

(a) Compressive forces

In analysing the lateral vibrations of simply supported beams subjected to concentrated end-forces and cantilevers under their uniformly distributed weight, Timoshenko [1928] took into account the effect of axial forces. His results showed that the axial compressive forces reduce the value of the frequencies of natural vibration by the factor of

$$\sqrt{1 - \frac{F}{F_{cr}}}, \qquad (3.79)$$

where F is the magnitude of the axial load and F_{cr} is the corresponding critical load. The reduction factor is considered exact for simply supported beams subjected to concentrated end-forces and can be used as reasonable approximation for cantilevers under uniformly distributed axial load.

Timoshenko's results can be generalized and the same approach can be applied to the *coupled* vibrations of thin walled cantilevers of open cross-section. Taking into account the effects of the uniformly distributed axial forces, the relevant formulae can be modified. This leads to the following formulae for the first basic frequencies

$$f_X = \frac{0.56 r_f}{H^2} \sqrt{\frac{EI_Y}{\rho A}\left(1 - \frac{N}{N_{cr,X}}\right)}, \quad f_Y = \frac{0.56 r_f}{H^2} \sqrt{\frac{EI_X}{\rho A}\left(1 - \frac{N}{N_{cr,Y}}\right)}, \qquad (3.80)$$

$$f_\varphi = \frac{\eta r_f}{i_p H^2} \sqrt{\frac{EI_\omega}{\rho A}\left(1 - \frac{N}{N_{cr,\varphi}}\right)}, \qquad (3.81)$$

which now take into consideration the effect of the axial forces. Critical loads $N_{cr,X}$, $N_{cr,Y}$ and $N_{cr,\varphi}$ are the basic critical loads and are obtained from the stability analysis as described in section 3.1.1.

(b) Shear deformation
When relatively short bracing elements (compared to the cross-sectional dimensions) are analysed and higher frequencies are also needed, the effect of the shear deformations can be of considerable magnitude. The deformations increase, resulting in a reduction in the frequencies. Various sources offer excellent treatment of this problem. Detailed mathematical background is given by Bishop and Price [1977] and Capuani, Savoiva and Laudiero [1992]. Closed form solutions are presented by Huang [1969], Timoshenko, Young and Weaver [1974] and Capuani, Savoiva and Laudiero [1992]. According to the investigations, the shear deformation may have considerable effect on the higher frequencies but only slightly modifies the fundamental frequency. As the effect of shear deformation is considered of secondary importance, it can be neglected when the fundamental frequency is calculated.

(c) Damping
Most frequency analyses including the one presented here ignore the fact that due to movements of a structure, energy absorption occurs through friction, air resistance and viscous behaviour resulting in damping. The omission of these effects leads to overestimated frequencies. The inclusion of damping characteristics in the governing equations would make the analysis much more complex. The effect of damping in multistorey buildings is not significant, as a rule, as the natural frequencies are not highly affected by the degree of damping. The problem is therefore often sidestepped in structural engineering design by estimating the damping forces separately. This can be done either theoretically or using test data [Littler, 1993]. Different forecast models are available for damping and vibration periods of buildings [Lagomarsino, 1993]. Damping coefficients are also provided in papers and monographs [Fintel, 1974; Hart and Vasudevan, 1975; Irwin, 1984; MacLeod, 1990; Goschy, 1990], which then may be used for estimating the damped frequency. A simple estimation [Fintel, 1974] is obtained using the formula

$$f_d = f\sqrt{1-\beta^2} \qquad (3.82)$$

which is derived for single-degree-of-freedom systems. In formula (3.82), f is the frequency when damping is neglected and f_d represents the frequency when damping is taken into account. Coefficient β is the damping ratio relative to the critical value of the damping coefficient, i.e. to the value of damping which would just cause an initial displacement to

62 Stability and frequency analyses

decay to zero without any oscillation. The value of coefficient β ranges from about 0.02 to 0.20 for most civil engineering structures; the effect of damping is usually well under 10%.

(d) Approximate methods

The accuracy of some approximate formulae for pure torsional vibrations is investigated in detail in [Zalka, 1994b]. Many approximate methods have been published for predicting the fundamental natural frequency of buildings [Ellis, 1980; Goldberg, 1973; Goschy, 1990]. Detailed evaluation of their accuracy is available [Ellis, 1980; Jeary and Ellis, 1981] and it is not the intention of this section to do further research in this area.

However, having evaluated a number of numerical examples, it is worth emphasizing one important point. Most approximate methods are one-parameter (height of building) or two-parameter (height and width of building) methods offering a simple formula for the fundamental frequency. They do not take into consideration the nature of the mode of vibration and ignore the possibility of mode coupling. The well-known formula

$$f = \frac{46}{H} \tag{3.83}$$

is perhaps the most characteristic example. It produces surprisingly good approximations for lateral frequencies. However, when torsion (and/or mode coupling!) plays an important role, the formula may produce totally incorrect results.

The relative success of the one-parameter formulae is due to the fact that they are based on the height, one of the three most important dynamic characteristics. The mass and the bending stiffness of the building do not vary much in relative terms and can be represented fairly well by a single constant, e.g. 46, in formula (3.83).

The situation is different with torsion. The mass of the building is still easily predictable and the Saint-Venant torsional stiffness plays a very little role. However, the effect of the warping stiffness can be significant. Its value depends on the *arrangement* of the bracing elements to a great extent. Using the same size of bracing elements producing the same bending stiffness and the same mass, the warping stiffness can be increased, or decreased, by orders of magnitude. (Figure 4.13 in Chapter 4 shows an example for maximizing the warping stiffness of a bracing system with given bracing elements.)

4

Stress analysis: an elementary approach

When the bracing system under horizontal load is analysed, the most important and at the same time the most difficult task is the establishment of the distribution of the load among the elements of the bracing system. The objective of this chapter is to provide the structural designer with simple procedures to carry out this task. The closed-form solutions and guidelines make it possible to determine the load shares on the bracing elements and to calculate the maximum translations and rotation of the building in a very simple and efficient manner.

The behaviour of the whole structure is complex and the exact calculation normally represents a formidable task. In practical structural design, however, the complexity of the task can be considerably reduced by introducing some carefully chosen simplifying assumptions while keeping the accuracy of the results within a range acceptable in practice.

The assumptions used in this chapter in addition to those listed in Chapter 1 are as follows.

- The horizontal load is uniformly distributed over the height of the building.
- The torsional stiffness of the individual bracing elements is small and is therefore neglected.

Both the above assumptions are dropped in the next chapter where a more advanced – albeit slightly more complicated – method is presented for the analysis.

To help to understand the basic principles governing the 3-dimensional behaviour of buildings under horizontal loads, the derivations of the design formulae are also presented. The derivations are based on elementary static considerations and do not require any special (mathematical) background.

The spatial behaviour of the building is described in section 2.1.

4.1 HORIZONTAL LOAD

Although the magnitude of the horizontal load must always be established according to the relevant Code of Practice regulations, some simple methods are given here in the following sections, which can also be used for

- comparing the different types of horizontal loads,
- checking the result of more sophisticated calculations.

4.1.1 Wind

Wind is probably the most important horizontal load. Design for wind can be carried out, in most cases, applying an equivalent static wind pressure on the building acting normal to the surface. The resulting wind force on the structure may be determined by means of the global force [Eurocode 1, 1995]. The global force is obtained from

$$F_w = c_e(z_e) c_d c_f A_{ref} q_{ref}, \qquad (4.1)$$

where

$c_e(z_e)$	is the exposure coefficient,
c_d	is the dynamic coefficient,
c_f	is the force coefficient,
A_{ref}	is the reference area for c_f

and

$$q_{ref} = \frac{\rho}{2} v_{ref}^2 \qquad (4.2)$$

is the reference mean wind velocity pressure with

ρ	air density ($\rho = 1.25$ kg/m^3, unless otherwise specified),
v_{ref}	reference wind velocity.

The reference wind velocity in formula (4.2) is obtained from

$$V_{ref} = C_{DIR} C_{TEM} C_{ALT} V_{ref,0}, \qquad (4.3)$$

where

c_{DIR} is the direction factor ($c_{DIR} = 1$, unless otherwise specified),
c_{TEM} is the temporary (seasonal) factor ($c_{TEM} = 1$, unless otherwise specified),
c_{ALT} is the altitude factor ($c_{ALT} = 1$, unless otherwise specified),
$v_{ref,0}$ is the basic value of the wind velocity given by means of wind maps.

For vertical cantilevered structures with a slenderness ratio height/width > 2 and with nearly constant cross-section over the height, the global force is calculated from

$$F_{wj} = c_e(z_j) c_d c_{fj} A_j q_{ref}, \qquad (4.4)$$

where

A_j is the incremental area,
z_j is the height of the centre of gravity of incremental area A_j,
c_{fj} is the force coefficient for incremental area A_j.

Information regarding specific values and the application of coefficients c_e, z_e, z_j, c_d, c_f, c_{fj}, A_{ref}, A_j, ρ, c_{DIR}, c_{TEM}, c_{ALT} and $v_{ref,0}$ is given in Part 2.4 in Eurocode 1 [1995].

Although code provisions (and the values of coefficients) may change from time to time, formula (4.1) shows the basic principle: the value of the global force basically depends on the size of the reference area and the wind pressure. Additional modifying effects (exposure, direction, altitude, shape, terrain, etc) are accounted for by using coefficients. Comprehensive treatment of and guidelines to wind loading of building structures are given by Cook [1985 and 1990].

4.1.2 Seismic load

Under certain circumstances, Code of Practice regulations in some European countries and American states allow the use of equivalent static load for seismic analyses. For a simplified analysis, for example, the global seismic force (the resultant of the equivalent static load) can be calculated from

$$S = QK\beta, \qquad (4.5)$$

where

Q is the total dead weight of the building [kN],
K is the seismic constant,
β is the dynamic coefficient.

The total dead weight of the building is conveniently calculated by using a spreadsheet covering all the different types of material used for the construction. Quick results and – for most conventional buildings – good estimates can be obtained by using an equivalent floor load for the calculation. For example, an equivalent floor load of $q = 10$ kN/m² per unit area, which also covers the weight of the vertical structures, normally represents a conservative estimate.

Values for seismic constant K are given in Table 4.1, according to the Mercalli–Sieberg–Cancani (MSC) scale, where the seismic zones in the first column should be obtained using official seismic zone maps.

Table 4.1 Values for seismic constant K

MSC	K
5	0.005
6	0.010
7	0.025
8	0.050

The value of dynamic coefficient β (which is in fact the fundamental frequency of the building) is calculated from the relationship

$$0.8 \le \beta = \frac{1}{T} \le 3.0, \qquad (4.6)$$

where T is the natural period of the building. Its values are given in Table 4.2, as a function of soil characteristics, where n is the number of storeys.

Table 4.2 Values for natural period T

Soil	T
stiff, rocky	$n/15$
medium	$n/20$
soft, loose (clay, sand)	$n/30$

Experience and comparisons to other Code of Practice regulations show that formula (4.5) results in reliable estimates for the seismic load for the analysis of relatively stiff buildings [Gergely, 1975]. However, formula (4.5) tends to lead to very conservative solutions for flexible bracing

Horizontal load 67

systems and therefore more accurate methods may be needed for the final analysis of flexible systems. Simple methods are given in section 3.2 for calculating the fundamental frequency and the natural period of the building.

Knowing the resultant of the total seismic force, the intensity of the horizontal seismic load of uniform distribution is obtained from

$$p_s = \frac{S}{LH}, \qquad (4.7)$$

where L is the width of the building in the perpendicular direction and H is the height of the building.

Formula (4.5) shows the basic principle: the value of an equivalent static load basically depends of the weight of the building, its fundamental frequency and the seismic zone it is in. According to different national Code of Practice regulations, additional effects can be accounted for by introducing more coefficients into the formula (e.g. foundation flexibility coefficient, damping factor, importance factor, mode factor, etc).

Several monographs are available for seismic analysis and earthquake resistant design offering theoretical background, simple approximate and more advanced methods and practical guidelines [Newmark and Rosenblueth, 1971; Key, 1988; Scarlat, 1996].

4.1.3 Construction misalignment

Additional horizontal forces may arise due to the inappropriate placing of the vertical load bearing elements. The following procedure offers a simple formula for the resulting horizontal load.

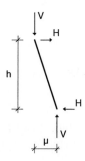

Fig. 4.1 Wall misalignment.

An out-of-plane vertical element in an inclined position (Fig. 4.1) develops the horizontal force

$$H = V\frac{\mu}{h}, \qquad (4.8)$$

where V [kN/m] is the vertical load on the element, h [m] is the storey height and μ [m] is the misalignment.

The value of the total horizontal force due to misalignment depends on the number of elements on one floor level (n_h) and the number of elements above each other, i.e. the number of storeys (n). The value of the resulting total horizontal load is not proportional to the number of vertical elements, since misalignment can occur in both directions and some of the resulting horizontal forces may cancel each other out. This can be approximately taken into account by introducing the probability factor $0.5(n_h n)^{0.5}$, which leads to the formula for the total horizontal force as

$$F_m = V\frac{\mu}{h}\frac{1}{2}\sqrt{n_h n}. \qquad (4.9)$$

This horizontal force represents the uniformly distributed horizontal load p_m [kN/m²]:

$$p_m = \frac{V\mu}{2h^2}\sqrt{n_h n}. \qquad (4.10)$$

The vertical load V [kN/m] in the above formula is the average vertical load of the vertical load bearing elements on a floor level.

4.1.4 Comparisons

The three types of horizontal load are quite different in nature. The magnitude of the wind load is in direct proportion to the size of the building and, for cross-wall system buildings, this leads to quite different wind loads in the two directions that are normally considered for the structural analysis since the wider the building, the greater the wind load (Fig. 4.2/a). Accidentally, this may go in parallel with a stronger bracing system since a wider building normally has more bracing elements.

The situation is different with the seismic load. Its magnitude is primarily determined by which seismic zone the building is in and the value is proportional to the mass of the building. It follows that the seismic load does not depend on the size of the area of the facade and it is of similar magnitude in every direction [cf. formula (4.5)] (Fig. 4.2/b). This fact should be borne in mind in seismic design when in many cases an 'ordinary' bracing system (one created for wind resistance) may easily be inadequate in the direction perpendicular to the cross walls.

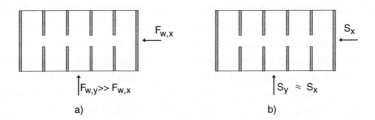

Fig. 4.2 Cross-wall system. a) Wind forces, b) seismic forces.

The comparison of the different types of horizontal load reveals interesting characteristics. The magnitude of the seismic force on a building in zone 8 by the MSC scale can be even ten times greater than that of the greater wind force. The two forces are normally of the same order of magnitude in zones MSC 5 and MSC 6. The magnitude of the horizontal load due to misalignment is normally smaller than that of the wind load; assuming a misalignment of $\mu = 0.015$ m, it is in the range of 5–40%.

4.2 BUILDINGS BRACED BY PARALLEL WALLS

Bracing low-rise building by using frameworks is normal practice but as the number of storeys increases, the building may develop deformations of unacceptable magnitude or it may lead to uneconomic structural solutions. Some or all of the frameworks are replaced by shear walls, resulting in the widely used cross-wall system for buildings subjected to wind load. When the system comprises shear walls and frameworks, the frameworks are often neglected in the structural analysis for horizontal loads as a conservative approximation. This approximation is justified as their lateral stiffness is indeed negligible compared to that of the shear walls. The bracing system is represented by a system of parallel shear walls, which can be investigated in a simple manner. This is shown in this section where the second moment of area of the walls perpendicular to their plane as well as their torsional constant are neglected as a conservative approximation.

4.2.1 Basic principles

Two basic principles are used for the analysis: the translational stiffness and the shear centre.

(a) Translational stiffness

Translational stiffness k is defined as the uniformly distributed load of intensity p and resultant F which, acting on a cantilever of length H and bending stiffness EI, develops unit translation at the free end.

The top translation of the cantilever with the above characteristics is obtained using the principle of virtual work in the usual way (Fig. 4.3):

$$c = \frac{1}{EI}\int M_p M_Q \, dz = \frac{1}{EI}\frac{pH^2}{2}\frac{H}{3}\frac{3H}{4} = \frac{pH^4}{8EI} = \frac{FH^3}{8EI}. \qquad (4.11)$$

The translational stiffness is obtained when this translation is of unit magnitude:

$$k = \frac{8EI}{H^3}. \qquad (4.12)$$

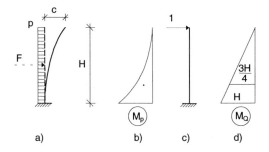

Fig. 4.3 Top translation of cantilever.

Using the translational stiffness, the maximum translation of the cantilever is obtained from

$$c = \frac{1}{k}F. \qquad (4.13)$$

(b) Shear centre

The shear centre is an important geometrical characteristic of the bracing system. By definition, an external force passing through the shear centre only develops translation but not rotation.

The shear centre is the centroid of the translational stiffnesses and is also called centre of stiffness. Its location is calculated using the rules of determining the centroid. The distance of the shear centre from an arbitrary axis is obtained by dividing the 'static moment' of the translational stiffnesses related to the axis with the sum of the stiffnesses. The x coordinate of shear centre O of the system of parallel walls in the coordinate system $\bar{x}-\bar{y}$ (Fig. 4.4) is obtained from

$$\bar{x}_o = \frac{\sum_{1}^{n} k_i \bar{x}_i}{\sum_{1}^{n} k_i}, \qquad (4.14)$$

where $\bar{x}_i$ is the distance of the ith wall from axis $\bar{y}$, k_i is the translational stiffness of the ith wall defined by formula (4.12) and n is the number of walls.

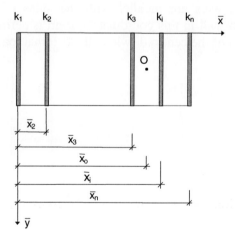

Fig. 4.4 Shear centre for parallel walls.

When all the walls are of the same material and height, the above formula simplifies to

$$\bar{x}_o = \frac{\sum_1^n I_i \bar{x}_i}{\sum_1^n I_i}. \tag{4.14a}$$

4.2.2 Load distribution

Consider the bracing system of the cross-wall system building whose layout is shown in Fig. 4.5/a. The resultant of the external horizontal load (F) passes through the geometrical centre of the layout (C). The floor slabs transmit forces F_i to the shear walls, which are then balanced by reaction forces $-F_i$. The objective is to determine these forces. (For the sake of simplicity, only the external forces are shown in the figures in this section.)

The shear centre is located first. Knowing the geometrical and stiffness characteristics of the walls, formula (4.14) yields the coordinate of the shear centre from axis $\bar{y}$. The investigation is then carried out in the new coordinate system x-y whose origin coincides with the shear centre and whose coordinate axes are parallel with the sides of the layout. The location of the ith wall in this coordinate system is defined by coordinate x_i and the location of the centroid (where the external load passes through) by x_c.

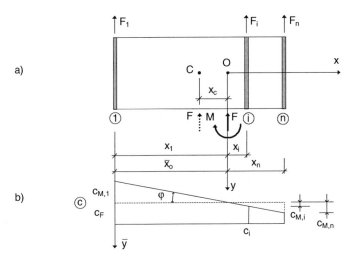

Fig. 4.5 Building braced by parallel walls. a) Layout, b) displacements.

The resultant of the lateral load (F) passing through the centroid is replaced by force F passing through the shear centre and the couple $M = Fx_c$ around the shear centre. Force F passing through the shear centre makes the building translate but not rotate. All walls translate by c_F. Couple M makes the building rotate around the shear centre, developing translations $c_{M,i}$ of the walls. The translation of the ith wall (Fig. 4.5/b) therefore is

$$c_i = c_F + c_{M,i}. \tag{4.15}$$

Making use of formulae (4.13) and (4.15), the load on the ith wall is given by

$$F_i = k_i(c_F + c_{M,i}). \tag{4.16}$$

According to Fig. 4.5/b,

$$c_{M,i} = x_i \tan\varphi, \tag{4.17}$$

where φ is the rotation of the building. Combining formulae (4.16) and (4.17) results in the load on the ith wall as

$$F_i = k_i(c_F + x_i \tan\varphi). \tag{4.18}$$

This equation contains three unknowns: F_i, c_F and φ. The two more equations needed to solve equation (4.18) are obtained using equilibrium considerations.

The equilibrium of forces in direction y is expressed by

$$F = \sum_1^n F_i \tag{4.19}$$

which, after substituting for F_i, assumes the form

$$F = \sum_1^n k_i(c_F + x_i \tan\varphi). \tag{4.20}$$

Rearranging equation (4.20) leads to

$$F = c_F \sum_1^n k_i + \tan\varphi \sum_1^n k_i x_i. \tag{4.21}$$

74 Elementary stress analysis

The second term on the right-hand side is zero since $\Sigma k_i x_i$ represents the 'static moment' of the stiffnesses with respect to the shear centre (i.e. to their centroid). The translation caused by force F is therefore obtained from equation (4.21) as

$$c_F = \frac{F}{\sum_{1}^{n} k_i}. \qquad (4.22)$$

Moment equilibrium with respect to the shear centre is expressed by

$$Fx_c = \sum_{1}^{n} M_i = \sum_{1}^{n} F_i x_i \qquad (4.23)$$

which, after substituting formula (4.18) for F_i, assumes the form

$$Fx_c = \sum_{1}^{n} k_i x_i (c_F + x_i \tan\varphi). \qquad (4.24)$$

Rearrangement leads to

$$Fx_c = c_F \sum_{1}^{n} k_i x_i + \tan\varphi \sum_{1}^{n} k_i x_i^2. \qquad (4.25)$$

The first term on the right-hand side is zero because $\Sigma k_i x_i$ (the 'static moment' with respect to the centroid) is zero. Thus, the rotation of the building is obtained as

$$\tan\varphi = \frac{Fx_c}{\sum_{1}^{n} k_i x_i^2}. \qquad (4.26)$$

Substituting for c_F and $\tan\varphi$ in equation (4.18) finally results in the load share of the total horizontal load on the ith wall:

$$F_i = \frac{k_i}{\sum_{1}^{n} k_i} F + \frac{k_i x_i}{\sum_{1}^{n} k_i x_i^2} Fx_c. \qquad (4.27)$$

In the practical case when all the walls are of the same height and material, H and E in the formula of translational stiffness can be taken out and formula (4.27) simplifies to

$$F_i = \frac{I_i}{\sum_1^n I_i} F + \frac{I_i x_i}{\sum_1^n I_i x_i^2} F x_c. \tag{4.28}$$

The first term in formula (4.28) demonstrates that of the total external load (F) each wall takes a load share proportional to their second moment of area. The second term shows that, due to the torsional moment of the external load (Fx_c), the walls also take additional load shares. The magnitude of this load share can be significant; its value depends on, in addition to their second moment of area, the distance of the walls from the shear centre (x_i). This geometrical property is called the 'torsion arm' of the wall. The term $\Sigma I_i x_i^2$ [m^6] in the denominator of the second term plays an important role in the torsional behaviour of the bracing system and is called bending torsional constant, or warping constant:

$$I_\omega = \sum_1^n I_i x_i^2. \tag{4.29}$$

Formula (4.28) for the load distribution is effectively identical to the one presented by Pearce and Matthews [1971], Dowrick [1976] and Irwin [1984] for cross-wall systems.

4.2.3 Deformations

When the load shares on the walls are known, the maximum top translation of the walls is obtained using formula (4.13):

$$c_i = \frac{1}{k_i} F_i = \frac{H^3}{8EI_i} F_i. \tag{4.30}$$

According to Fig. 4.5/b, the maximum translation occurs at one end of the building. If there is a wall there, then formula (4.30) can be used for the calculation of the maximum translation of the building. If there is no wall where the maximum translation occurs, then formula (4.30) cannot be used. In that case, formulae (4.15) and (4.17) should be combined:

76 Elementary stress analysis

$$c_{max} = c_F + c_{M,max} = c_F + x_{max}\tan\varphi, \qquad (4.31)$$

where x_{max} is the location of the maximum translation measured from the shear centre. Substituting for c_F and $\tan\varphi$ in formula (4.31), the formula for the maximum translation assumes the form:

$$c_{max} = \frac{H^3}{8E}\left(\frac{F}{\sum_1^n I_i} + \frac{Fx_c}{\sum_1^n I_i x_i^2} x_{max}\right). \qquad (4.32)$$

Assuming small angles when the approximation $\tan\varphi \approx \varphi$ holds and making use of formulae (4.26) and (4.12) the rotation of the building is calculated from

$$\varphi = \frac{H^3}{8E}\frac{Fx_c}{\sum_1^n I_i x_i^2}. \qquad (4.33)$$

The rotations obtained from the above formulae are given in radian. According to formula (5.33) in section 5.2.2, their value in degree is φ [degree] = 57.3φ [radian].

4.3 BUILDINGS BRACED BY PERPENDICULAR WALLS

Buildings with a square layout (or with plan width and breadth of similar size) under wind load and buildings of any layout geometry in seismic zones need bracing elements in both directions. The spatial stiffness of such systems is maximized when the perpendicular walls are built together. However, in many practical cases the walls are not built together (e.g. they are not located near each other), or they are built together but the vertical joints are not constructed to transfer shear. This type of independent perpendicular wall system is investigated in this section. The principals and the treatment are identical to those applied earlier to the system of parallel walls.

4.3.1 Load distribution

The procedure is presented for buildings with perpendicular walls, under horizontal load F_y in direction y (Fig. 4.6/a). It is assumed that the height and modulus of elasticity are identical for each wall.

The first step is to determine the location of the shear centre. This is done by using the same procedure as in the previous section [which led to formulae (4.14) and (4.14a)] but because there are walls in both the x and y directions, the calculation is carried out in both directions resulting in both coordinates of the shear centre:

$$\bar{x}_o = \frac{\sum_{1}^{n} I_{x,i} \bar{x}_i}{\sum_{1}^{n} I_{x,i}}, \qquad \bar{y}_o = \frac{\sum_{1}^{n} I_{y,i} \bar{y}_i}{\sum_{1}^{n} I_{y,i}}. \qquad (4.34)$$

In the above formulae $I_{x,i}$ and $I_{y,i}$ are the second moments of area of the ith wall with respect to its centroidal axes and n is the number of walls. The origin of the coordinate system is now transferred to shear centre O.

The external horizontal load (F_y) passing through the geometrical centre (C) is replaced by F_y acting in and moment $M = F_y x_c$ acting around the shear centre. Under F_y in the shear centre, the building develops a translation in direction y but no rotation occurs. This translation, identical for each wall, is denoted by c_{yF}. Under moment M, the building develops rotation φ around the shear centre. Because of this rotation, the walls undergo additional translations. These translations are denoted by $c_{xM,i}$ and $c_{yM,i}$ in directions x and y. The translation of the ith wall in direction y is expressed by

$$c_{y,i} = c_{yF} + c_{yM,i}. \qquad (4.35)$$

The two terms on the right-hand side show the effect of the horizontal force passing through the shear centre and that of the moment around the shear centre (Fig. 4.6/b).

In making use of equations (4.13) and (4.35), the load share on the ith wall is obtained from

$$F_{y,i} = k_{x,i}(c_{yF} + c_{yM,i}), \qquad (4.36)$$

where

78 Elementary stress analysis

$$k_{x,i} = \frac{8EI_{x,i}}{H^3}. \qquad (4.36a)$$

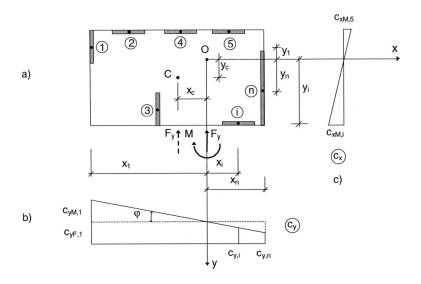

Fig. 4.6 Building braced by perpendicular walls. a) Layout, b)–c) displacements.

According to the diagram of translations in direction y (Fig. 4.6/b),

$$c_{yM,i} = x_i \tan\varphi. \qquad (4.37)$$

Substituting this formula for $c_{yM,i}$ in equation (4.36) leads to

$$F_{y,i} = k_{x,i}(c_{yF} + x_i \tan\varphi). \qquad (4.38)$$

There is no external load on the building in direction x, but due to moment M, the walls develop translations in direction x. These translations are proportional to the distance of the walls from the shear centre (Fig. 4.6/c):

$$c_{xM,i} = y_i \tan\varphi. \qquad (4.39)$$

Making use of formula (4.13), formula (4.39) can be rearranged as

$$\frac{1}{k_{y,i}} F_{x,i} = y_i \tan\varphi, \qquad (4.40)$$

from which the load on the ith wall in direction x, due to the rotation of the building, is obtained:

$$F_{x,i} = k_{y,i} y_i \tan\varphi, \qquad (4.41)$$

where

$$k_{y,i} = \frac{8EI_{y,i}}{H^3}. \qquad (4.41a)$$

The actual values of the load shares cannot be calculated from formulae (4.38) and (4.41) as they contain the two deformations c_{yF} and φ, yet unknown. They can be determined using equilibrium considerations.

After substituting formula (4.38) for $F_{y,i}$, the equilibrium equation of the forces in direction y

$$F_y = \sum_1^n F_{y,i} \qquad (4.42)$$

assumes the form

$$F_y = \sum_1^n k_{x,i}(c_{yF} + x_i \tan\varphi) \qquad (4.43)$$

or, after rearrangement

$$F_y = c_{yF} \sum_1^n k_{x,i} + \tan\varphi \sum_1^n k_{x,i} x_i. \qquad (4.44)$$

Because of $\Sigma k_{x,i} x_i = 0$, the second term on the right-hand side is zero and therefore the translation of the building in direction y, due to the external force, is

$$c_{yF} = \frac{F_y}{\sum_1^n k_{x,i}}. \qquad (4.45)$$

Moment equilibrium of the internal and external forces around the shear centre is expressed by

80 Elementary stress analysis

$$\sum_{1}^{n} M_i = F_y x_c = \sum_{1}^{n} F_{y,i} x_i + \sum_{1}^{n} F_{x,i} y_i . \tag{4.46}$$

After substituting formulae (4.38) and (4.41) for $F_{y,i}$ and $F_{x,i}$ and some rearrangement, equation (4.46) assumes the form

$$F_y x_c = c_{yF} \sum_{1}^{n} k_{x,i} x_i + \tan\varphi \left(\sum_{1}^{n} k_{x,i} x_i^2 + \sum_{1}^{n} k_{y,i} y_i^2 \right). \tag{4.47}$$

It is now the first term on the right-hand side which is zero and the rotation of the building is obtained from the equation as

$$\tan\varphi = \frac{F_y x_c}{\sum_{1}^{n} k_{x,i} x_i^2 + \sum_{1}^{n} k_{y,i} y_i^2} . \tag{4.48}$$

Substituting formulae (4.45) and (4.48) of the deformations for c_{yF} and $\tan\varphi$ in equations (4.38) and (4.41) gives the loads on the individual walls. Of the total external horizontal load F_y, the ith wall takes

$$F_{y,i} = \frac{k_{x,i}}{\sum_{1}^{n} k_{x,i}} F_y + \frac{k_{x,i} x_i}{\sum_{1}^{n} k_{x,i} x_i^2 + \sum_{1}^{n} k_{y,i} y_i^2} F_y x_c \tag{4.49}$$

in direction y, and

$$F_{x,i} = -\frac{k_{y,i} y_i}{\sum_{1}^{n} k_{x,i} x_i^2 + \sum_{1}^{n} k_{y,i} y_i^2} F_y x_c \tag{4.50}$$

in direction x. The negative sign in formula (4.50) accounts for the fact that a positive twisting moment develops negative translation in the x direction (as opposed to the positive translation it develops in the y direction) – cf. Fig. 5.7.

As all the walls are of the same material and height, the above formulae simplify to

$$F_{y,i} = \frac{I_{x,i}}{\sum_{1}^{n} I_{x,i}} F_y + \frac{I_{x,i} x_i}{I_\omega} F_y x_c \tag{4.51}$$

in direction y, and

$$F_{x,i} = -\frac{I_{y,i} y_i}{I_\omega} F_y x_c \qquad (4.52)$$

in direction x, where

$$I_\omega = \sum_1^n I_{x,i} x_i^2 + \sum_1^n I_{y,i} y_i^2 \qquad (4.53)$$

is the warping constant of the bracing system.

Formulae (4.51) to (4.53) are in line with those presented by Beck and Schäfer [1969], König and Liphardt [1990] and MacLeod [1990].

4.3.2 Deformations

Formula (4.13) makes it possible to calculate the maximum top translation of any wall if its load share is known. The maximum translation of the ith wall due to its load share in direction y is

$$c_{y,i} = \frac{H^3}{8EI_{x,i}} F_{y,i}. \qquad (4.54)$$

When the load acts in direction x the translation is

$$c_{x,i} = \frac{H^3}{8EI_{y,i}} F_{x,i}. \qquad (4.55)$$

If the maximum translation of the building occurs at a location where there is no wall, the above formulae cannot be used. In such cases the maximum translation can be calculated using equations (4.35) and (4.37):

$$c_{y,\max} = c_{yF} + x_{\max} \tan\varphi, \qquad (4.56)$$

where $x_{\max}$ is the location of the maximum translation – a corner of the building. After substituting formulae (4.45) and (4.48) for c_{yF} and $\tan\varphi$ and also making use of formulae (4.36a) and (4.41a), the general formula for the maximum translation is obtained as

$$c_{y,\max} = \frac{H^3}{8E} \left(\frac{F_y}{\sum_{1}^{n} I_{x,i}} + \frac{F_y x_c}{I_\omega} x_{\max} \right). \tag{4.57}$$

According to Fig. 4.6/c, the building under horizontal load in direction y also develops translations in direction x. If there is no wall at the location of the maximum translation in direction x (at one side of the building parallel with axis x), then formula (4.39) should be used, which, after substituting formula (4.48) for $\tan\varphi$ and making use of formulae (4.36a) and (4.41a), assumes the form

$$c_{x,\max} = \frac{H^3}{8E} \frac{F_y x_c}{I_\omega} y_{\max}. \tag{4.58}$$

Assuming small rotations when the approximation $\tan\varphi \approx \varphi$ holds and making use of formulae (4.36a) and (4.41a), the rotation of the building is calculated using formula (4.48):

$$\varphi = \frac{H^3}{8E} \frac{F_y x_c}{I_\omega}. \tag{4.59}$$

In formulae (4.57) and (4.59) I_ω is the warping constant given by formula (5.53).

4.4 BUILDINGS BRACED BY FRAMEWORKS

Low-rise buildings can be provided with adequate lateral stiffness using frameworks. In addition to the assumptions listed in the introduction, it is also assumed in this section that the frameworks of the bracing system have the same geometrical and stiffness characteristics. Both symmetrical and unsymmetrical arrangements are considered.

4.4.1 Frameworks in a symmetrical arrangement

Consider the building braced by frameworks of the same stiffness in a symmetrical arrangement (Fig. 4.7). The resultant of the horizontal load F

passes through the geometrical centre of the layout which, because of the symmetrical arrangement, coincides with the shear centre.

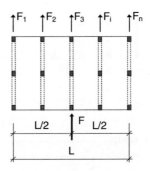

Fig. 4.7 Frameworks in a symmetrical arrangement.

The floor slabs assumed to be stiff in their plane make the frameworks develop the same translations and therefore all the frameworks take the same amount of the total horizontal load

$$F_i = \frac{F}{n}, \qquad (4.60)$$

where F_i is the load share on the ith framework and n is the number of frameworks. (The more general formula (4.28) also leads to the above formula, if it is taken into consideration that each bracing element has the same lateral stiffness and, because of the concurrent shear centre and centroid, x_c is zero.)

4.4.2 Frameworks in an asymmetrical arrangement

A practical case is shown in Fig. 4.8/a. A building of plan width L is braced by parallel frameworks of identical lateral stiffness and a single wall at the side of the building. It is assumed that the stiffness of the wall is infinitely great compared to the stiffness of the frameworks. The system can be modelled by a rigid beam of length L on one fixed support (the wall) and several flexible supports (the frameworks). The beam effectively models the floor slabs of the building. Each floor slab moves as a rigid body with a translation characterized by a triangular diagram.

84 Elementary stress analysis

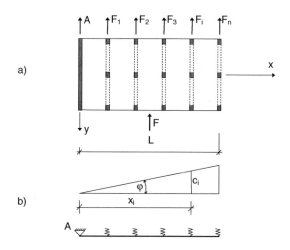

Fig. 4.8 Frameworks in an asymmetrical arrangement. a) Layout, b) displacements.

The translation is zero at the wall (which is assumed to be infinitely stiff compared to the frameworks) and then increases at the frameworks in proportion to their distance from the wall (Fig. 4.8/b):

$$c_i = x_i \tan\varphi, \tag{4.61}$$

where x_i is the distance of the ith framework from the wall.

The ith wall takes the load share

$$F_i = k c_i, \tag{4.62}$$

proportional to its translation, where k represents the lateral stiffness of the frameworks. Substituting the translation of the ith framework for c_i, the load share on the framework assumes the form

$$F_i = k x_i \tan\varphi. \tag{4.63}$$

In addition to stiffness k, the rotation of the building (φ) is also an unknown quantity in this formula. Its value can be determined by investigating the moment equilibrium of the internal and external forces with respect to point 'A' defining the location of the wall:

$$\sum_1^n F_i x_i = F\frac{L}{2}. \tag{4.64}$$

Substituting formula (4.63) for F_i leads to

$$\sum_1^n k x_i^2 \tan\varphi = F\frac{L}{2}, \tag{4.65}$$

from which the angle of rotation is obtained as

$$\tan\varphi = \frac{FL}{2k\sum_1^n x_i^2}. \tag{4.66}$$

Knowing the rotation, the load share on the ith framework is obtained using formula (4.63):

$$F_i = \frac{x_i}{\sum_1^n x_i^2}\frac{FL}{2}. \tag{4.67}$$

The load share on the wall is also needed. Its value is obtained using the equilibrium of the external and internal forces:

$$A = F - \sum_1^n F_i. \tag{4.68}$$

The load share on the first framework is

$$F_1 = \frac{x_1}{\sum_1^n x_i^2}\frac{FL}{2} \tag{4.69}$$

and when the frameworks are evenly spaced the load share on the subsequent frameworks is easily calculated from

$$F_i = iF_1, \tag{4.70}$$

where $i = 2, 3, \ldots n$. The formula for the load share on the wall is also simplified:

$$A = F - F_1(1 + 2 + \ldots + n). \tag{4.71}$$

It is noted here that the deformations of the building (c_i and φ) can only be calculated when the value of the stiffness of the frameworks (k) is known.

4.5 MAXIMUM BENDING MOMENTS IN THE BRACING ELEMENTS

When the load shares on the bracing elements (walls and frameworks) are known, the next step is to ensure that there is not too much load on the bracing elements. For this purpose, the maximum bending moment in the element in question is needed. The maximum bending moment in a shear wall develops at the bottom and assumes the value

$$M_{max} = \frac{pH^2}{2}, \qquad (4.72)$$

where p is the intensity of the horizontal load.

It is more difficult to find the maximum bending moment in the beams and columns of multistorey frameworks but the following simple procedure offers a quick way of doing the calculation.

The framework can be replaced by an equivalent column and, by distributing the horizontal load and the supporting effect of the beams, a continuum model can be set up (Fig. 4.9).

The bending moments on the equivalent column are expressed by

$$M = \frac{pz^2}{2} - \int_0^z \overline{m}\,dz, \qquad (4.73)$$

where

$$p = \frac{F}{h} \qquad (4.74)$$

is the horizontal load, uniformly distributed over the height,

$$\overline{m} = -Ky' \qquad (4.75)$$

is the supporting effect of the beams, distributed over the height,

$$K = 2\sum_{1}^{n-1} \frac{6EI_{b,i}}{hl_i} \qquad (4.76)$$

is the sum of the stiffnesses of the beams, distributed over the height,

E is the modulus of elasticity,
$I_{b,i}$ is the second moment of area of the beams,

l_i is the length of the beams,
h is the height of storey,
n is the number of columns.

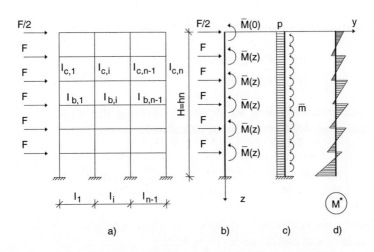

Fig. 4.9 Framework under horizontal load. a) Framework, b) equivalent column, c) continuum model, d) moments on the equivalent column.

The governing differential equation of the equivalent column [Beck, 1956; Csonka, 1965a] assumes the form

$$M'' - \frac{K}{EI_c}M = p, \quad (4.77)$$

where

$$I_c = \sum_1^n I_{c,i} \quad (4.78)$$

is the sum of the second moments of area of the columns and M is the bending moment on the column.

The boundary conditions for differential equation (4.77) express that the bending moment is zero at the top of the equivalent column and the tangent to the column is vertical at the bottom:

$$M(0) = 0, \quad (4.79a)$$

$$y'(H) = 0. \quad (4.79b)$$

The solution of equation (4.77) yields

$$M = \frac{p}{\alpha^2}\left(\frac{\alpha H - \sinh \alpha H}{\cosh \alpha H}\sinh \alpha z + \cosh \alpha z - 1\right) \quad (4.80)$$

which, after differentiating formula (4.73) once, leads to

$$\overline{m} = -\frac{p}{\alpha}\left(\frac{\alpha H - \sinh \alpha H}{\cosh \alpha H}\cosh \alpha z + \sinh \alpha z\right) + pz. \quad (4.81)$$

The concentrated bending moments representing the supporting effect of the beams (Fig. 4.9/b) can now be calculated:

$$\overline{M}(0) = \int_0^{\frac{h}{2}} \overline{m}\,dz = -\frac{p}{\alpha^2}\left(\frac{\alpha H - \sinh \alpha H}{\cosh \alpha H}\sinh \frac{\alpha h}{2} + \cosh \frac{\alpha h}{2} - 1\right) + \frac{ph^2}{8} \quad (4.82)$$

and

$$\overline{M}(z) = \int_{z-\frac{h}{2}}^{z+\frac{h}{2}} \overline{m}\,dz =$$

$$= -\frac{2p}{\alpha^2}\left(\frac{\alpha H - \sinh \alpha H}{\cosh \alpha H}\cosh \alpha z + \sinh \alpha z\right)\sinh \frac{\alpha h}{2} + phz, \quad (4.83)$$

where

$$\alpha = \sqrt{\frac{K}{EI_c}}. \quad (4.84)$$

Bending moments M^* on the equivalent column (Fig. 4.9/d) can now be calculated in the usual manner as the loads on the statically determined cantilever (horizontal forces F and concentrated supporting moments $\overline{M}$) are now known quantities.

Finally, the moments in the columns and beams of the framework are calculated. The moments in the *i*th column of the framework are obtained by distributing the moment on the equivalent column in proportion to its stiffness to the total stiffness of the columns:

$$M_i^* = \frac{I_{c,i}}{I_c} M^*. \tag{4.85}$$

The moments in the beams are calculated in the same manner:

$$\overline{M}_i = \frac{K_i}{K} \overline{M}. \tag{4.86}$$

The above method offers good approximations for medium-rise (4–25 storey high) frameworks.

It is useful to know the maximum values of the bending moments. The maximum bending moment on the equivalent column develops at the bottom. Its value can be obtained by making use of formula (4.73) and (4.81). After carrying out the necessary integration, a closed-form solution can be produced. In most practical cases, when $\alpha H > 5$ holds, the solution simplify considerably [Csonka, 1965b] and the formula for the maximum bending moment on the equivalent column assumes the form:

$$M_{max}^* = \frac{p}{\alpha^2} \left(\frac{\alpha H}{e^{\alpha h/2}} - 1 \right) + \frac{p}{2} \left(Hh - \frac{h^2}{4} \right). \tag{4.87}$$

The formula for the maximum bending moments also simplifies:

$$\overline{M}_{max} = ph \left(H - \frac{1}{\alpha}(1 + \ln \alpha H) \right). \tag{4.88}$$

Again, to obtain the actual moments in the columns and the beams, these moments should be distributed according to the relevant stiffnesses as shown in formulae (4.85) and (4.86).

The location of the maximum bending moment in the beams is obtained from

$$z_{max} = H - \frac{1}{\alpha} \ln \alpha H. \tag{4.89}$$

When the bending moments on the bracing elements are known, the last step in the design process is to calculate the stresses and to size the bracing elements. Section 5.5 deals with the calculation of the normal and shear stresses in shear walls and cores.

4.6 WORKED EXAMPLES

Three examples show how the formulae can be used in practice.

4.6.1 Example 1: building braced by parallel walls

Determine the load share F_1 on wall No. 1, the rotation and the maximum translation of the 7-storey building shown in Fig. 4.10. The magnitude of the horizontal load is $F = 1000$ kN, the modulus of elasticity is $E = 23$ kN/mm², the thickness of the walls is $t = 0.25$ m and the storey height is $h = 3$ m.

(a) Stiffness characteristics
Second moments of area of walls No. 1, 2 and 3:

$$I_1 = I_3 = \frac{0.25 \times 10.25^3}{12} = 22.435 \text{ m}^4, \quad I_2 = \frac{0.25 \times 5.125^3}{12} = 2.804 \text{ m}^4.$$

Sum of second moments of area:

$$\sum_{1}^{3} I_i = 47.675 \text{ m}^4.$$

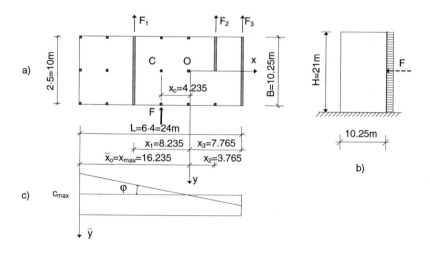

Fig. 4.10 Building for Example 1. a) Layout, b) elevation, c) displacements.

The location of the shear centre is calculated from formula (4.14a):

$$\bar{x}_o = \frac{22.435 \times 8 + 2.804 \times 20 + 22.435 \times 24}{47.675} = 16.235 \text{ m}.$$

(b) Load share
The load share on wall No. 1 is calculated using formula (4.28):

$$F_1 = -\frac{22.435}{47.675} 1000 - \frac{22.435 \times 8.235 \times 1000 \times 4.235}{22.435(8.235^2 + 7.765^2) + 2.804 \times 3.765^2} =$$

$$= -470.58 - 268.51 = -739.1 \text{ kN}.$$

The denominator of the second term in the formula represents the warping constant:

$$I_\omega = 22.435(8.235^2 + 7.765^2) + 2.804 \times 3.765^2 = 2913.9 \text{ m}^6.$$

(c) Maximum translation and rotation
The building develops maximum translation at the left-hand side (Fig. 4.10/c). There is no bracing wall there so formula (4.32) should be used:

$$c_{max} = \frac{-21^3 \times 10^3}{8 \times 23 \times 10^6} \left(\frac{1}{47.675} + \frac{4.235 \times 16.235}{2913.9} \right) =$$

$$= -(1.05 + 1.19)10^{-3} \text{m} = -2.24 \text{ mm}.$$

The rotation of the building is given by formula (4.33):

$$\varphi = \frac{21^3 \times 10^3 \times 4.235}{8 \times 23 \times 10^6 \times 2913.9} = 7.315 \times 10^{-5} \text{rad} = 0.0042°.$$

4.6.2 Example 2: building braced by perpendicular walls

Calculate the load share on walls No. 1 and 5, the rotation and the maximum translation of the 7-storey building shown in Fig. 4.11. The resultant of the horizontal load is F_y = 1000 kN, the modulus of elasticity

92 Elementary stress analysis

is $E = 23$ kN/mm^2, the thickness of the walls is $t = 0.25$ m and the storey height is $h = 3$ m.

This building only differs from the one shown in Fig. 4.10 in that two bracing walls in direction x (Nos 4 and 5) are added to the bracing system. This makes it possible to make an interesting comparison.

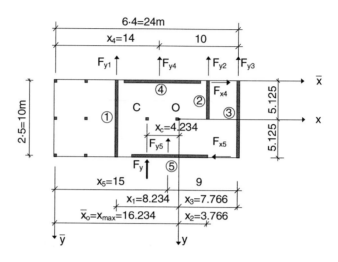

Fig. 4.11 Building for Example 2.

(a) Geometrical and stiffness characteristics
Second moments of area of walls No. 1 and 3:

$$I_{x,1} = I_{x,3} = \frac{0.25 \times 10.25^3}{12} = 22.435 \text{ m}^4,$$

$$I_{y,1} = I_{y,3} = \frac{10.25 \times 0.25^3}{12} = 0.0133 \text{ m}^4.$$

Second moments of area of wall No. 2:

$$I_{x,2} = \frac{0.25 \times 5.125^3}{12} = 2.804 \text{ m}^4, \qquad I_{y,2} = \frac{5.125 \times 0.25^3}{12} = 0.0067 \text{ m}^4.$$

Second moments of area of walls No. 4 and 5:

$$I_{x,4} = I_{x,5} = \frac{10 \times 0.25^3}{12} = 0.013 \text{ m}^4, \quad I_{y,4} = I_{y,5} = \frac{0.25 \times 10^3}{12} = 20.833 \text{ m}^4.$$

Sum of second moments of area:

$$\sum_1^5 I_{x,i} = 47.701 \text{ m}^4, \qquad \sum_1^5 I_{y,i} = 41.700 \text{ m}^4.$$

The coordinates of the shear centre are calculated using formulae (4.34):

$$\bar{x}_o = \frac{22.435(8+24) + 2.804 \times 20 + 0.013(14+15)}{47.701} = 16.234 \text{ m}$$

and

$$\bar{y}_o = \frac{2 \times 0.013 \times 5.125 + 0.0067 \times 2.5625 + 20.833(10.25 + 0.125)}{41.700} = 5.125 \text{ m}.$$

The warping constant is calculated using formula (4.53):

$$I_\omega = 22.435(8.234^2 + 7.766^2) + 2.804 \times 3.766^2 + 0.0133(2.234^2 + 1.234^2) +$$
$$+ 0.0067 \times 2.5625^2 + 20.833(5^2 + 5^2) = 2914.0 + 1041.7 = 3955.7 \text{m}^6.$$

(b) Load share

Loads on walls No. 1 and 5 in the x and y directions are calculated using formulae (4.51) and (4.52):

$$F_{y,1} = \frac{-22.435}{47.701} 1000 - \frac{22.435 \times 8.234}{3955.7} 1000 \times 4.234 = -470 - 198 = -668 \text{kN},$$

$$F_{x,1} = 0,$$

$$F_{y,5} = \frac{-0.013}{47.701} 1000 - \frac{0.013 \times 1.234}{3955.7} 1000 \times 4.234 = -0.27 - 0.02 = -0.29 \text{kN},$$

$$F_{x,5} = -\frac{20.833 \times 5}{3955.7} 1000 \times 4.234 = -111.5 \text{kN}.$$

94 Elementary stress analysis

(c) Deformations
The building develops maximum translation at the left-hand side. Its value is calculated from formula (4.57):

$$c_{y,max} = \frac{-21^3 \times 10^3}{8 \times 23 \times 10^6}\left(\frac{1}{47.701} + \frac{4.234 \times 16.234}{3955.7}\right) =$$

$$= -(1.05 + 0.88)10^{-3}\,m = -1.93\,mm.$$

The rotation of the building is calculated from formula (4.59):

$$\varphi = \frac{21^3 \times 10^3 \times 4.234}{8 \times 23 \times 10^6 \times 3955.7} = 5.39 \times 10^{-5}\,rad = 0.0031°.$$

4.6.3 Comparison

The comparison of the two numerical examples shows interesting results. Adding the two walls with negligible stiffness in the direction of the external load to the bracing system considerably changes the results: the magnitude of force $F_{y,1}$, the maximum translation and the rotation decreased by 10.1%, 14.9% and 28.9%, respectively. This follows from the fact that torsion plays a very important role in the behaviour of the bracing system; the longer the distance between the shear centre and the centroid of the layout, the more important this role is. Although walls No. 4 and 5 have negligible stiffnesses in the direction of the external load, their contribution to the warping constant is significant, as demonstrated in the examples above. It also follows that, because of the rotation of the building, walls No. 4 and 5 also take forces in the x direction.

4.6.4 Example 3: building braced by frameworks and a single wall

Calculate the load share on the frameworks and the wall of the 7-storey building shown in Fig. 4.12. The magnitude of the horizontal load is $F = 1000$ kN. The frameworks are evenly spaced so formulae (4.69) to (4.71) can be used for the calculation.

The load on framework No.1 is obtained from formula (4.69):

$$F_1 = \frac{4}{1456} \frac{1000 \times 24}{2} = 32.97 \text{ kN},$$

where the denominator is calculated as

$$\sum_{1}^{6} x_i^2 = 4^2 + 8^2 + 12^2 + 16^2 + 20^2 + 24^2 = 1456 \text{m}^2.$$

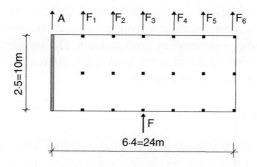

Fig. 4.12 Building for Example 3.

The load share on the subsequent frameworks is easily calculated from formula (4.70):

$$F_2 = 2 \times 32.97 = 65.93 \text{ kN},$$

$$F_3 = 3 \times 32.97 = 98.90 \text{ kN},$$

$$F_4 = 4 \times 32.97 = 131.87 \text{ kN},$$

$$F_5 = 5 \times 32.97 = 164.84 \text{ kN},$$

$$F_6 = 6 \times 32.97 = 197.80 \text{ kN}.$$

The load share on the wall is obtained from formula (4.71):

$$A = 1000 - 32.97(1 + 2 + 3 + 4 + 5 + 6) = 307.69 \text{ kN}.$$

96 *Elementary stress analysis*

4.7 DISCUSSION

The behaviour of the building under horizontal load is determined by its translations and rotation. This fact is demonstrated, for example, by the first and second terms on the right-hand side of formulae (4.28) and (4.51) as well as formulae (4.32) and (4.57). The derivation of these formulae clearly shows the role that the translations and the rotation of the building play in determining the performance of the building.

When the external load passes through the shear centre, the building only develops translations (in the direction of the external load). This is the optimum case, both statically and economically, when the bracing system has a doubly symmetrical arrangement. The building develops the smallest translations and the external load is distributed among the bracing elements according to their stiffnesses.

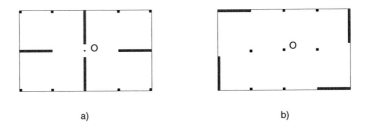

Fig. 4.13 Torsional resistance. a) Small, b) great.

Because of functional restrictions, for example, the optimum (doubly symmetrical) arrangement cannot be achieved in many practical cases. The geometrical centre – through which the external load passes – and the shear centre of the bracing system normally do not coincide and consequently the building also develops rotation. This rotation results in additional translations and also additional forces on the bracing elements. As these additional translations and loads are due to the rotation of the building, the means to reduce these unfavourable effects are also linked to the rotation. The evaluation of the second term (which is associated with torsion) in equations (4.28), (4.51), (4.32) and (4.57) indicates two possibilities for improving the performance of the building: reducing the perpendicular distance of the line of action of the horizontal load from the shear centre and increasing the value of the warping constant. First, by reducing the perpendicular distance between the line of action of the horizontal load and the shear centre, the magnitude of the external torsional moment can be reduced. The optimum case emerges when the distance is eliminated altogether, i.e. when the system is doubly symmetrical – see above. Second, the efficiency of the system can be increased by increasing the magnitude of the warping constant. This is most effectively done by creating an arrangement of the bracing elements when the perpendicular distance between the shear centre and the bracing elements (their 'torsion arm') is the greatest.

Figure 4.13 shows simple examples for this approach. The torsional resistance of the bracing system in Fig. 4.13/a is very small. On the contrary, the torsional resistance of the bracing system in Fig. 4.13/b is considerably greater although the system consists of the same four walls. The favourable change is due to the fact that the perpendicular distances between the walls and the shear centre are increased.

The spatial stiffness of bracing systems comprising perpendicular walls can be increased by constructing some of the walls together along their vertical edges (Fig. 4.14/a). However, the increase in stiffness can only be realized if the construction is such that the walls can actually work together, i.e. they can take the shear forces along the edges. In addition to the second moments of area I_x and I_y and the Saint-Venant constant I_t, neglected for the analysis is this chapter, such systems also normally have the product of inertia I_{xy}.

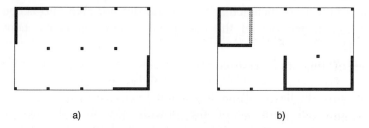

Fig. 4.14 Increasing 3-dimensional stiffness. a) Built-up walls, b) cores.

The spatial stiffness can be further increased by building more walls together and using bracing elements of closed or partially closed cross-section, e.g. the elevator and staircase shaft in the left-hand corner in Fig. 4.14/b. Such elements normally have significant Saint-Venant stiffness and they also have warping stiffness. The procedure presented here cannot be applied directly to such systems but similar, albeit more complex, formulae and design diagrams are presented in the next chapter for a more advanced analysis. In fact, most of the formulae given in this chapter can be considered as simplified versions of those general formulae in Chapter 5.

5

Stress analysis: an advanced approach

When the building is subjected to a horizontal load of trapezoidal distribution and/or the bracing elements have (Saint-Venant and warping) torsional stiffnesses, simple static considerations alone are no longer sufficient to investigate the 3-dimensional behaviour and the methods given in the previous chapter may not be applicable or would result in inaccurate solutions. The governing differential equations of the bracing system are needed to be set up and solved to enable the spatial analysis to be carried out.

In appraising the design of box-frame structures, Pearce and Matthews [1971] presented a simple approximate method for the load distribution in shear wall structures. The method was only applicable to cross-wall systems (with no perpendicular walls) and did not take into account the torsional characteristics of the individual bracing elements. Dowrick's method [1976] handled a system of perpendicular shear walls but ignored the Saint-Venant and warping torsional characteristics of the individual bracing elements. Irwin [1984] included the warping stiffness of the bracing cores in the analysis but did not take into account the Saint-Venant torsional stiffness of the bracing system. All the methods mentioned above assumed a uniformly distributed horizontal load.

The method to be presented in the following can be considered as an extension of the above methods, with additional contributions in three areas: a) horizontal load of trapezoidal distribution is considered, b) exact load and stress distributions are given on the individual bracing elements, originated from torsion, c) the location of the maximum of the Saint-Venant torsional moments is given.

5.1 THE EQUIVALENT COLUMN AND ITS LOAD

The building is replaced by the equivalent column for the first part of the analysis. The stiffness characteristics are calculated using the relevant formulae in section 2.2. The load on the equivalent column is also needed for the analysis. The building is subjected to a horizontal load of trapezoidal distribution with the intensity of

$$q(z) = q_0 + q_1 \frac{z}{H} = q_0\left(1 + \mu \frac{z}{H}\right), \tag{5.1}$$

where q_0 is the intensity of the uniform part of the load, $(q_0 + q_1)$ is the intensity of the load at the top of the structure (Fig. 5.1) and H is the height of the building.

The slope of the load function is defined by coefficient μ:

$$\mu = \frac{q_1}{q_0}. \tag{5.2}$$

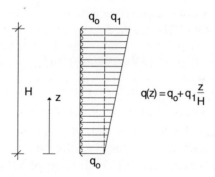

Fig. 5.1 Horizontal load of trapezoidal distribution.

A load system of such distribution covering a wide range of distributions between the uniform and triangular cases can be effectively used for the global static and dynamic analyses of building structures subjected to horizontal (wind or seismic) load.

The resultant of the horizontal load normally does not pass through the shear centre of the bracing system of the building. When the bracing system is replaced by the equivalent column for the analysis, the horizontal load is transferred to the shear centre axis, where it is decomposed into load components q_x and q_y parallel to coordinate axes x

and y (Fig. 5.2). (For buildings with a rectangular layout, coordinate axes x and y are conveniently set parallel with the sides of the layout.) The external load is defined by the two load components (q_x and q_y) and the torsional moment (m_z):

$$q_x(z) = q_{0x} + q_{1x}\frac{z}{H} = q_{0x}\left(1+\mu\frac{z}{H}\right), \qquad (5.3)$$

$$q_y(z) = q_{0y} + q_{1y}\frac{z}{H} = q_{0y}\left(1+\mu\frac{z}{H}\right) \qquad (5.4)$$

and

$$m_z(z) = eq(z) = eq_0\left(1+\mu\frac{z}{H}\right) = m_{z0}\left(1+\mu\frac{z}{H}\right) \qquad (5.5)$$

or, using the load components

$$m_z(z) = q_x y_c + q_y x_c = (q_{0x}y_c + q_{0y}x_c)\left(1+\mu\frac{z}{H}\right). \qquad (5.6)$$

In formulae (5.5) and (5.6)

$$m_{z0}(z) = eq_0 = q_{0x}y_c + q_{0y}x_c \qquad (5.7)$$

is the torsional moment of the uniform part of the load and

$$x_c = \frac{L}{2} - \bar{x}_o, \qquad y_c = \frac{B}{2} - \bar{y}_o \qquad (5.8)$$

are the coordinates of the centroid of the building layout in the coordinate system whose origin is in the shear centre. In the above formulae

- e is the perpendicular distance between the line of action of the horizontal load and the shear centre,
- q_{0x}, q_{0y} are the components of the uniform part of the load,
- q_{1x}, q_{1y} are the components of the triangular part of the load (Fig. 5.1).

Load components q_x and q_y are considered positive in the positive x and y directions and m_z is positive in the clockwise direction. The load function defined by formula (5.1) represents a uniformly distributed load when $\mu = 0$ holds. This special case is dealt with in section 5.7.

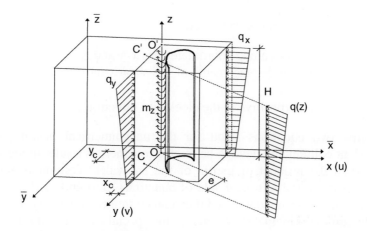

Fig. 5.2 The equivalent column and its external load.

Another load case, concentrated top load, applicable to single storey buildings and to multistorey building models when the load is only applied on top floor level, can also be treated in a similar way as the trapezoidal distribution. Closed-form solutions for this load case are given in section 5.6.

After modelling the building by the equivalent column, the structure can be conveniently analysed by the continuum method, with the stiffnesses of the structure being uniformly distributed over the height. The model and the method are simple and at the same time give a realistic picture of the behaviour of the structure. The resulting closed-form formulae for the global deformations, load distribution among the bracing elements, shear forces, bending and torsional moments are accurate enough for everyday use in design offices.

The governing differential equations defining the unsymmetrical bending and torsion of the equivalent column assume the following form in the left-handed coordinate system x-y-z (Fig. 5.2) [Vlasov, 1940]:

$$EI_y \frac{d^4u}{dz^4} + EI_{xy} \frac{d^4v}{dz^4} = q_x(z), \qquad (5.9)$$

$$EI_x \frac{d^4v}{dz^4} + EI_{xy} \frac{d^4u}{dz^4} = q_y(z), \qquad (5.10)$$

$$EI_\omega \frac{d^4\varphi}{dz^4} - GJ\frac{d^2\varphi}{dz^2} = m_z(z), \tag{5.11}$$

where the following notation is used:

- x, y, z coordinates along coordinate axes x, y and z,
- u, v horizontal displacements of the shear centre in directions x and y,
- φ rotation (positive in the clockwise direction).

The first two equations stand for the unsymmetrical bending of the equivalent column and have to be treated together, as a rule. In the special case when the product of inertia of the equivalent column is zero, the second term on the left-hand side of equations (5.9) and (5.10) vanishes. The two equations uncouple and they can be treated separately. Directions x and y are the principal directions in this special case. The third equation characterizes the torsion of the equivalent column.

Apart from some special cases, building structures develop a combination of lateral displacements and torsion. The nature of the behaviour (and the extent of the combination) depend on the relative position of the shear centre of the bracing system O and the line of action of the horizontal load (which passes through the centroid of the layout C).

Boundary conditions complete the simultaneous differential equations. Both equations (5.9) and (5.10) characterize bending equilibrium so the boundary conditions are of the same nature. The third equation defines torsional equilibrium and it should be taken into consideration when the corresponding boundary conditions are established that warping is completely prevented at the built-in end and can freely develop at the free upper end. The coordinate system is fixed at the bottom of the cantilever (Fig. 5.2) and the boundary conditions are given as follows.

The displacements and rotation is zero at the built-in end of the cantilever:

$$u(0) = v(0) = \varphi(0) = 0. \tag{5.12}$$

The slope of the deflection curve is equal to zero and no warping develops at the built-in end:

$$u'(0) = v'(0) = \varphi'(0) = 0. \tag{5.13}$$

The bending moments and the warping stresses are zero at the free end of the cantilever:

$$u''(H) = v''(H) = \varphi''(H) = 0. \tag{5.14}$$

The shear forces and the total torque also vanish at the free end:

$$u'''(H) = v'''(H) = EI_\omega \varphi'''(H) - GJ\varphi'(H) = 0. \tag{5.15}$$

5.2 DEFORMATIONS OF THE EQUIVALENT COLUMN

Under horizontal load, the equivalent column normally develops both lateral displacements and rotation. They can be determined using the governing differential equations (5.9) to (5.11) and the corresponding boundary conditions (5.12) to (5.15).

5.2.1 Horizontal displacements

In the general case, equations (5.9) and (5.10) of unsymmetrical bending are coupled and have to be treated together as both contain the horizontal displacements u and v. After combining the equations, integrating four times and using the relevant boundary conditions, the horizontal displacements in directions x and y are obtained as:

$$u(z) = \frac{\overline{q}_x}{E}(Z_1 + \mu Z_2) \tag{5.16}$$

and

$$v(z) = \frac{\overline{q}_y}{E}(Z_1 + \mu Z_2), \tag{5.17}$$

where

$$Z_1 = \frac{z^4}{24} - \frac{Hz^3}{6} + \frac{H^2 z^2}{4}, \qquad Z_2 = \frac{z^5}{120H} - \frac{Hz^3}{12} + \frac{H^2 z^2}{6} \tag{5.18}$$

are auxiliary functions and

$$\bar{q}_x = \frac{I_x q_{0x} - I_{xy} q_{0y}}{I_x I_y - I_{xy}^2}, \qquad \bar{q}_y = \frac{I_y q_{0y} - I_{xy} q_{0x}}{I_x I_y - I_{xy}^2} \qquad (5.19)$$

are auxiliary load functions.

Figure 5.3 shows the characteristic horizontal displacements for $\mu = 0$ (uniformly distributed load), $\mu = 0.2$, $\mu = 0.4$, $\mu = 0.6$, $\mu = 0.8$, $\mu = 1.0$ and $\mu = 2$.

Maximum displacements develop at the top and are obtained by substituting H for z in formulae (5.16) to (5.18):

$$u_{max} = u(H) = \frac{\bar{q}_x}{E}\left(\frac{1}{8} + \mu\frac{11}{120}\right)H^4 \qquad (5.20)$$

and

$$v_{max} = v(H) = \frac{\bar{q}_y}{E}\left(\frac{1}{8} + \mu\frac{11}{120}\right)H^4. \qquad (5.21)$$

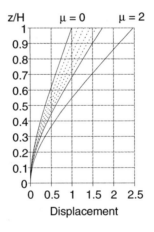

Fig. 5.3 Horizontal displacements of the equivalent column.

5.2.2 Rotations

The fourth order, linear, inhomogeneous differential equation (5.11) characterizes the rotation of the equivalent column. Its general solution is obtained in the usual way in the form

$$\varphi(z) = C_1 + C_2 z + C_3 \sinh\frac{k}{H}z + C_4 \cosh\frac{k}{H}z - \frac{m_{z0}}{6EI_\omega}(3z^2 + \frac{\mu}{H}z^3)\frac{H^2}{k^2}, \quad (5.22)$$

where C_1, C_2, C_3 and C_4 are constants of integration and k is the well-known torsion parameter defined by formula (3.19).

The first four terms in equation (5.22) represent the solution of the homogeneous part of equation (5.11) and the last term is a particular solution of the corresponding inhomogeneous equation. Constants of integration C_1, C_2, C_3 and C_4 are obtained using the relevant parts of boundary conditions (5.12) to (5.15):

$$C_1 = -\frac{m_{z0}H^2}{kGJ\cosh k}\left(\frac{1+\mu}{k} + (1+\frac{\mu}{2}-\frac{\mu}{k^2})\sinh k\right), \quad (5.23)$$

$$C_2 = \frac{m_{z0}H}{GJ}\left(1+\frac{\mu}{2}-\frac{\mu}{k^2}\right), \quad (5.24)$$

$$C_3 = -\frac{m_{z0}H^2}{kGJ}\left(1+\frac{\mu}{2}-\frac{\mu}{k^2}\right), \quad (5.25)$$

$$C_4 = \frac{m_{z0}H^2}{kGJ\cosh k}\left(\frac{1+\mu}{k} + (1+\frac{\mu}{2}-\frac{\mu}{k^2})\sinh k\right). \quad (5.26)$$

After substituting for the constants of integration in equation (5.22), the rotation of the equivalent column is obtained as:

$$\varphi(z) = \frac{m_{z0}H^2}{k^2 GJ\cosh k}\left\{(1+\mu)\left(\cosh\frac{kz}{H}-1\right) + (1+\frac{\mu}{2}-\frac{\mu}{k^2})k \times \right.$$
$$\left. \times \left(\sinh(k-\frac{kz}{H}) - \sinh k\right) + \frac{zk^2}{H^2}\cosh k\left(H - \frac{z}{2} + \mu(\frac{H}{2} - \frac{H}{k^2} - \frac{z^2}{6H})\right)\right\}. \quad (5.27)$$

This solution clearly demonstrates that no rotation develops when the horizontal load passes through the shear centre (i.e. when $m_z = 0$).

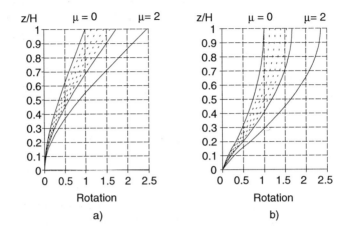

Fig. 5.4 Rotations of the equivalent column. a) $k < 0.1$, b) $k > 30$.

Figure 5.4 shows the characteristic rotations for the ranges $k < 0.1$ and $k > 30$ with $\mu = 0$ (uniformly distributed load), $\mu = 0.2$, $\mu = 0.4$, $\mu = 0.6$, $\mu = 0.8$, $\mu = 1.0$ and $\mu = 2$.

The case $k < 0.1$ represents torsion when the effect of the warping stiffness is dominant over that of the Saint-Venant stiffness and $k > 30$ shows the torsional deformations when torsion is dominated by the Saint-Venant torsional stiffness. It is interesting to note that the nature of the displacement curve (Fig. 5.3) and that of the rotation curve for small values of k (Fig. 5.4/a) are similar. This follows from the fact that both phenomena are associated with the bending of the elements of the bracing system.

Rotation assumes maximum value at the top at $z = H$:

$$\varphi_{max} = \frac{m_{z0}H^2}{GJ}\left(\frac{(1+\mu)(\cosh k - 1)}{k^2 \cosh k} - (1 + \frac{\mu}{2} - \frac{\mu}{k^2})\frac{\tanh k}{k} + \frac{1}{2} + \frac{\mu}{3} - \frac{\mu}{k^2}\right). \tag{5.28}$$

Two special cases are worth considering here: columns with Saint-Venant torsional stiffness only (and with no warping stiffness) and columns whose Saint-Venant torsional stiffness is negligibly small compared to their warping stiffness.

The denominator in formula (3.19) of k vanishes when the equivalent column has no warping stiffness and the formulae derived for the rotations cannot be used. This situation may emerge when a building is braced by a single column whose warping stiffness is zero. Figure 5.5 shows cross-sections with no or negligible warping stiffness.

Fig. 5.5 Cross-sections with no or negligible warping stiffness.

In such cases, the original differential equation of torsion (5.11) has to be solved again, but this time with $I_\omega = 0$. After making use of the relevant part of boundary condition (5.13), the rotation of the equivalent column emerges as

$$\varphi(z) = \frac{m_{z0} z}{GJ}\left(H - \frac{z}{2} + \mu(\frac{H}{2} - \frac{z^2}{6H}) \right) \tag{5.29}$$

and the maximum rotation at $z = H$ is calculated from

$$\varphi_{max} = \varphi(H) = \frac{m_{z0} H^2}{2GJ}\left(1 + \frac{2\mu}{3}\right). \tag{5.30}$$

When the contribution of the warping stiffness to the overall torsional resistance is much greater than that of the Saint-Venant torsional stiffness, the latter can be safely neglected as a conservative approximation [Vlasov, 1940]. In such cases, instead of formula (5.27), the much simpler formula

$$\varphi(z) = \frac{m_{z0}}{EI_\omega}(Z_1 + \mu Z_2) \tag{5.31}$$

can be used where Z_1 and Z_2 are defined by formulae (5.18). The maximum rotation in such cases is obtained from

$$\varphi_{max} = \varphi(H) = \frac{m_{z0}}{EI_\omega}\left(\frac{1}{8} + \mu \frac{11}{120}\right) H^4. \tag{5.32}$$

108 Advanced stress analysis

The similarity indicated by the comparison of the horizontal displacement curve (Fig. 5.3) and the rotation curve for $k < 0.1$ (Fig. 5.4/a) now becomes identity: the corresponding curves and the structures of the corresponding formulae are identical. This is clearly shown by comparing formulae (5.16) and (5.31) as well as formulae (5.20) and (5.32).

The rotations obtained from the above formulae are given in radian; they can be more conveniently used in degrees in practical applications. The conversion is given by

$$\varphi \text{ [degree]} = \varphi \text{ [radian]} \frac{180°}{\pi} \approx 57.3\varphi \text{ [radian]}. \tag{5.33}$$

5.3 DEFORMATIONS OF THE BUILDING

When the deformations of the building are needed, only the formulae related to the rotations can be used directly. As for the calculation of the displacements of the building, one has to remember that formulae (5.16), (5.17) and (5.20), (5.21) give the displacements of the *equivalent column*, i.e. the displacements of the shear centre axis of the building. If the building also develops rotations, then the additional displacements, due to the rotation around the shear centre, also have to be taken into account. The horizontal displacements of the corner points of the building are of practical importance as the building develops the maximum displacement at one of them. In making use of the deformations of the equivalent column and the size of the layout, these can be easily calculated. When the location of the shear centre and that of the resultant of the horizontal load are known, common sense usually shows which corner points of the building develop the greatest displacement (Fig. 5.6):

$$v_{\text{building}} = v_A = v + x_A \tan\varphi \quad \text{or} \quad u_{\text{building}} = u_B = u + y_B \tan\varphi. \tag{5.34}$$

In the above formula u and v are the displacements of the shear centre (i.e. the equivalent column), x_A and y_B are the coordinates of corner points A and B and φ is the rotation. If it is not obvious which corner point develops the maximum displacement, the calculation has to be repeated for more (maybe for all four) corner points.

An alternative method is also available for the determination of the maximum displacements when a bracing element is situated (at the corner point) where the maximum displacement is expected to develop. The first step in this case is the determination of the load acting on the bracing element in question by using the procedure to be given later. Having calculated load shares $q_{x,k}$ and $q_{y,k}$ (assuming that it is the kth bracing element where the maximum displacement develops), formulae (5.20) and (5.21) directly yield the maximum displacements, where the moments of inertia to be used are those of the kth element.

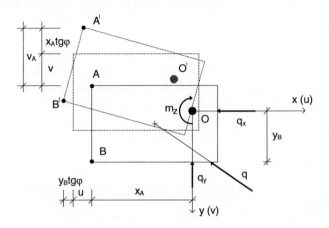

Fig. 5.6 Maximum displacements at corner points A and B.

Maximum displacements are restricted in multistorey building structures in order to ensure the comfort of the occupants and to avoid damage to structural and non-structural elements and mechanical systems. In certain cases, it is also advantageous to limit second-order effects in this way. The commonly accepted range for maximum horizontal displacements is from 0.0016 to 0.0035 times the height of the building, depending on the building height and the magnitude of the wind pressure. In accordance with the recommendation of the Committee on Wind Bracing of the American Society of Civil Engineers, for example, the bracing system is normally considered adequate if the conditions

$$u_{max} \le \frac{H}{500} \quad \text{and} \quad v_{max} \le \frac{H}{500} \tag{5.35}$$

are fulfilled, where H is the height of the building [Schueller, 1977].

5.4 LOAD DISTRIBUTION AMONG THE BRACING ELEMENTS

The bracing elements can only fulfil their primary task to provide the building with adequate lateral and torsional stiffnesses if they safely transmit the external horizontal load to the foundation. The stresses in the bracing elements must not exceed their limit values. The magnitude of the stresses in the individual bracing elements basically depends on how much the element in question takes of the total external load. The establishment of the load distribution is therefore an important part of the design procedure if the individual bracing elements are to be sized properly.

The equivalent column in the shear centre of the bracing system, subjected to q_x, q_y and m_z, is only a fictitious column and in reality its load is transmitted to the bracing elements by the floor slabs by making use of their in-plane stiffness. Due to the rigid-body translation and rotation of the floor slabs, shear forces are transmitted to each bracing element. In addition to these shear forces that develop bending in the bracing elements, the rotation of the floor slabs also develops torsion in the bracing elements. These two effects are investigated separately in the next two sections.

5.4.1 Shear forces and bending moments

In addition to horizontal displacements u and v, identical to those of the shear centre, the ith bracing element also develops additional horizontal displacements φy_i and φx_i (Fig. 5.7), due to the rotation around the shear centre of the bracing system.

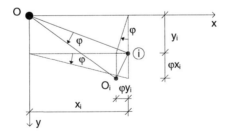

Fig. 5.7 Additional horizontal displacement of the ith bracing element.

Thus the total horizontal displacements of the ith element in directions x and y are:

$$v_i = v + x_i \varphi \quad \text{and} \quad u_i = u - y_i \varphi, \tag{5.36}$$

where coordinates x_i and y_i are the coordinates of the shear centre of the ith bracing element in the coordinate system whose origin is in the shear centre of the bracing system.

In making use of u_i and v_i and applying equations (5.9) and (5.10) to the ith bracing element, and after four differentiations and some rearrangement, the horizontal load intensities on the ith element are obtained as

$$q_{x,i} = (I_{y,i}\bar{q}_x + I_{xy,i}\bar{q}_y)(1 + \mu \frac{z}{H}) - \frac{m_{z0}}{I_\omega}(I_{y,i}y_i - I_{xy,i}x_i)\eta_q, \tag{5.37}$$

$$q_{y,i} = (I_{x,i}\bar{q}_y + I_{xy,i}\bar{q}_x)(1 + \mu \frac{z}{H}) + \frac{m_{z0}}{I_\omega}(I_{x,i}x_i - I_{xy,i}y_i)\eta_q, \tag{5.38}$$

where

$$\eta_q = \frac{1}{\cosh k}\left(\cosh\frac{kz}{H}(1+\mu) + k\sinh(k-\frac{kz}{H})(1 + \frac{\mu}{2} - \frac{\mu}{k^2})\right) \tag{5.39}$$

is the load factor.

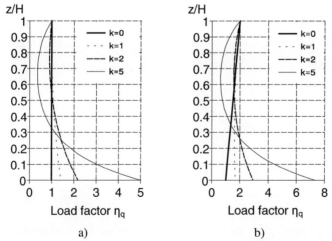

Fig. 5.8 Load factor η_q: Load on the ith bracing element, due to rotation. a) $\mu = 0$, b) $\mu = 1$.

112 Advanced stress analysis

The first term in formulae (5.37) and (5.38) represents the external load share on the ith bracing element, due to the bending of the bracing system in directions x and y. The intensity of this load is of trapezoidal distribution, just like that of the external load. The second term in the formulae stands for the load share that is due to the rotation of the bracing system around the shear centre. Its distribution is characterized by the distribution of load factor η_q, shown in Fig. 5.8, as a function of parameter k, for $k = 0$, $k = 1$, $k = 2$ and $k = 5$, with $\mu = 0$ and $\mu = 1$. In the special case of $k = 0$ (no Saint-Venant stiffness) and $\mu = 0$ (uniformly distributed load) formulae (5.37) and (5.38) simplify and become identical to those given by König and Liphardt [1990] in Beton-Kalender.

The diagrams in Fig. 5.8 show that as the value of parameter k increases, i.e. as the effect of the Saint-Venant stiffness becomes greater, the nature of the distribution of the load on the ith bracing element, due to rotation, changes. The linear distribution for $k < 0.1$ representing dominant warping stiffness becomes a variable distribution characterized by a curve for $k > 0.1$ representing more contribution from the Saint-Venant stiffness. The variation is such that the bracing element is subjected to more load in the lower region and less in the upper region and the 'centroid' of the load moves downwards over the height.

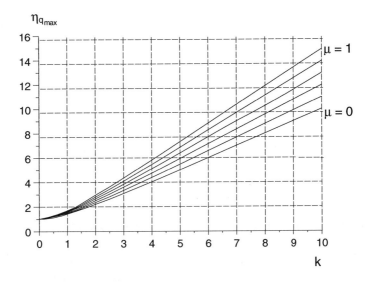

Fig. 5.9 Load factor $\eta_{q_{max}}$ for maximum load at bottom, due to rotation.

In parallel with this downward shift of the centroid of the load, the intensity of the load increases at ground floor level where it assumes local maximum. The value of this local maximum is given by

$$\eta_{q_{max}} = \eta_q(0) = \frac{1}{\cosh k}\left(1+\mu+(1+\frac{\mu}{2}-\frac{\mu}{k^2})k\sinh k\right). \qquad (5.40)$$

Factor η_{qmax} is shown in Fig. 5.9 as a function of parameter k and μ.

The shear forces in the bracing elements are obtained in the usual way by integrating the formulae for the external load given by equations (5.37) and (5.38), taking into consideration the corresponding boundary conditions defined by formulae (5.15):

$$T_{x,i} = \left(H-z-\mu(\frac{z^2}{2H}-\frac{H}{2})\right)(I_{y,i}\bar{q}_x+I_{xy,i}\bar{q}_y)- \\ -\frac{I_{y,i}y_i-I_{xy,i}x_i}{I_\omega}m_{z0}H\eta_T, \qquad (5.41)$$

$$T_{y,i} = \left(H-z-\mu(\frac{z^2}{2H}-\frac{H}{2})\right)(I_{x,i}\bar{q}_y+I_{xy,i}\bar{q}_x)+ \\ +\frac{I_{x,i}x_i-I_{xy,i}y_i}{I_\omega}m_{z0}H\eta_T, \qquad (5.42)$$

where

$$\eta_T = \frac{1}{k\cosh k}\left\{(1+\frac{\mu}{2}-\frac{\mu}{k^2})\left(\cosh(k-\frac{kz}{H})-1\right)k- \\ -(1+\mu)(\sinh\frac{kz}{H}-\sinh k)\right\} \qquad (5.43)$$

is the shear force factor representing the effect of rotation around the shear centre.

The first term in formulae (5.41) and (5.42) stands for shear forces due to the horizontal displacement of the shear centre axis of the bracing system. As with the standard case of statically determinate cantilevers, they only depend on the height, the intensity of the horizontal load and the bending stiffness of the bracing element (and of the whole system), and

114 Advanced stress analysis

have a linear distribution for $\mu = 0$. The second term represents the effect of the rotation around the shear centre, with η_T being the shear force factor characterizing the distribution of this part of the shear force. When parameter k assumes zero and $\mu = 0$ holds, the distribution of the shear forces due to rotation is also linear. As the value of k – and the effect of the Saint-Venant stiffness – increases, the shape of η_T changes: the value of the shear forces in the upper region of the column decreases. The shear force factor representing the distribution of the shear force is shown in Fig. 5.10, as a function of k and μ.

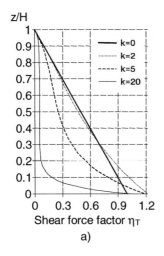
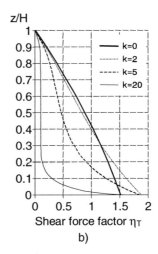

Fig. 5.10 Shear force factor η_T: shear forces on the ith bracing element, due to rotation. a) $\mu = 0$, b) $\mu = 1$.

The maximum shear forces develop at the bottom of the bracing elements:

$$T_{x,i}^{max} = T_{x,i}(0) = H(1+\frac{\mu}{2})(I_{y,i}\bar{q}_x + I_{xy,i}\bar{q}_y) - \frac{I_{y,i}y_i - I_{xy,i}x_i}{I_\omega} m_{z0} H \eta_{T_{max}}, \quad (5.44)$$

$$T_{y,i}^{max} = T_{y,i}(0) = H(1+\frac{\mu}{2})(I_{x,i}\bar{q}_y + I_{xy,i}\bar{q}_x) + \frac{I_{x,i}x_i - I_{xy,i}y_i}{I_\omega} m_{z0} H \eta_{T_{max}}. \quad (5.45)$$

In formulae (5.44) and (5.45)

$$\eta_{T_{max}}(k,\mu) = (1+\frac{\mu}{2}-\frac{\mu}{k^2})(1-\frac{1}{\cosh k})+(1+\mu)\frac{\tanh k}{k} \qquad (5.46)$$

is the maximum shear force factor whose values are given in Fig. 5.11 and in Table 5.1, as a function of k and μ.

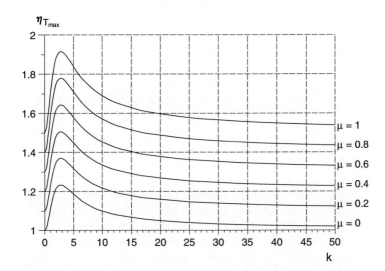

Fig. 5.11 Shear force factor η_{Tmax} for maximum shear force, due to rotation.

The distribution of the shear forces over the height is drastically affected by the value of parameter k, which is in accordance with the change in the distribution of the load shown in Fig. 5.8. The maximum value of the shear force is also affected considerably: the increase in the value of the maximum shear force is 23.3% at $k = 2.8$, compared to the case $k = 0$ and $\mu = 0$.

Figure 5.11 and Table 5.1 show that the value of the maximum shear force factor is always greater than 1.0 demonstrating that the value of the shear forces, due to rotation, always exceeds the one obtained in the 'standard' case, i.e. when the load distribution is defined by a straight line. It is in the region $0 < k < 5$ where the deviation from $\eta_T = 1.0$ is the greatest: the very region where most practical cases fall.

Table 5.1 Maximum shear force factor $\eta_{T\max}$

k	$\eta_{T\max}$						
	$\mu=0.0$	$\mu=0.2$	$\mu=0.4$	$\mu=0.6$	$\mu=0.8$	$\mu=1.0$	$\mu=2.0$
0.0	1.000	1.100	1.200	1.300	1.400	1.500	2.000
0.2	1.007	1.108	1.209	1.309	1.410	1.511	2.016
0.4	1.025	1.129	1.232	1.336	1.440	1.544	2.062
0.6	1.052	1.159	1.267	1.375	1.483	1.590	2.129
0.8	1.082	1.195	1.307	1.420	1.532	1.644	2.206
1.0	1.114	1.231	1.348	1.465	1.582	1.699	2.285
1.2	1.142	1.264	1.385	1.507	1.629	1.750	2.358
1.4	1.167	1.293	1.418	1.544	1.669	1.794	2.421
1.6	1.188	1.317	1.445	1.574	1.702	1.831	2.474
1.8	1.204	1.335	1.467	1.598	1.729	1.860	2.516
2.0	1.216	1.349	1.482	1.616	1.749	1.882	2.547
2.2	1.225	1.359	1.494	1.628	1.763	1.897	2.570
2.4	1.230	1.365	1.501	1.636	1.772	1.907	2.585
2.6	1.233	1.369	1.505	1.641	1.777	1.913	2.594
2.8	1.233	1.370	1.506	1.642	1.779	1.915	2.597
3.0	1.232	1.369	1.505	1.642	1.778	1.914	2.596
3.5	1.225	1.361	1.496	1.632	1.768	1.903	2.581
4.0	1.213	1.347	1.482	1.616	1.750	1.885	2.556
4.5	1.200	1.333	1.465	1.598	1.730	1.863	2.526
5.0	1.187	1.317	1.448	1.579	1.710	1.840	2.494
6.0	1.162	1.289	1.416	1.544	1.671	1.798	2.435
7.0	1.141	1.265	1.390	1.514	1.638	1.763	2.384
8.0	1.124	1.246	1.368	1.490	1.612	1.733	2.342
9.0	1.111	1.231	1.350	1.470	1.590	1.710	2.308
10	1.100	1.218	1.336	1.454	1.572	1.690	2.280
15	1.067	1.179	1.292	1.404	1.516	1.629	2.191
20	1.050	1.160	1.269	1.379	1.488	1.598	2.145
30	1.033	1.140	1.246	1.353	1.459	1.566	2.098
40	1.025	1.130	1.235	1.340	1.445	1.549	2.074
50	1.020	1.124	1.228	1.332	1.436	1.540	2.059
100	1.010	1.112	1.214	1.316	1.418	1.520	2.030
>100	1.000	1.100	1.200	1.300	1.400	1.500	2.000

The bending moments in the bracing elements are obtained by integrating formulae (5.41) and (5.42) for the shear forces, with taking into consideration the corresponding boundary conditions defined by formulae (5.14):

$$M_{x,i} = -\left(\frac{(z-H)^2}{2} + \mu(\frac{z^3}{6H} - \frac{zH}{2} + \frac{H^2}{3})\right)(I_{y,i}\bar{q}_x + I_{xy,i}\bar{q}_y) +$$
$$+ \frac{I_{y,i}y_i - I_{xy,i}x_i}{I_\omega}\frac{m_{z0}H^2}{2}\eta_M, \qquad (5.47)$$

$$M_{y,i} = -\left(\frac{(z-H)^2}{2} + \mu(\frac{z^3}{6H} - \frac{zH}{2} + \frac{H^2}{3})\right)(I_{x,i}\bar{q}_y + I_{xy,i}\bar{q}_x) -$$
$$- \frac{I_{x,i}x_i - I_{xy,i}y_i}{I_\omega}\frac{m_{z0}H^2}{2}\eta_M, \qquad (5.48)$$

where

$$\eta_M = \frac{2}{k^2 \cosh k}\left\{(1+\mu)(\cosh\frac{kz}{H} - \frac{kz}{H}\sinh k - \cosh k + k\sinh k) + \right.$$
$$\left. + (1 + \frac{\mu}{2} - \frac{\mu}{k^2})\left(\sinh(k - \frac{kz}{H}) + \frac{kz}{H} - k\right)k\right\} \qquad (5.49)$$

is the bending moment factor representing the effect of rotation around the shear centre.

The first term in formulae (5.47) and (5.48) stands for bending moments due to the unsymmetrical bending of the shear centre axis of the bracing system. As with the standard case of statically determinate cantilevers, they only depend on the height, the intensity of the horizontal load and the bending stiffness of the bracing element (and of the whole system), and have a distribution of a third order parabola over the height.

The second term represents the effect of the rotation around the shear centre, with η_M being the bending moment factor and characterizing the distribution of this part of the bending moment. The bending moment factor is shown in Fig. 5.12, as a function of k and μ. When parameter k assumes zero value, the bending moments due to rotation have the same distribution as that of the bending moments due to the unsymmetrical bending of the shear centre axis (i.e. the first term in the formulae). As the value of parameter k increases, representing the growing effect of the Saint-Venant torsional stiffness, the value of the bending moments in the bracing element – after a small initial increase – decreases. This is due to the fact that with the growing value of k the 'centroid' of the load on the bracing element shifts downwards (Fig. 5.8) resulting in a decreasing moment arm.

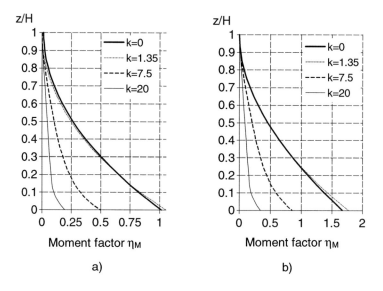

Fig. 5.12 Bending moment factor η_M: bending moments in the ith bracing element, due to rotation. a) $\mu = 0$, b) $\mu = 1$.

The maximum bending moments develop at the bottom of the bracing elements:

$$M_{x,i}^{max} = M_{x,i}(0) = -\frac{H^2}{2}(1+\frac{2\mu}{3})(I_{y,i}\bar{q}_x + I_{xy,i}\bar{q}_y) + \frac{I_{y,i}y_i - I_{xy,i}x_i}{I_\omega}\frac{m_{z0}H^2}{2}\eta_{M_{max}},$$

(5.50)

$$M_{y,i}^{max} = M_{y,i}(0) = -\frac{H^2}{2}(1+\frac{2\mu}{3})(I_{x,i}\bar{q}_y + I_{xy,i}\bar{q}_x) - \frac{I_{x,i}x_i - I_{xy,i}y_i}{I_\omega}\frac{m_{z0}H^2}{2}\eta_{M_{max}},$$

(5.51)

where

$$\eta_{M_{max}} = 2\left\{(1+\mu)\left(\frac{1}{k^2 \cosh k} - \frac{1}{k^2} + \frac{\tanh k}{k}\right) + \right.$$
$$\left. +(1+\frac{\mu}{2}-\frac{\mu}{k^2})\left(\frac{\tanh k}{k} - \frac{1}{\cosh k}\right)\right\} \quad (5.52)$$

is the maximum bending moment factor.

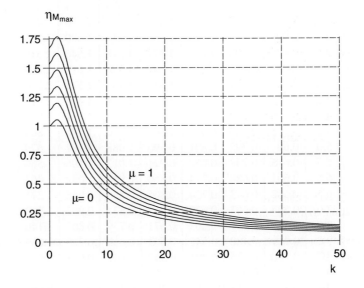

Fig. 5.13 Bending moment factor η_{Mmax} for maximum bending moment, due to rotation.

Values for the maximum bending moment factor are given in Table 5.2 and in Fig. 5.13, as a function of k and μ.

The value of maximum bending moment factor η_{Mmax} increases in the range $0 < k < 1.35$; its value is greater than 1.0 up to $k = 2.346$; then its value decreases rapidly. It peaks at $k = 1.35$ representing a 5.4% increase in the value of the maximum bending moment in the bracing element, due to rotation around the shear centre, compared to the theoretical case of $k = 0$ (and $\mu = 0$), when the behaviour of the bracing system is not affected by the Saint-Venant torsional stiffness. When the value of k exceeds 40, then the value of the maximum bending moment factor drops below 0.10 (with $\mu = 0$) and then becomes practically negligible.

Table 5.2 Maximum bending moment factor η_{Mmax}

k	η_{Mmax}						
	$\mu=0.0$	$\mu=0.2$	$\mu=0.4$	$\mu=0.6$	$\mu=0.8$	$\mu=1.0$	$\mu=2.0$
0.0	1.000	1.133	1.267	1.400	1.533	1.667	2.333
0.5	1.018	1.154	1.290	1.427	1.563	1.699	2.380
1.0	1.046	1.188	1.329	1.470	1.611	1.752	2.458
1.5	1.053	1.196	1.339	1.482	1.625	1.768	2.484
2.0	1.029	1.170	1.311	1.452	1.593	1.734	2.439
2.2	1.013	1.153	1.292	1.431	1.570	1.710	2.406
2.4	0.995	1.132	1.269	1.406	1.543	1.680	2.365
2.6	0.974	1.108	1.243	1.377	1.512	1.646	2.319
2.8	0.952	1.083	1.215	1.347	1.478	1.610	2.269
3.0	0.928	1.057	1.185	1.314	1.443	1.572	2.216
3.5	0.867	0.988	1.109	1.230	1.351	1.472	2.077
4.0	0.806	0.919	1.032	1.145	1.258	1.371	1.937
4.5	0.748	0.853	0.959	1.064	1.170	1.276	1.804
5.0	0.694	0.793	0.891	0.990	1.088	1.187	1.679
5.5	0.645	0.737	0.829	0.921	1.013	1.106	1.566
6.0	0.601	0.688	0.774	0.860	0.946	1.032	1.463
6.5	0.562	0.643	0.724	0.805	0.886	0.966	1.371
7.0	0.527	0.603	0.679	0.755	0.831	0.907	1.288
8.0	0.467	0.535	0.603	0.671	0.739	0.807	1.146
9.0	0.419	0.480	0.542	0.603	0.664	0.725	1.031
10	0.380	0.435	0.491	0.547	0.602	0.658	0.936
12	0.319	0.366	0.413	0.460	0.507	0.554	0.789
14	0.276	0.316	0.357	0.398	0.438	0.479	0.682
16	0.242	0.278	0.314	0.350	0.386	0.421	0.601
18	0.216	0.248	0.280	0.312	0.344	0.376	0.536
20	0.195	0.224	0.253	0.282	0.311	0.340	0.485
30	0.131	0.151	0.170	0.190	0.209	0.229	0.327
40	0.099	0.113	0.128	0.143	0.158	0.172	0.246
50	0.079	0.091	0.103	0.115	0.127	0.138	0.198
100	0.040	0.046	0.052	0.058	0.064	0.070	0.099
150	0.027	0.031	0.035	0.039	0.043	0.046	0.066
200	0.020	0.023	0.026	0.029	0.032	0.035	0.050
300	0.013	0.015	0.017	0.019	0.021	0.023	0.033
500	0.008	0.009	0.010	0.012	0.013	0.014	0.020
>500	0	0	0	0	0	0	0

There seems to be an analogy in the behaviour of an element of the bracing system subjected to uniformly distributed horizontal load ($\mu = 0$)

and a cantilever subjected to uniformly distributed vertical load, both with Saint-Venant and warping torsional stiffnesses. Certain similarities emerge in the stress analysis of the former case and the stability analysis of the latter case as the value of torsion parameter k increases. The torsional buckling analysis of the cantilever shows that the axis of the cantilever does not develop rotation in the upper regions but a sudden twist develops near the bottom (Fig. 5.14). In the former case, the load and the shear force, due to the rotation of the bracing system, tend to decrease significantly in the upper section of the bracing element, with the 'centroid' of the load shifting downwards. In both cases, 'activities' on the column (load, shear force, rotation) are limited to the bottom section of the column as the value of the Saint-Venant torsional stiffness increases.

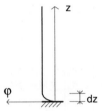

Fig. 5.14 Pure torsional buckling of a cantilever with dominant GJ.

The fact that this analogy is only one in a series of analogies makes the situation even more interesting. The same phenomenon can be observed with cantilevers developing both bending and shear deformations. Furthermore, as far as bending and shear deformations are concerned, similar phenomenon develops during the full-height buckling of frameworks [Zalka and Armer, 1992].

5.4.2 Torsional moments

According to equations (5.5) and (5.11), unless the external load passes through the shear centre of the bracing system, the equivalent column and the whole building it represents undergo rotation. The floor slabs make the individual bracing elements rotate and consequently they develop torsional moments.

The torsional resistance of the bracing system is provided by two sources: the Saint-Venant torsional stiffness and the warping (bending torsional) stiffness. Consequently, the torsional moments in the individual bracing elements develop according to their corresponding torsional stiffnesses:

$$M_{t,i}(z) = \frac{I_{t,i}}{I_t} M_t(z) \tag{5.53}$$

and

$$M_{\omega,i}(z) = \frac{I_{\omega,i}}{I_\omega} M_\omega(z), \tag{5.54}$$

where $M_t(z)$ is the Saint-Venant torsional part and $M_\omega(z)$ is the warping torsional part of the total torque $M_z(z)$:

$$M_z(z) = M_t(z) + M_\omega(z). \tag{5.55}$$

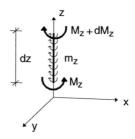

Fig. 5.15 Torsional equilibrium of a differential element.

The equilibrium of an elementary section of the equivalent column (Fig. 5.15) is expressed by

$$-M_z + m_z dz + (M_z + dM_z) = 0, \tag{5.56}$$

which leads to

$$m_z = -\frac{dM_z}{dz}. \tag{5.57}$$

The Saint-Venant and warping torsional parts of the total torque are now obtained from equation (5.11) as:

$$M_t(z) = GJ\varphi'(z) \tag{5.58}$$

and

$$M_\omega(z) = -EI_\omega\varphi'''(z). \tag{5.59}$$

After differentiating the function of rotation defined by formula (5.27) once and three times, respectively, and rearranging the resulting expressions, the solutions for the Saint-Venant and warping torsional moments assume the form:

$$M_t = m_{z0}\left\{H - z + \mu\left(\frac{H}{2} - \frac{z^2}{2H} - \frac{H}{k^2}\right) + \sinh\frac{kz}{H}\left((1+\frac{\mu}{2} - \frac{\mu}{k^2})\sinh k + \right.\right.$$
$$\left.\left. + \frac{1+\mu}{k}\right)\frac{H}{\cosh k} - \left(1+\frac{\mu}{2} - \frac{\mu}{k^2}\right)H\cosh\frac{kz}{H}\right\} \tag{5.60}$$

and

$$M_\omega = m_{z0}\left\{\frac{\mu H}{k^2} - \sinh\frac{kz}{H}\left((1+\frac{\mu}{2} - \frac{\mu}{k^2})\sinh k + \frac{1+\mu}{k}\right)\frac{H}{\cosh k} + \right.$$
$$\left. + \left(1+\frac{\mu}{2} - \frac{\mu}{k^2}\right)H\cosh\frac{kz}{H}\right\}. \tag{5.61}$$

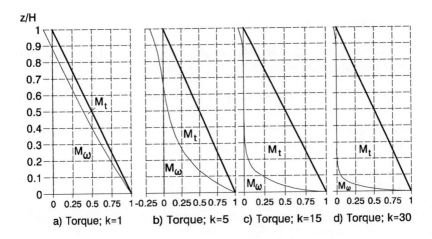

Fig. 5.16 Typical torsional moment distribution for $\mu = 0$ (UDL).

In making use of formulae (5.55), (5.60) and (5.61), the total torque is obtained from

$$M_z(z) = m_{z0}\left[H - z + \mu(\frac{H}{2} - \frac{z^2}{2H})\right].\qquad(5.62)$$

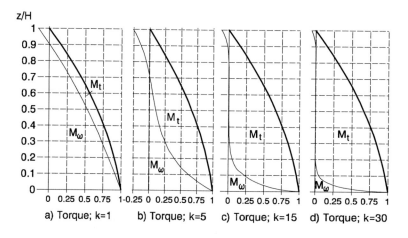

Fig. 5.17 Typical torsional moment distribution for $\mu = 4$.

Typical distributions of the torsional moments are shown in Fig. 5.16 and Fig. 5.17, as a function of k and μ. The warping torsional moment $M_\omega(z)$ always assumes maximum at ground floor level:

$$M_{\omega,\max} = M_\omega(0) = m_{z0}H\left(1 + \frac{\mu}{2}\right).\qquad(5.63)$$

The Saint-Venant torsional moment $M_t(z)$ is zero at ground floor level. The location of its maximum value over the height depends on the values of parameters k and μ and – after differentiating function (5.60) once – is obtained as the solution in z of the equation

$$\cosh\frac{kz}{H}\left(\tanh k(1+\frac{\mu}{2} - \frac{\mu}{k^2})k + \frac{1+\mu}{\cosh k}\right) - k(1+\frac{\mu}{2} - \frac{\mu}{k^2})\sinh\frac{kz}{H} - \frac{\mu z}{H} - 1 = 0.\qquad(5.64)$$

Table 5.3 Location of the maximum Saint-Venant torsional moment $M_{t,\max}$

k	z/H						
	$\mu=0.0$	$\mu=0.2$	$\mu=0.4$	$\mu=0.6$	$\mu=0.8$	$\mu=1.0$	$\mu=2.0$
0.00	1.000	1.000	1.000	1.000	1.000	1.000	1.000
0.05	0.999	0.999	0.999	0.999	0.999	0.999	0.999
0.10	0.997	0.997	0.997	0.997	0.997	0.997	0.997
0.20	0.987	0.987	0.988	0.988	0.988	0.989	0.989
0.30	0.971	0.972	0.973	0.974	0.974	0.975	0.976
0.40	0.950	0.952	0.953	0.954	0.955	0.956	0.958
0.50	0.925	0.928	0.930	0.931	0.933	0.934	0.936
0.60	0.897	0.901	0.903	0.905	0.907	0.908	0.912
0.70	0.867	0.871	0.875	0.877	0.879	0.881	0.886
0.80	0.835	0.840	0.844	0.848	0.850	0.852	0.858
0.90	0.803	0.809	0.814	0.817	0.820	0.822	0.829
1.0	0.772	0.779	0.784	0.787	0.791	0.793	0.801
1.2	0.712	0.720	0.726	0.730	0.733	0.736	0.745
1.4	0.659	0.667	0.673	0.677	0.681	0.684	0.694
1.6	0.612	0.620	0.626	0.631	0.635	0.638	0.649
1.8	0.572	0.580	0.586	0.591	0.595	0.598	0.609
2.0	0.537	0.545	0.551	0.556	0.560	0.563	0.574
2.5	0.471	0.479	0.484	0.489	0.493	0.496	0.507
3.0	0.426	0.432	0.438	0.442	0.446	0.449	0.459
4.0	0.367	0.373	0.378	0.382	0.385	0.388	0.398
5.0	0.329	0.335	0.340	0.343	0.347	0.350	0.359
6.0	0.301	0.307	0.311	0.315	0.318	0.321	0.331
7.0	0.279	0.284	0.289	0.292	0.296	0.298	0.308
8.0	0.260	0.265	0.270	0.273	0.277	0.279	0.289
9.0	0.244	0.249	0.253	0.257	0.260	0.263	0.272
10	0.230	0.235	0.239	0.243	0.246	0.248	0.257
20	0.150	0.153	0.156	0.158	0.161	0.162	0.170
30	0.113	0.116	0.118	0.120	0.122	0.123	0.129
40	0.092	0.094	0.096	0.097	0.099	0.100	0.105
50	0.078	0.080	0.081	0.083	0.084	0.085	0.089
60	0.068	0.070	0.071	0.072	0.073	0.074	0.077
70	0.061	0.062	0.063	0.064	0.065	0.066	0.069
80	0.055	0.056	0.057	0.058	0.058	0.059	0.062
90	0.050	0.051	0.052	0.053	0.053	0.054	0.057
100	0.046	0.047	0.048	0.048	0.049	0.050	0.052

Figure 5.18 and Table 5.3 give values for z/H defining the location of $M_{t,\max}$ as a function of torsion parameter k and parameter μ which

characterizes the slope of the load function of trapezoidal distribution. It is interesting to note that the location of the Saint-Venant torsional moment hardly depends on the slope of the load function.

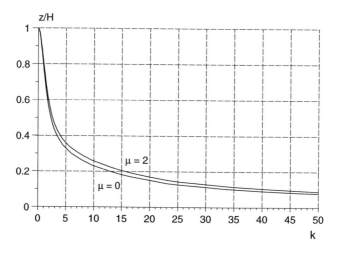

Fig. 5.18 Location of the maximum Saint-Venant torsional moment $M_{t,max}$ over the height.

5.5 STRESSES IN THE BRACING ELEMENTS

Knowing the shear forces, bending moments, Saint-Venant torsional and warping torsional moments in the ith bracing element, the last step of the analysis is the determination of the normal and shear stresses. The calculation is carried out using the formulae derived for combined unsymmetrical bending and mixed torsion [Kollbrunner and Basler, 1969; Murray, 1984; Zbirohowski-Koscia, 1967]. A local x-y coordinate system is used whose origin is in the centroid of the ith bracing element and whose axes are parallel with the global axes x and y.

Normal stresses are caused by bending moments and the warping torsional moments:

$$\sigma_{z,i}(z) = M_{x,i}(z)\frac{I_{y,i}y - I_{xy,i}x}{I_{x,i}I_{y,i} - I_{xy,i}^2} + M_{y,i}(z)\frac{I_{x,i}x - I_{xy,i}y}{I_{x,i}I_{y,i} - I_{xy,i}^2} + \frac{W_i(z)}{I_{\omega,i}}\omega, \quad (5.65)$$

where

$$W_i(z) = \int M_{\omega,i}(z)dz \qquad (5.66)$$

is the bimoment and ω is the sectorial coordinate. The sectorial coordinates represent triangular areas, multiplied by two, defined by three points: the shear centre of the cross-section and two points along the middle line of the cross-section defining the section to which the sectorial coordinate belongs. The sign of the sectorial coordinate is positive if the advancement along the middle line in creating the area represents an anticlockwise rotation around the shear centre.

The first two terms in formula (5.65) represent the effect of unsymmetrical bending and the third term stands for warping torsion. Both unsymmetrical bending and warping torsion develop maximum normal stresses on ground floor level. Consequently, the maximum value of the normal stresses is obtained by setting $z = 0$. Formulae (5.50), (5.51) and (5.63) can be used for the practical calculation.

It should be borne in mind in the structural design of the bracing elements that the vertical load on the bracing elements also develops normal stresses and formula (5.65) has to be supplemented by the term

$$\frac{F_i}{A_i}, \qquad (5.67)$$

where F_i is the vertical load on the ith element, resulting from its own weight and the share of dead and live vertical load it takes, and A_i is its cross-sectional area.

The shear stresses originate from three sources:

$$\tau_i(z) = \tau_i^b(z) + \tau_i^t(z) + \tau_i^\omega(z), \qquad (5.68)$$

where the first, second and third term on the right-hand side represent shear stresses caused by unsymmetrical bending, Saint-Venant torsion and warping torsion, respectively. The shear stresses resulting from unsymmetrical bending are obtained using Jourawski's well-known formula [Timoshenko, 1955]

$$\tau_i^b(z) = \frac{1}{t_i}\left(T_{x,i}(z)\frac{I_{x,i}S'_{y,i} - I_{xy,i}S'_{x,i}}{I_{x,i}I_{y,i} - I_{xy,i}^2} + T_{y,i}(z)\frac{I_{y,i}S'_{x,i} - I_{xy,i}S'_{y,i}}{I_{x,i}I_{y,i} - I_{xy,i}^2}\right), \qquad (5.69)$$

128 Advanced stress analysis

where t_i is the wall thickness and $S'_{x,i}$ and $S'_{y,i}$ are the first (statical) moments of area with respect to the neutral axis. The area to be considered is the one which is cut off from the cross-section by a line parallel to the neutral axis, where the shear stress is needed, i.e. above (or below) the point considered.

Shear stresses develop from the Saint-Venant torsion as well. The formula for the kth wall element of an open cross-section is

$$\tau^t_{i,k}(z) = \frac{M_{t,i}(z)}{J_i} t_{i,k}, \tag{5.70}$$

where

$$J_i = \frac{1}{3}\sum_k t^3_{i,k} h_{i,k} \tag{5.71}$$

is the Saint-Venant torsional constant of the cross-section and $t_{i,k}$ and $h_{i,k}$ are the wall thickness and length of the middle lane of the kth wall element of the cross-section of the ith bracing element.

When the ith element has a closed cross-section, the formula for the shear stresses is

$$\tau^t_{i,s}(z) = \frac{M_{t,i}(z)}{2A_o t(s)}, \tag{5.72}$$

where s is the arc length, A_o is the area enclosed by the middle line of the wall and $t(s)$ is the wall thickness at s (Fig. 5.19).

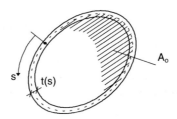

Fig. 5.19 Closed cross-section.

Finally, warping torsion also develops shear stresses:

$$\tau_i^\omega(z) = \frac{M_{\omega,i}(z)S_{\omega,i}}{I_{\omega,i}t(s)}, \qquad (5.73)$$

where

$$S_{\omega,i} = \int \omega\, dA \qquad (5.74)$$

is the sectorial statical moment.

Two of the three components of the shear stress, τ_i^b and τ_i^ω, defined by formulae (5.69) and (5.73), respectively, assume maximum on ground floor level. However, the situation with τ_i^t defined by formula (5.70) for open cross-sections and by formula (5.72) for closed cross-sections is different. The location of the maximum of τ_i^t varies over the height as a function of k and μ and can be calculated by using Table 5.3 and Fig. 5.18. The fact that the different shear stress components develop maximum at different locations over the height has to be taken into account when the maximum shear stress is calculated.

Comprehensive theoretical background to the calculation of sectorial properties is presented in Vlasov's monograph [1940] and step-by-step instructions for practical calculations are also available [Kollbrunner and Basler, 1969; Zbirohowski-Koscia, 1967].

5.6 CONCENTRATED FORCE AT TOP LEVEL; SINGLE-STOREY BUILDINGS

The equations and formulae presented in sections 5.1 to 5.4 are valid for buildings subjected to a horizontal load of trapezoidal distribution and can be used for the structural analysis of multistorey buildings. The situation with single-storey buildings is somewhat different in that the external horizontal load is transmitted through the facade to the bracing system as a single concentrated force at the floor level. This load case is therefore of practical importance and is dealt with in this section.

The formulae for this case may also be useful for investigating model structures where, because of practical considerations, a single lateral force of eccentricity e on top of the structure is applied, instead of distributed lateral load over the height.

The design formulae for this load case are obtained from the more general ones in sections 5.2 to 5.4 by taking into account the fact that the

130 Advanced stress analysis

shear force is constant over the height. The solution of the differential equations

$$EI_y \frac{d^3u}{dz^3} + EI_{xy} \frac{d^3v}{dz^3} = F_x, \qquad (5.75a)$$

$$EI_x \frac{d^3v}{dz^3} + EI_{xy} \frac{d^3u}{dz^3} = F_y \qquad (5.75b)$$

with the boundary conditions

$$u(0) = v(0) = 0, \qquad (5.76a)$$

$$u'(0) = v'(0) = 0, \qquad (5.76b)$$

$$u''(H) = v''(H) = 0 \qquad (5.76c)$$

in the usual manner leads to the lateral displacements of the equivalent column:

$$u(z) = \frac{F_x I_x - F_y I_{xy}}{E(I_x I_y - I_{xy}^2)} \left(\frac{Hz^2}{2} - \frac{z^3}{6} \right), \qquad (5.77a)$$

$$v(z) = \frac{F_y I_y - F_x I_{xy}}{E(I_x I_y - I_{xy}^2)} \left(\frac{Hz^2}{2} - \frac{z^3}{6} \right), \qquad (5.77b)$$

where F_x and F_y are the components of the concentrated top load F in directions x and y.

Maximum displacement develops at $z = H$:

$$u_{max} = u(H) = \frac{(F_x I_x - F_y I_{xy}) H^3}{3E(I_x I_y - I_{xy}^2)}, \qquad (5.78a)$$

$$v_{max} = v(H) = \frac{(F_y I_y - F_x I_{xy}) H^3}{3E(I_x I_y - I_{xy}^2)}. \qquad (5.78b)$$

The rotations of the equivalent column are obtained from the solution of the governing differential equation

$$EI_\omega \frac{d^3\varphi}{dz^3} - GJ \frac{d\varphi}{dz} = M_z, \qquad (5.79)$$

where

$$M_z = Fe = F_x y_c + F_y x_c \qquad (5.80)$$

is the torque at the top of the equivalent column. In making use of the boundary conditions

$$\varphi(0) = 0, \qquad (5.81a)$$

$$\varphi'(0) = 0, \qquad (5.81b)$$

$$\varphi''(H) = 0, \qquad (5.81c)$$

the rotation is obtained as

$$\varphi(z) = \frac{M_z H}{GJk}\left(\frac{kz}{H} - \sinh\frac{kz}{H} + (\cosh\frac{kz}{H} - 1)\tanh k\right). \qquad (5.82)$$

Maximum rotation develops at $z = H$ and, in the general case, is obtained from

$$\varphi_{max} = \varphi(H) = \frac{M_z H}{GJ}(1 - \frac{\tanh k}{k}). \qquad (5.82a)$$

When the warping stiffness dominates over the Saint-Venant torsional stiffness, the simpler formulae

$$\varphi(z) = \frac{M_z}{EI_\omega}\left(\frac{Hz^2}{2} - \frac{z^3}{6}\right) \qquad (5.82b)$$

and

$$\varphi_{max} = \varphi(H) = \frac{M_z H^3}{3EI_\omega} \qquad (5.82c)$$

can be used instead of formulae (5.82) and (5.82a).

132 Advanced stress analysis

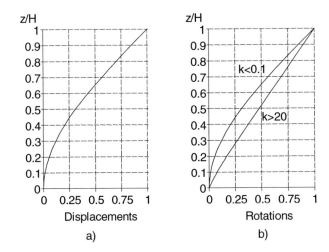

Fig. 5.20 Deformations of the equivalent column. a) Displacements, b) rotations.

Formulae (5.82), (5.82a), (5.82b) and (5.82c) cannot be used when the warping stiffness is zero (e.g. for thin-walled closed cross-sections). In such cases, the rotation is calculated from

$$\varphi(z) = \frac{M_z z}{GJ} \qquad (5.82d)$$

and the maximum rotation at $z = H$ is

$$\varphi_{max} = \varphi(H) = \frac{M_z H}{GJ}. \qquad (5.82e)$$

Figure 5.20 shows the displacements and rotations of the equivalent column.

The expressions for the shear forces in the ith bracing element are

$$T_{x,i} = I_{y,i}\overline{F}_x + I_{xy,i}\overline{F}_y - \frac{I_{y,i}y_i - I_{xy,i}x_i}{I_\omega} M_z \eta_{T'}, \qquad (5.83a)$$

$$T_{y,i} = I_{x,i}\overline{F}_y + I_{xy,i}\overline{F}_x + \frac{I_{x,i}x_i - I_{xy,i}y_i}{I_\omega} M_z \eta_{T'}. \qquad (5.83b)$$

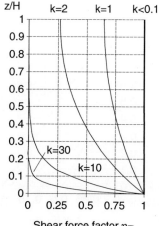

Fig. 5.21 Shear force factor $\eta_{T'}$: shear forces in the ith bracing element, due to rotation.

The notation used in formulae (5.83a) and (5.83b) is

$$\overline{F}_x = \frac{I_x F_x - I_{xy} F_y}{I_x I_y - I_{xy}^2}, \qquad \overline{F}_y = \frac{I_y F_y - I_{xy} F_x}{I_x I_y - I_{xy}^2}, \qquad (5.84)$$

$$\eta_{T'} = \cosh\frac{kz}{H} - \tanh k \sinh\frac{kz}{H}. \qquad (5.85)$$

Parameter $\eta_{T'}$ is the shear force factor representing the effect of rotation around the shear centre. Figure 5.21 shows typical shear force distributions as a function of k. The maximum shear force, due to rotation, develops at $z = 0$ and always assumes $\eta_{T'\text{max}} = \eta_{T'}(0) = 1$ so the maximum shear forces in the ith bracing element are

$$T_{x,i}^{\max} = I_{y,i}\overline{F}_x + I_{xy,i}\overline{F}_y - \frac{I_{y,i}y_i - I_{xy,i}x_i}{I_\omega}M_z, \qquad (5.86a)$$

$$T_{y,i}^{\max} = I_{x,i}\overline{F}_y + I_{xy,i}\overline{F}_x + \frac{I_{x,i}x_i - I_{xy,i}y_i}{I_\omega}M_z. \qquad (5.86b)$$

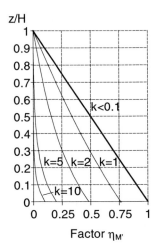

Fig. 5.22 Bending moment factor $\eta_{M'}$: bending moments in the ith bracing element, due to rotation.

The bending moments in the ith bracing element are obtained from formulae (5.83a) and (5.83b) by integration:

$$M_{x,i} = (I_{y,i}\overline{F}_x + I_{xy,i}\overline{F}_y)(z - H) + \frac{I_{y,i}y_i - I_{xy,i}x_i}{I_\omega} M_z H \eta_{M'}, \qquad (5.87a)$$

$$M_{y,i} = (I_{x,i}\overline{F}_y + I_{xy,i}\overline{F}_x)(z - H) - \frac{I_{x,i}x_i - I_{xy,i}y_i}{I_\omega} M_z H \eta_{M'}, \qquad (5.87b)$$

where

$$\eta_{M'} = \frac{1}{k}\left(\tanh k \cosh\frac{kz}{H} - \sinh\frac{kz}{H}\right) \qquad (5.88)$$

is the bending moment factor representing the effect of rotation around the shear centre. The bending moment parameter is shown in Fig. 5.22, as a function of k.

Maximum bending moments develop at the bottom of the bracing elements:

$$M_{x,i}^{max} = -(I_{y,i}\overline{F}_x + I_{xy,i}\overline{F}_y)H + \frac{I_{y,i}y_i - I_{xy,i}x_i}{I_\omega} M_z H \eta_{M'_{max}}, \quad (5.89)$$

$$M_{y,i}^{max} = -(I_{x,i}\overline{F}_y + I_{xy,i}\overline{F}_x)H - \frac{I_{x,i}x_i - I_{xy,i}y_i}{I_\omega} M_z H \eta_{M'_{max}}, \quad (5.90)$$

where

$$\eta_{M'_{max}} = \frac{\tanh k}{k}. \quad (5.91)$$

Values for $\eta_{M'_{max}}$ are given in Table 5.4 and in Fig. 5.23.

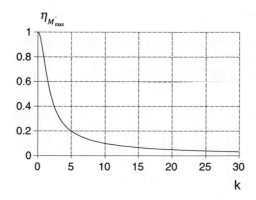

Fig. 5.23 Bending moment factor $\eta_{M'_{max}}$: maximum bending moments in the ith bracing element, due to rotation.

The external torque $M_z = Fe$ is balanced by the torsional moments

$$M_t(z) = M_z \left(1 - \cosh\frac{kz}{H} + \tanh k \sinh\frac{kz}{H}\right) \quad (5.92)$$

and

$$M_\omega(z) = M_z \left(\cosh\frac{kz}{H} - \tanh k \sinh\frac{kz}{H}\right). \quad (5.93)$$

136 Advanced stress analysis

Table 5.4 Maximum bending moment factor $\eta_{M'_{max}}$

k	$\eta_{M'_{max}}$	k	$\eta_{M'_{max}}$	k	$\eta_{M'_{max}}$	k	$\eta_{M'_{max}}$	k	$\eta_{M'_{max}}$
0.0	1.000	0.8	0.830	1.8	0.526	4.0	0.250	7	0.143
0.1	0.997	1.0	0.762	2.0	0.482	4.5	0.222	8	0.125
0.2	0.987	1.2	0.695	2.5	0.395	5.0	0.200	9	0.111
0.4	0.950	1.4	0.632	3.0	0.332	5.5	0.182	10	0.100
0.6	0.895	1.6	0.576	3.5	0.285	6.0	0.167	>10	1/k

Typical torsional moment distributions are shown in Fig. 5.24 as a function of k.

The Saint-Venant torsional moment always assumes maximum at $z = H$:

$$M_t^{max}(z) = M_t(H) = M_z\left(1 - \frac{1}{\cosh k}\right). \tag{5.94}$$

The maximum of the warping torsional moment is at $z = 0$:

$$M_\omega^{max}(z) = M_\omega(0) = M_z. \tag{5.95}$$

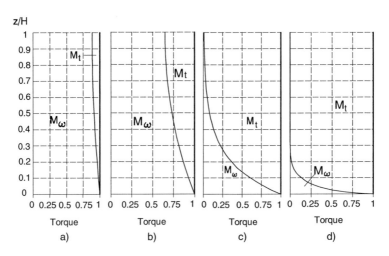

Fig. 5.24 Typical torsional moment distributions. a) $k=0.5$, b) $k=1$, c) $k=5$, d) $k=20$.

5.7 BUILDINGS WITH $I_{xy} = 0$, SUBJECTED TO UNIFORMLY DISTRIBUTED HORIZONTAL LOAD

In the special case when the bracing elements have no product of inertia and the building is subjected to a uniformly distributed horizontal load, the design formulae given in section 2.2 and earlier in this chapter simplify considerably. The simplified versions are given in this section.

Formulae (2.1) and (2.2) of the coordinates of the shear centre simplify to

$$\bar{x}_o = \frac{\sum_{1}^{n} I_{x,i} \bar{x}_i}{I_x}, \qquad \bar{y}_o = \frac{\sum_{1}^{n} I_{y,i} \bar{y}_i}{I_y}. \tag{5.96}$$

Formula (2.5) of the warping constant also assumes a simpler form:

$$I_\omega = \sum_{1}^{n} (I_{\omega,i} + I_{x,i} x_i^2 + I_{y,i} y_i^2). \tag{5.97}$$

The maximum lateral displacements of the equivalent column are obtained from formulae (5.20) and (5.21):

$$u_{max} = u(H) = \frac{q_x H^4}{8EI_y}, \qquad v_{max} = v(H) = \frac{q_y H^4}{8EI_x}. \tag{5.98}$$

In the general case, the maximum top rotation is given by formula (5.28):

$$\varphi_{max} = \varphi(H) = \frac{m_z H^2}{GJ}\left(\frac{\cosh k - 1}{k^2 \cosh k} - \frac{\tanh k}{k} + \frac{1}{2}\right). \tag{5.99}$$

When the warping stiffness dominates over the Saint-Venant torsional stiffness, the much simpler formula

$$\varphi_{max} = \varphi(H) = \frac{m_z H^4}{8EI_\omega} \tag{5.99a}$$

can be used instead of formula (5.99).

Formulae (5.99) and (5.99a) cannot be used when the warping stiffness is zero (thin-walled closed cross-sections). In such cases, the maximum rotation is calculated from

$$\varphi_{max} = \varphi(H) = \frac{m_z H^2}{2GJ}. \tag{5.99b}$$

The maximum shear force factor [formula (5.46)] and the maximum bending moment factor [formula (5.52)] simplify considerably:

$$\eta_{T_{max}} = 1 - \frac{1}{\cosh k} + \frac{\tanh k}{k} \tag{5.100}$$

and

$$\eta_{M_{max}} = \frac{2}{k^2 \cosh k} - \frac{2}{k^2} + \frac{4 \tanh k}{k} - \frac{2}{\cosh k}. \tag{5.101}$$

The maximum shear forces in the bracing elements are obtained from formulae (5.44) and (5.45):

$$T_{x,i}^{max} = T_{x,i}(0) = q_x H \frac{I_{y,i}}{I_y} - m_z H \frac{I_{y,i} y_i}{I_\omega} \eta_{T_{max}}, \tag{5.102}$$

$$T_{y,i}^{max} = T_{y,i}(0) = q_y H \frac{I_{x,i}}{I_x} + m_z H \frac{I_{x,i} x_i}{I_\omega} \eta_{T_{max}}. \tag{5.103}$$

The maximum bending moments are calculated using formulae (5.50) and (5.51):

$$M_{x,i}^{max} = M_{x,i}(0) = -\frac{q_x H^2}{2} \frac{I_{y,i}}{I_y} + \frac{m_z H^2}{2} \frac{I_{y,i} y_i}{I_\omega} \eta_{M_{max}}, \tag{5.104}$$

$$M_{y,i}^{max} = M_{y,i}(0) = -\frac{q_y H^2}{2} \frac{I_{x,i}}{I_x} - \frac{m_z H^2}{2} \frac{I_{x,i} x_i}{I_\omega} \eta_{M_{max}}. \tag{5.105}$$

Formula (5.60) for the Saint-Venant torsional moments simplifies to

$$M_t = m_z \left\{ H - z + \sinh \frac{kz}{H} \left(\sinh k + \frac{1}{k} \right) \frac{H}{\cosh k} - H \cosh \frac{kz}{H} \right\} \tag{5.106}$$

and the location of its maximum is obtained using Table 5.3.

The maximum warping torsional moment at $z = 0$ is obtained from formula (5.63):

$$M_{\omega,\max} = M_\omega(0) = m_z H. \tag{5.107}$$

5.8 WORKED EXAMPLE: A 6-STOREY BUILDING IN LONDON

To illustrate the use of the method, it is applied to the 6-storey building in Brook Street, London W1, whose layout is shown in Fig. 5.25. The modulus of elasticity and the shear modulus are $E = 2\times10^4$ MN/m^2 and $G = 8.33\times10^3$ MN/m^2, respectively. The total height of the building is 22.8 m. The maximum deformations and the rotation of the building as well as the maximum shear forces, bending moments and torsional moments in the bracing elements are to be determined.

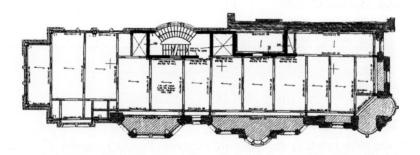

Fig. 5.25 6-storey building in Brook Street, London W1.

The external load is given by a uniformly distributed horizontal load in direction y, whose intensity, for the sake of simplicity, is assumed 1.0 kN/m^2. This leads to the horizontal load as

$q_{0y} = 33.0$ kN/m   or $F_y = 752.4$ kN.

Two investigations will be carried out. It is assumed for the first analysis that the bracing shear walls are not built together and they act independently of each other. For the second analysis, it is assumed that some of the shear walls are built together in such a manner that the reinforcement along the vertical joints makes the adjacent walls work together resulting in 3-dimensional bracing elements (cores).

5.8.1 Model: individual shear walls

The bracing system consists of four shear walls in the y and five shear walls in the x directions (Fig. 5.26), whose geometrical and stiffness characteristics are summarized in Table 5.5, where h and t are the width and thickness of the shear walls. The product of inertia $(I_{xy,i})$ and the warping constant $(I_{\omega,i})$ of all the elements are zero.

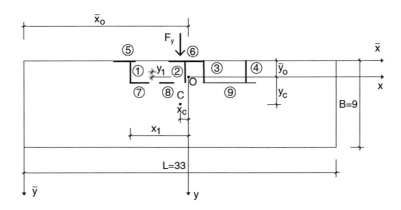

Fig. 5.26 Model: individual shear walls.

As the equivalent column is situated concurrent with the shear centre, the first step is to locate the shear centre. Its coordinates are obtained from formulae (5.96) in section 5.7:

$$\bar{x}_o = \frac{17.102}{0.9639} = 17.742 \text{ m}, \qquad \bar{y}_o = \frac{6.5766}{3.8315} = 1.716 \text{ m}.$$

When the location of the shear centre is known, the coordinate system is moved to the shear centre and the warping constant is obtained from formula (5.97):

$$I_\omega = 18.80 + 4.00 = 22.80 \text{ m}^6.$$

Table 5.5 Geometrical and stiffness characteristics

Element	$\bar{x}_i$	$\bar{y}_i$	h	t	$I_{x,i}$	$I_{y,i}$	x_i	y_i	$I_{x,i}x_i^2$	$I_{y,i}y_i^2$	J_i
1	11.30	1.125	2.25	0.250	0.2373	0.0029	-6.442	-0.591	9.848	0.001	0.0117
2	17.10	1.125	2.25	0.250	0.2373	0.0029	-0.642	-0.591	0.098	0.001	0.0117
3	19.10	1.125	2.25	0.250	0.2373	0.0029	1.358	-0.591	0.437	0.001	0.0117
4	23.60	1.125	2.25	0.250	0.2373	0.0029	5.858	-0.591	8.143	0.001	0.0117
5	10.80	0.125	2.20	0.250	0.0029	0.2218	-6.942	-1.591	0.140	0.562	0.0115
6	17.35	0.125	3.50	0.250	0.0046	0.8932	-0.392	-1.591	0.001	2.262	0.0182
7	12.20	2.375	2.00	0.250	0.0026	0.1667	-5.542	0.659	0.080	0.072	0.0104
8	15.05	2.375	1.50	0.250	0.0020	0.0703	-2.692	0.659	0.014	0.031	0.0078
9	21.75	2.375	5.50	0.178	0.0026	2.4679	4.008	0.659	0.042	1.070	0.0103
Σ					0.9639	3.8315			18.803	4.001	0.1050

Details of this calculation are shown in Table 5.5, where x_i and y_i are coordinates of the shear centres of the bracing shear walls in the coordinate system whose origin coincides with the shear centre. The value of torsion parameter k is obtained from formula (3.19):

$$k = 22.8 \sqrt{\frac{0.8333 \times 0.105}{2 \times 22.80}} = 0.999.$$

The coordinates of the geometrical centre of the plan of the building in the coordinate system whose origin is in the shear centre are obtained from (2.10) in section 2.2:

$$x_c = \frac{33}{2} - 17.742 = -1.242 \text{ m}, \qquad y_c = \frac{9}{2} - 1.716 = 2.784 \text{ m}.$$

The torque around the shear centre is obtained from formula (5.6) where μ is zero as the horizontal load is uniformly distributed:

$$m_z(z) = m_{z0} = q_{0y}x_c = -33 \times 1.242 = -40.987 \text{ kNm/m}.$$

The auxiliary load parameters are calculated from formulae (5.19):

$$\bar{q}_x = 0, \qquad \bar{q}_y = \frac{33}{0.9639} = 34.236.$$

The maximum top translation of the equivalent column is obtained from formula (5.98):

142 Advanced stress analysis

$$v_{max} = v(H) = \frac{33 \times 22.8^4}{8 \times 2 \times 10^7 \times 0.9639} = 0.0578 \, m = 57.8 \, mm.$$

The maximum top rotation is calculated from formula (5.99):

$$\varphi_{max} = \frac{-40.987 \times 22.8^2}{8.333 \times 10^6 \times 0.105} \left(\frac{\cosh 0.999 - 1}{0.999^2 \cosh 0.999} - \frac{\tanh 0.999}{0.999} + \frac{1}{2} \right) =$$

$$= -2.19 \times 10^{-3} \, rad.$$

Formula (5.33) converts radians into degrees:

$$\varphi_{max} = -57.3 \times 2.19 \times 10^{-3} = -0.126°.$$

The rotation of the building is equal to the rotation of the equivalent column. As for the maximum horizontal displacements of the building, however, it should be borne in mind that, due to the rotation around the shear centre, the characteristic (corner) points of the building develop additional displacements. The maximum horizontal displacement of the building is obtained using formula (5.34):

$$v_{building} = 57.8 + 17742 \times 2.19 \times 10^{-3} = 96.6 \, mm.$$

Table 5.6 Maximum shear forces

Bracing element	$T_{x,i}$ $\{q_x\}$	$T_{x,i}$ $\{m_z\}$	$T_{x,i}$	$T_{y,i}$ $\{q_y\}$	$T_{y,i}$ $\{m_z\}$	$T_{y,i}$
1	0	-0.1	-0.1	185.2	69.8	255.0
2	0	-0.1	-0.1	185.2	7.0	192.2
3	0	-0.1	-0.1	185.2	-14.7	170.5
4	0	-0.1	-0.1	185.2	-63.4	121.8
5	0	-16.1	-16.1	2.3	0.9	3.2
6	0	-64.9	-64.9	3.6	0.1	3.7
7	0	5.0	5.0	2.0	0.7	2.7
8	0	2.2	2.2	1.6	0.2	1.8
9	0	74.2	74.2	2.1	-0.6	1.5
Σ	0	0.0	0.0	752.4	0.0	752.4

The maximum shear forces in the bracing elements are obtained from formulae (5.102) and (5.103). The shear forces in the first bracing element, for example, are

$$T_{x,1}^{max} = -\frac{0.0029 \times 0.591}{22.80} 40.987 \times 22.8 \times 1.113 = -0.078 \text{kN},$$

$$T_{y,1}^{max} = \frac{33 \times 22.8 \times 0.237}{0.9639} + \frac{0.237 \times 6.442}{22.80} 40.987 \times 22.8 \times 1.113 = 255 \text{kN},$$

where

$$\eta_{T_{max}} = 1 - \frac{1}{\cosh 0.999} - 1 + \frac{\tanh 0.999}{0.999} = 1.113$$

is the maximum shear force factor [formula (5.100)].

Details for all the nine walls are given in Table 5.6.

The maximum bending moments are calculated using formulae (5.104) and (5.105). The bending moments in the first shear wall are:

$$M_{x,1}^{max} = \frac{0.0029 \times 0.591}{22.80} \frac{40.987 \times 22.8^2}{2} 1.046 = 0.838 \text{kNm},$$

$$M_{y,1}^{max} = -\frac{33 \times 22.8^2 \times 0.237}{2 \times 0.9639} - \frac{0.237 \times 6.442}{22.80} \frac{40.987 \times 22.8^2}{2} 1.046 =$$

$$= -2859 \text{ kNm},$$

where – from formula (5.101) – the maximum bending moment factor is

$$\eta_{M_{max}} = \frac{2}{0.999^2 \cosh 0.999} - \frac{2}{0.999^2} + \frac{4 \times \tanh 0.999}{0.999} - \frac{2}{\cosh 0.999} = 1.046.$$

Details for all the nine walls are given in Table 5.7.

The above formulae for the shear forces and bending moments consist of two parts. The first part represents the shear force/bending moment share which is due to the external load components q_x and q_y acting at the shear centre of the building, developing horizontal displacements in directions x and y. The second part represents the shear force/bending moment share which is due to the external torque m_z developing rotation around the shear centre and, consequently, additional displacements. The two parts are marked $\{q_x\}$, $\{q_y\}$ and $\{m_z\}$ in the Tables.

144 Advanced stress analysis

Table 5.7 Maximum bending moments

Bracing element	$M_{x,i}$ $\{q_x\}$	$M_{x,i}$ $\{m_z\}$	$M_{x,i}$	$M_{y,i}$ $\{q_y\}$	$M_{y,i}$ $\{m_z\}$	$M_{y,i}$
1	0	0.8	0.8	-2112	-747	-2859
2	0	0.8	0.8	-2112	-74	-2186
3	0	0.8	0.8	-2112	158	-1954
4	0	0.8	0.8	-2112	680	-1432
5	0	172.5	172.5	-25.8	-9.8	-35.6
6	0	694.8	694.8	-40.9	-0.9	-41.8
7	0	-53.7	-53.7	-23.1	-7.1	-30.2
8	0	-22.6	-22.6	-17.8	-2.6	-20.4
9	0	-794.4	-794.4	-23.1	5.1	-18.0

As a function of k ($k = 0.999$), Table 5.3 gives the location of $M_{t,\max}$ at $z = 0.772H = 17.60$ m and, according to formula (5.106), the maximum value is

$$M_t = -40.987 \left\{ 22.8 - 17.6 + \sinh\frac{0.999 \times 17.6}{22.8} \left(\sinh 0.999 + \frac{1}{0.999} \right) \times \right.$$

$$\left. \times \frac{22.8}{\cosh 0.999} - 22.8 \cosh\frac{0.999 \times 17.6}{22.8} \right\} = -106.8 \,\text{kNm}.$$

The warping torsional moment assumes maximum at $z = 0$, which is obtained from formula (5.107) as

$$M_{\omega,\max} = M_\omega(0) = -40.987 \times 22.8 = -934.5 \,\text{kNm}.$$

5.8.2 Model: built-up shear walls and cores

Assuming that some of the shear walls are built together resulting in Z, T and U shaped cores, the building is now braced by four bracing elements (Fig. 5.27), whose geometrical and stiffness characteristics are summarized in Table 5.8. The maximum deformations and the rotation of the building as well as the maximum shear forces, bending moments and torsional moments in the bracing elements are to be determined.

The auxiliary parameters [formulae (5.19)] assume the value

$$\bar{q}_x = \frac{-1.073 \times 33}{2.736 \times 8.493 - 1.073^2} = -1.602,$$

$$\bar{q}_y = \frac{8.493 \times 33}{2.736 \times 8.493 - 1.073^2} = 12.683.$$

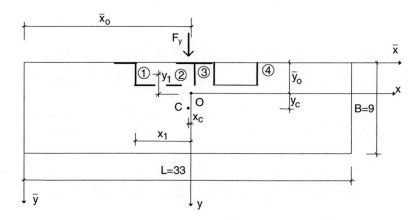

Fig. 5.27 Model: built-up shear walls and cores.

The coordinates of the shear centre are obtained from formulae (2.1) and (2.2):

$$\bar{x}_o = \frac{1.073 \times 8.048 + 8.493 \times 42.553}{2.736 \times 8.493 - 1.073^2} = 16.758\,\text{m},$$

$$\bar{y}_o = \frac{2.736 \times 8.048 + 1.073 \times 42.553}{2.736 \times 8.493 - 1.073^2} = 3.073\,\text{m}.$$

Table 5.8 Geometrical and stiffness characteristics

Element	$\bar{x}_i$	$\bar{y}_i$	$I_{x,i}$	$I_{y,i}$	$I_{xy,i}$	x_i	y_i	$I_{x,i}x_i^2$	$I_{y,i}y_i^2$	$I_{xy,i}x_iy_i$	$I_{\omega,i}$	J_i
1	11.300	1.125	1.217	1.492	1.073	-5.458	-1.948	36.264	5.660	11.402	0.735	0.032
2	15.050	2.125	0.002	0.070	0	-1.708	-0.948	0.006	0.063	0	0	0.008
3	17.350	0.125	0.573	0.896	0	0.592	-2.948	0.201	7.785	0	0.008	0.029
4	21.225	3.033	0.944	6.035	0	4.467	-0.040	18.848	0.010	0	3.015	0.030
Σ			2.736	8.493	1.073			55.319	13.517	11.402	3.758	0.099

146 Advanced stress analysis

The warping constant is calculated using formula (2.5):

$$I_\omega = 3.76 + 55.32 + 13.52 - 2 \times 11.40 = 49.78 \text{ m}^6.$$

Details of this calculation are shown in Table 5.8. The value of torsion parameter is obtained from formula (3.19):

$$k = 22.8 \sqrt{\frac{0.8333 \times 0.099}{2 \times 49.78}} = 0.654.$$

The coordinates of the geometrical centre of the plan of the building in the coordinate system whose origin is in the shear centre are obtained from (2.10):

$$x_c = \frac{33}{2} - 16.758 = -0.258, \qquad y_c = \frac{9}{2} - 3.073 = 1.427.$$

The torque around the shear centre is obtained from formula (5.6):

$$m_z(z) = m_{z0} = q_{0y} \times x_c = -33 \times 0.258 = -8.514 \text{ kNm/m}.$$

The maximum top translation of the equivalent column is obtained using formulae (5.19) and (5.21):

$$v_{max} = v(H) = \frac{33 \times 8.493 \times 22.8^4}{8 \times 2 \times 10^7 (2.736 \times 8.493 - 1.073^2)} = 0.0214 \text{ m} = 21.4 \text{ mm}.$$

The maximum top rotation is calculated from formula (5.28):

$$\varphi_{max} = \frac{-8.514 \times 22.8^2}{8.333 \times 10^6 \times 0.099} \left(\frac{\cosh 0.655 - 1}{0.655^2 \cosh 0.655} - \frac{\tanh 0.655}{0.655} + \frac{1}{2} \right) =$$

$$= -2.48 \times 10^{-4} \text{ rad}.$$

Formula (5.33) converts radians into degrees:

$$\varphi_{max} = -57.3 \times 2.48 \times 10^{-4} = -0.0142°.$$

The rotation of the building is equal to the rotation of the equivalent column. As for the maximum translation of the building, however, in addition to the translation of the shear centre, additional translations due to the rotation of the building have to be taken into account. The maximum horizontal displacement of the building is obtained using formula (5.34):

$$v_{building} = 21.4 + 16758 \times 2.48 \times 10^{-4} = 25.6 \text{ mm},$$

showing a drastic reduction, compared to the previous case (cf. $v_{building} = 96.6$ mm on page 142).

The maximum shear force factor is given by formula (5.46):

$$\eta_{T_{max}} = 1 - \frac{1}{\cosh 0.655} + \frac{\tanh 0.655}{0.655} = 1.060.$$

The maximum shear forces in the bracing elements are obtained from formulae (5.44) and (5.45). The shear forces in the first bracing element are

$$T_{x,1}^{max} = 22.8(-1.492 \times 1.602 + 1.073 \times 12.683) +$$

$$+ \frac{-1.492 \times 1.948 + 1.073 \times 5.458}{49.78} 8.514 \times 22.8 \times 1.06 = 267.8 \text{ kN},$$

$$T_{y,1}^{max} = 22.8(1.217 \times 12.683 - 1.073 \times 1.602) -$$

$$- \frac{-1.217 \times 5.458 + 1.073 \times 1.948}{49.78} 8.514 \times 22.8 \times 1.06 = 331.7 \text{ kN}.$$

Details for all the four bracing elements are given in Table 5.9.

Table 5.9 Maximum shear forces

Bracing element	$T_{x,i}$ $\{q_x\}$	$T_{x,i}$ $\{m_z\}$	$T_{x,i}$	$T_{y,i}$ $\{q_y\}$	$T_{y,i}$ $\{m_z\}$	$T_{y,i}$
1	255.7	12.1	267.8	312.9	18.8	331.7
2	-2.6	-0.2	-2.8	0.6	0.0	0.6
3	-32.7	-10.9	-43.6	165.8	-1.4	164.4
4	-220.4	-1.0	-221.4	273.1	-17.4	255.7
Σ	0.0	0.0	0.0	752.4	0.0	752.4

The maximum bending moments are calculated using formulae (5.50) and (5.51). The bending moments in the first shear wall are:

$$M_{x,1}^{max} = -\frac{22.8^2}{2}(-1.492 \times 1.602 + 1.073 \times 12.683) -$$

$$-\frac{-1.492 \times 1.948 + 1.073 \times 5.458}{49.79} \frac{8.51 \times 22.8^2}{2} 1.028 = -3049 \text{ kNm},$$

$$M_{y,1}^{max} = -\frac{22.8^2}{2}(1.217 \times 12.683 - 1.073 \times 1.602) -$$

$$-\frac{1.217 \times 5.458 - 1.073 \times 1.948}{49.79} \frac{8.51 \times 22.8^2}{2} 1.028 = -3775 \text{ kNm},$$

where

$$\eta_{M_{max}} = \frac{2}{0.654^2 \cosh 0.654} - \frac{2}{0.654^2} + \frac{4 \times \tanh 0.654}{0.654} - \frac{2}{\cosh 0.654} = 1.028.$$

Details relating to all the four walls are given in Table 5.10.

Table 5.10 Maximum bending moments

Bracing element	$M_{x,i}$ $\{q_x\}$	$M_{x,i}$ $\{m_z\}$	$M_{x,i}$	$M_{y,i}$ $\{q_y\}$	$M_{y,i}$ $\{m_z\}$	$M_{y,i}$
1	-2915	-134	-3049	-3567	-208	-3775
2	29.3	3.0	32.3	-6.6	-0.1	-6.7
3	372.9	120.6	493.5	-1890	15	-1875
4	2513	11.0	2524	-3114	193	-2921

As a function of k ($k = 0.654$), Table 5.3 gives the location of $M_{t,max}$ at $z = 0.892 \times H = 20.34$ m and, according to formula (5.60), the maximum value is

$$M_t = -8.51 \left\{ 22.8 - 20.34 + \sinh\frac{0.654 \times 20.34}{22.8} \left(\sinh 0.654 + \frac{1}{0.654} \right) \times \right.$$

$$\left. \times \frac{22.8}{\cosh 0.654} - 22.8 \cosh\frac{0.654 \times 20.34}{22.8} \right\} = -11.6 \text{ kNm}.$$

The warping torsional moment assumes maximum at $z = 0$, which is obtained from formula (5.63) as

$$M_{\omega,\max} = M_{\omega}(0) = -8.51 \times 22.8 = -194.0 \text{ kNm}.$$

5.9 SUPPLEMENTARY REMARKS

Although the formulae presented in this chapter are applicable to bracing systems consisting of solid shear walls and cores on rigid foundation, developing predominantly bending deformations, the method can be extended to cover other types of bracing elements and support conditions as well.

5.9.1 Frameworks and coupled shear walls

When frameworks and/or coupled shear walls are also included among the elements of the bracing system, then the determination of the stiffnesses of these elements needed for the establishment of the equivalent column requires special considerations. They do not have their own warping torsional stiffness and their Saint-Venant torsional stiffness and product of inertia can safely be neglected. This leaves the lateral stiffnesses EI_x and EI_y. As with most planar bracing elements, their stiffness perpendicular to their plane is small and may be neglected. Their in-plane stiffness can be easily calculated by using one of the procedures given in Chapter 9. When the lateral stiffnesses of the frameworks and coupled shear walls have been determined, they can be replaced by fictitious solid walls of the same stiffnesses and the spatial analysis, starting with the establishment of the equivalent column, can be carried out.

The above procedure offers good approximation for X-braced frameworks and infilled frames, and to some extent for coupled shear walls with relatively small openings and with stiff wall strips and flexible lintels, as they develop predominantly bending deformations. The situation is different with sway frames and with coupled shear walls with flexible wall strips and stiff lintels as they tend to develop predominantly shear type deformations. Replacing such bracing elements with fictitious walls can lead to unacceptable approximations, particularly as far as the location of the shear centre is concerned, and has to be considered very carefully.

5.9.2 Bracing systems with shear or a mixture of shear and bending deformations

When the elements of the bracing system develop significant shear-type deformation, the formulae given in section 2.2 for the coordinates of the shear centre cannot be used. The wall strips of a built-up bracing element develop shear deformations independently from each other. A simple, albeit approximate, procedure and closed form solutions are presented in [Kollár and Póth, 1994] for the determination of the location of the shear centre.

The situation is even more complicated when the bracing system has some elements which develop bending deformations while other elements develop shear deformations (e.g. sway frames and shear walls). The shear centre of such systems also depend on the external load and may vary over the height [Stüssi, 1965]. Stafford Smith and Vézina [1985] developed a 2-dimensional model for the determination of the shear centre and introduced the notion 'centre of resistance'. Their method also takes into consideration the distribution of the external lateral load over the height. The special case of bracing systems containing parallel plane members subjected to uniformly distributed load was investigated by Manninger and Kollár [1998]. They showed that even in this relatively simple case the notion 'shear centre' should be replaced by the notion 'shear centre curve' reflecting the fact that the position of the shear centre changes over the height of the building.

5.9.3 Special cases – scope for simplification

The closed form solutions given in this chapter are already simple enough for quick checks on maximum stresses and deformations. However, the treatment further simplifies in some special cases.

When the product of inertia of the equivalent column is zero (either because the elements have no product of inertia or they have but cancel each other out), most of the formulae considerably simplify and the amount of calculation involved is significantly reduced. (The simplified formulae for $I_{xy} = 0$ – and for uniformly distributed load – are given in section 5.7.) When only a few elements have a product of inertia, they can be neglected when the number and arrangement of the bracing elements are considered. However, when the bending and shear stresses are calculated, the products of inertia of the individual bracing elements should not be neglected as their contribution can be significant and neglecting them can lead to unacceptable approximations [Kollár and Póth, 1994].

The contribution of the warping stiffnesses of the individual elements to the global warping stiffness is usually very small in most practical cases – first term in formula (2.5) – and it can be neglected when the torsional stiffnesses of the equivalent column are determined. However, as with the product of inertia, the warping characteristics always have to be taken into account when the bending and shear stresses in the individual bracing elements are calculated.

The formulae also simplify considerably when the contribution of the Saint-Venant torsional stiffness to the overall torsional resistance is negligible, compared to that of the warping stiffness. Research into structural elements of thin-walled, open cross-section shows that neglecting the Saint-Venant torsional stiffness for cross-sections with $k < 0.02$ does not result in appreciable error [Vlasov, 1940]. Furthermore, the evaluation of values for the shear force factor and for the bending moment factor given in section 5.4.1 shows that the error committed when the Saint-Venant torsional stiffness for equivalent columns with $k < 0.1$ is neglected in the calculation of shear forces and bending moments due to rotation around the shear centre is smaller than 1%.

5.9.4 Second-order effects

When the formulae for the deformations and stresses were derived, only the effects of lateral loads were taken into account. However, building structures are always subjected to vertical loads which develop compression in the vertical load bearing elements, including the bracing elements. Because of the simultaneous effects of the lateral and vertical loads, the deformations and stresses increase. This effect can be taken into consideration in a simple way by applying the magnification factor

$$\frac{1}{1-v} \qquad (5.108)$$

to the deformations and stresses calculated considering the lateral loads only. In formula (5.108)

$$v = \frac{N}{N_{cr}} \qquad (5.109)$$

152 Advanced stress analysis

is the global critical load ratio with N being the total vertical load and N_{cr} being the global critical load of the building. (Chapter 7 deals with the critical load ratio in detail.) Formula (5.108), derived by Timoshenko and Gere [1961] for the stability analysis of beam-columns, gives good approximation; the error for $v < 0.6$ is less than 2%.

5.9.5 Soil-structure interaction

The formulae given in this chapter (and in Chapter 4) assume rigid foundations for the elements of the bracing system. This is not always the case in practice. Typical examples show that foundation flexibility can have an important effect on the behaviour of the bracing system. The flexible foundation affects mainly the stresses at the lower portion of shear walls and cores and, in general, has negligible effect on the stresses at the upper portion of the structure [Nadjai and Johnson, 1996]. For practical purposes, foundations built on hard rock or dense sand can be considered rigid.

Due to the rotation of the foundation, the deformations of the building also increase. Under a horizontal load of trapezoidal distribution, for example, the top translation of a column on flexible support (i.e. the equivalent column), due to the rotation of the foundation, can be determined using simple static considerations.

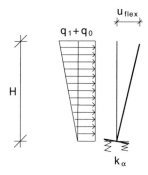

Fig. 5.28 Top translation of rigid column on flexible support.

The moment equilibrium of the column (Fig. 5.28) is expressed by

$$\frac{u_{flex} k_\alpha}{H} - \frac{q_0 H^2}{2} - \frac{q_1 H}{2} \frac{2H}{3} = 0, \qquad (5.110)$$

where

$$k_\alpha = k'_\alpha I_f \qquad (5.111)$$

is the spring constant and

- k'_α is the spring coefficient for rocking motion,
- I_f is the second moment of the contact area between foundation and soil in the relevant direction.

Values for the spring constant and the spring coefficient are given in Tables 3.4 and 3.5 in section 3.1.5.

Equation (5.110) leads to the translations, due to flexible support, as

$$u_{flex} = \frac{q_{0x}}{k'_\alpha I_{fy}} \left(\frac{1}{2} + \frac{\mu}{3} \right) H^3 \qquad (5.112a)$$

and

$$v_{flex} = \frac{q_{0y}}{k'_\alpha I_{fx}} \left(\frac{1}{2} + \frac{\mu}{3} \right) H^3, \qquad (5.112b)$$

where $\mu = q_1/q_0$ is the slope of the variable part of the horizontal load.

The total top translation of the equivalent column is obtained by adding up the translation calculated assuming a flexible column on a rigid foundation [formulae (5.20) and (5.21)] and the translation obtained by assuming a rigid column on a flexible support [formulae (5.112a) and (5.112b)].

The flexibility of the foundation against torsional motions does not normally have a great effect on the behaviour. However, if the structure of the foundation is sensitive to torsional motions (e.g. isolated footings under the bracing elements or rigid box foundation under an eccentric bracing system), this effect may also have to be taken into account.

Discrete models have also been developed for the analysis of soil-structure interaction. Both simple models [Kaliszky, 1978] and more complex procedures [Spyrakos and Beskos, 1986; Nadjai and Johnson, 1996] are available for the 3-dimensional stress analysis. Different aspects of soil-structure interaction are discussed in [Council..., 1978c].

6

Illustrative example; Qualitative and quantitative evaluation

The application of the method developed for the global structural analysis is illustrated here by a series of examples. The illustration also makes it possible to show how the performance of the bracing system can be evaluated and, in the next chapter, to advocate the introduction of a 'performance indicator'.

The 8-storey building in the example is braced by four shear walls (Fig. 6.1/a). Three different arrangements of the same four shear walls are considered (Figs 6.1/b, 6.2 and 6.3) to show how the method can be used for assessing the efficiency of the bracing system. The critical loads, the fundamental frequencies, the maximum deformations and the rotation of the building as well as the maximum shear forces, bending moments and torsional moments in the bracing elements are determined. The second order effects are not taken into account.

It is assumed for the analysis that

- only the shear walls resist the horizontal load and the contribution of the columns to the lateral stiffness of the bracing system is negligible,
- the shear walls only develop bending deformation.

The modulus of elasticity and the shear modulus for the example are $E = 2 \times 10^4$ MN/m^2 and $G = 8.33 \times 10^3$ MN/m^2, respectively. The storey height is 3.0 m and the total height of the building is 24 m. The weight per unit volume of the building (for the dynamic analysis) is $\gamma = 2.50$ kN/m^3.

Assuming normal wind pressure of uniform distribution with an intensity of $q = 28.0$ kN/m, making 50° with axis x, the load components for a simplified static analysis are

$$q_x = -28.0 \times \cos 50° = -18.0 \text{ kN/m,}$$

$q_y = -28.0 \times \sin 50° = -21.45$ kN/m,

representing a total horizontal load of $F = -672$ kN, whose components are

$$F_x = -432 \text{ kN} \quad \text{and} \quad F_y = -514.8 \text{ kN}.$$

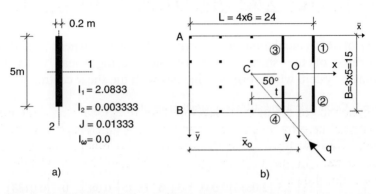

Fig. 6.1 8-storey building for the example. a) Bracing shear wall, b) 'Layout 1'.

6.1 CASE 1

In the first arrangement ('Layout 1') all four walls are placed in the right-hand side of the building (Fig. 6.1/b). The second moments of area of the shear walls are given in Fig. 6.1/a.

As the equivalent column is situated concurrent with the shear centre, the first step is to locate the shear centre. Its coordinates are obtained from formulae (5.96) which are simplified versions of formulae (2.1) and (2.2) when $I_{xy} = 0$ holds:

$$\bar{x}_o = \frac{2.0833(18+18+24+24)}{8.3333} = 21.0 \text{ m},$$

$$\bar{y}_o = \frac{0.0033(2.5+2.5+7.5+7.5)}{0.01333} = 7.5 \text{ m}.$$

Structural engineering common sense also predicts this result as the centre of stiffness of the four shear walls in the doubly symmetrical arrangement is at the geometrical centre of the wall system. The geometrical and

156 Illustrative example

stiffness characteristics needed for the calculation – after making use of formulae (2.3) and (2.4) – are given in Table 6.1.

When the location of the shear centre is known, the warping constant can be obtained from formula (5.97):

$$I_\omega = \sum_1^n (I_{\omega,i} + I_{x,i} x_i^2 + I_{y,i} y_i^2) = 75.33 \text{ m}^6.$$

Details of this calculation are shown in Table 6.1 where x_i and y_i are coordinates of the shear centres of the bracing shear walls in the coordinate system whose origin coincides with the shear centre.

Table 6.1 Geometrical and stiffness characteristics for 'Layout 1'

Element	$\bar{x}_i$	$\bar{y}_i$	$I_{x,i}$	$I_{y,i}$	x_i	y_i	$I_{x,i} x_i^2$	$I_{y,i} y_i^2$	$I_{\omega,i}$	J_i
1	24	2.5	2.0833	0.0033	3	-5	18.75	0.08	0	0.0133
2	24	12.5	2.0833	0.0033	3	5	18.75	0.08	0	0.0133
3	18	2.5	2.0833	0.0033	-3	-5	18.75	0.08	0	0.0133
4	18	12.5	2.0833	0.0033	-3	5	18.75	0.08	0	0.0133
Σ			8.3333	0.0133			75.00	0.33	75.33	0.0533

The coordinates of the geometrical centre of the plan of the building in the coordinate system whose origin is in the shear centre are obtained from formulae (2.10):

$$x_c = \frac{L}{2} - \bar{x}_o = 12 - 21 = -9 \text{ m}, \qquad y_c = \frac{B}{2} - \bar{y}_o = 7.5 - 7.5 = 0.$$

Formula (2.9) gives the distance between shear centre O and geometrical centre C:

$$t = \sqrt{x_c^2 + y_c^2} = 9.0 \text{ m},$$

with which the radius of gyration can be determined using formula (2.11):

$$i_p = \sqrt{\frac{L^2 + B^2}{12} + t^2} = \sqrt{\frac{24^2 + 15^2}{12} + 9^2} = \sqrt{147.75} = 12.16 \text{ m}.$$

6.1.1 Critical load

The critical loads for sway buckling when the building is under uniformly distributed floor load are calculated from formulae (3.11):

$$N_{cr,X} = \frac{7.84 r_s EI_Y}{H^2} = \frac{7.84 \times 0.834 \times 2 \times 10^4 \times 0.0133}{24^2} = 3.0 \text{ MN},$$

$$N_{cr,Y} = \frac{7.84 r_s EI_X}{H^2} = \frac{7.84 \times 0.834 \times 2 \times 10^4 \times 8.333}{24^2} = 1892 \text{ MN},$$

where modifier r_s (0.834) is obtained from Table 3.1.

Torsion parameter k is obtained from formula (3.19):

$$k = H\sqrt{\frac{GJ}{EI_\omega}} = 24\sqrt{\frac{8.333 \times 10^3 \times 0.05333}{2 \times 10^4 \times 75.33}} = 0.412.$$

With the above value of k, parameter k_s is obtained from (3.18) as

$$k_s = \frac{k}{\sqrt{r_s}} = \frac{0.412}{\sqrt{0.834}} = 0.451.$$

The critical load parameter (eigenvalue of pure torsional buckling) is obtained as a function of k_s, using Table 3.2 in section 3.1:

$$\alpha = 8.45.$$

The critical load for pure torsional buckling can now be calculated from formula (3.16):

$$N_{cr,\varphi} = \frac{\alpha r_s EI_\omega}{i_p^2 H^2} = \frac{8.45 \times 0.834 \times 2 \times 10^4 \times 75.33}{147.75 \times 24^2} = 124.8 \text{ MN}.$$

Interaction among the basic modes may be taken into account by cubic equation (3.22). (The fact that the system is monosymmetric is ignored in this worked example in order to demonstrate how the effect of the interaction is calculated in the general case.) The coefficients needed in the equation are obtained from formulae (3.23), (3.24) and (3.25):

158 Illustrative example

$$\tau_X = \frac{x_c}{i_p} = \frac{9.0}{12.16} = 0.74, \qquad \tau_Y = \frac{y_c}{i_p} = \frac{0.0}{12.16} = 0.0,$$

$$a_0 = \frac{N_{cr,X} N_{cr,Y} N_{cr,\varphi}}{1-\tau_X^2-\tau_Y^2} = \frac{3 \times 1892 \times 124.8}{1-0.74^2} = 1566467,$$

$$a_1 = \frac{N_{cr,X} N_{cr,Y} + N_{cr,X} N_{cr,\varphi} + N_{cr,\varphi} N_{cr,Y}}{1-\tau_X^2-\tau_Y^2} =$$
$$= \frac{3 \times 1892 + 124.8 \times 3 + 124.8 \times 1892}{1-0.74^2} = 535536,$$

$$a_2 = \frac{N_{cr,X}\tau_X^2 + N_{cr,Y}\tau_Y^2 - N_{cr,\varphi} - N_{cr,X} - N_{cr,Y}}{1-\tau_X^2-\tau_Y^2} =$$
$$= \frac{3 \times 0.74^2 - 124.8 - 3 - 1892}{1-0.74^2} = -4462.9.$$

In taking into account the coupling of the basic modes, the global critical load is obtained as the smallest root of equation

$$N^3 - 4462.9 N^2 + 535536 N - 1566467 = 0$$

as

$$N_1 = N_{cr} = 3.0 \text{ MN}.$$

The effect of the interaction of the basic modes could have been predicted offhand in this simple case. Because of the special arrangement of the bracing system (the shear centre lies on axis x), interaction only occurs between $N_{cr,Y} = 1892$ and $N_{cr,\varphi} = 124.8$. The global critical load is the minimum of the combining $N_{cr,Y}$ and $N_{cr,\varphi}$ and the independent $N_{cr,X}$, which, considering the magnitudes of the basic critical loads in question, is obviously

$$N_{cr} = N_{cr,X} = 3.0 \text{ MN}.$$

6.1.2 Fundamental frequency

The first natural frequencies for lateral vibrations in the principal directions are calculated from formulae (3.52):

$$f_X = \frac{0.56 r_f}{H^2}\sqrt{\frac{EI_Y}{\rho A}} = \frac{0.56 \times 0.892}{24^2}\sqrt{\frac{2\times 10^7 \times 0.0133 \times 9.81}{2.5 \times 24 \times 15}} = 0.047 \text{ Hz}$$

and

$$f_Y = \frac{0.56 r_f}{H^2}\sqrt{\frac{EI_X}{\rho A}} = \frac{0.56 \times 0.892}{24^2}\sqrt{\frac{2\times 10^7 \times 8.3333 \times 9.81}{2.5 \times 24 \times 15}} = 1.168 \text{ Hz},$$

where the value of modifier r_f (0.892) is obtained from Table 3.8.

The first natural frequency for pure torsional vibrations is calculated from formula (3.55):

$$f_\varphi = \frac{\eta r_f}{i_p H^2}\sqrt{\frac{EI_\omega}{\rho A}} = \frac{0.577 \times 0.892}{12.16 \times 24^2}\sqrt{\frac{2\times 10^7 \times 75.33 \times 9.81}{2.5 \times 24 \times 15}} = 0.298 \text{ Hz}.$$

The value of frequency parameter η (0.577) in the above formula is obtained from Table 3.9 in section 3.2.1 as a function of torsion parameter k.

The coupling of the basic modes can be taken into account using cubic equation (3.59). (The fact that the calculation can be simplified in this monosymmetrical case is once again ignored.) Formulae (3.60) and (3.61) yield the coefficients needed for the calculation:

$$a_0 = \frac{f_\varphi^2 f_X^2 f_Y^2}{1-\tau_X^2 - \tau_Y^2} = \frac{0.047^2 \times 1.168^2 \times 0.298^2}{1-0.74^2} = 0.000592,$$

$$a_1 = \frac{f_X^2 f_Y^2 + f_\varphi^2 f_X^2 + f_\varphi^2 f_Y^2}{1-\tau_X^2 - \tau_Y^2} = \frac{0.047^2 \times 1.168^2 + 0.298^2(0.047^2 + 1.168^2)}{1-0.74^2} =$$

$$= 0.275,$$

$$a_2 = \frac{f_X^2 \tau_X^2 + f_Y^2 \tau_Y^2 - f_\varphi^2 - f_X^2 - f_Y^2}{1-\tau_X^2 - \tau_Y^2} =$$

$$= \frac{0.047^2 \times 0.74^2 - 0.298^2 - 0.047^2 - 1.168^2}{1 - 0.74^2} = -3.215,$$

where $\tau_X = 0.74$ and $\tau_Y = 0$, already available from the stability analysis, are also used.

The value of the smallest coupled frequency is obtained from

$$(f^3)^2 - 3.215(f^2)^2 + 0.275 f^2 - 0.000592 = 0$$

as the smallest root:

$$f_1 = f = 0.047 \text{ Hz}.$$

Knowing the values of the basic frequencies and the fact that the arrangement is monosymmetrical, the above value could have been obtained without calculation – as was the case with the stability analysis.

6.1.3 Maximum stresses and deformations

The torque around the shear centre is obtained from formula (5.6) where μ is zero as the horizontal load is uniformly distributed:

$$m_z = q_x y_c + q_y x_c = 21.45 \times 9 = 193.05 \text{ kNm/m}.$$

As the procedure simplifies considerably for buildings subjected to uniformly distributed load and with zero product of inertia, the simplified versions of the relevant formulae given in section 5.7 will be used in the following.

The maximum top translations of the equivalent column are obtained from formulae (5.98):

$$u_{max} = \frac{-18 \times 24^4}{8 \times 2 \times 10^7 \times 0.0133} = -2.81 \text{m}, \quad v_{max} = \frac{-21.45 \times 24^4}{8 \times 2 \times 10^7 \times 8.333} = -0.005 \text{m}.$$

The maximum top rotation is calculated from formula (5.99):

$$\varphi_{max} = \frac{193.05 \times 24^2}{8.3333 \times 10^3 \times 0.0533} \left(\frac{\cosh(0.412) - 1}{0.412^2 \cosh(0.412)} - \frac{\tanh(0.412)}{0.412} + \frac{1}{2} \right) =$$

$$= 0.00499 \text{ rad}.$$

Formula (5.33) converts radians into degrees:

$$\varphi_{max} = 57.3 \times 0.00499 = 0.286°.$$

The rotation of the building is equal to the rotation of the equivalent column. As for the maximum horizontal displacements of the building, however, it should be borne in mind that, due to the rotation around the shear centre, the characteristic (corner) points of the building develop additional displacements. The maximum horizontal displacements of the building at corner points 'A' and 'B' (Fig. 6.1/b) are obtained using formulae (5.34):

$$v_{max} = v_A = v + x_A \tan\varphi = -0.005 - 21 \times \tan(0.286) = -0.11 \text{ m},$$

$$u_{max} = u_B = u + y_B \tan\varphi = -2.81 - 7.5 \times \tan(0.286) = -2.84 \text{ m}.$$

The maximum shear force factor [formula (5.100)] and the maximum bending moment factor [formula (5.101)] are needed for the calculation of the maximum shear forces and bending moments in the bracing shear walls:

$$\eta_{T_{max}} = 1 - \frac{1}{\cosh(0.412)} + \frac{\tanh(0.412)}{0.412} = 1.026,$$

$$\eta_{M_{max}} = \frac{2}{0.412^2 \cosh(0.412)} - \frac{2}{0.412^2} + \frac{4\tanh(0.412)}{0.412} - \frac{2}{\cosh(0.412)} =$$

$$= 1.012.$$

The maximum shear forces in the bracing elements are obtained from formulae (5.102) and (5.103). The shear forces in the first shear wall are

$$T_{x,1}^{max} = -18 \times 24 \frac{0.0033}{0.0133} + 193.05 \times 24 \frac{0.0033 \times 5}{75.33} 1.026 =$$

$$= -108.0 + 1.0 = -107 \text{ kN},$$

$$T_{y,1}^{max} = -21.45 \times 24 \frac{2.0833}{8.3333} + 193.05 \times 24 \frac{2.0833 \times 3}{75.33} 1.026 =$$

$$= -128.7 + 394.3 = 265.6 \text{ kN}$$

162 Illustrative example

and details for the four walls are given in Table 6.2.

Table 6.2 Maximum shear forces for 'Layout 1'

Bracing element	$T_{x,i}$ $\{q_x\}$	$T_{x,i}$ $\{m_z\}$	$T_{x,i}$	$T_{y,i}$ $\{q_y\}$	$T_{y,i}$ $\{m_z\}$	$T_{y,i}$
1	-108.0	1.0	-107.0	-128.7	394.3	265.6
2	-108.0	-1.0	-109.0	-128.7	394.3	265.6
3	-108.0	1.0	-107.0	-128.7	-394.3	-523.0
4	-108.0	-1.0	-109.0	-128.7	-394.3	-523.0
Σ	-432.0	0.0	-432.0	-514.8	0.0	-514.8

The maximum bending moments are calculated using formulae (5.104) and (5.105). The bending moments in the first shear wall are:

$$M_{x,1}^{max} = \frac{18 \times 24^2}{2} \frac{0.0033}{0.0133} - \frac{193.05 \times 24^2}{2} \frac{0.0033 \times 5}{75.33} 1.012 = 1283.5 \text{ kNm},$$

$$M_{y,1}^{max} = \frac{21.45 \times 24^2}{2} \frac{2.0833}{8.3333} - \frac{193.05 \times 24^2}{2} \frac{2.0833 \times 3}{75.33} 1.012 =$$

$$= -3123.8 \text{ kNm}.$$

Details relating to all the four walls are given in Table 6.3.

Table 6.3 Maximum bending moments for 'Layout 1'

Bracing element	$M_{x,i}$ $\{q_x\}$	$M_{x,i}$ $\{m_z\}$	$M_{x,i}$	$M_{y,i}$ $\{q_y\}$	$M_{y,i}$ $\{m_z\}$	$M_{y,i}$
1	1296.0	-12.5	1283.5	1544.4	-4668.2	-3123.8
2	1296.0	12.5	1308.5	1544.4	-4668.2	-3123.8
3	1296.0	-12.5	1283.5	1544.4	4668.2	6212.6
4	1296.0	12.5	1308.5	1544.4	4668.2	6212.6

The above formulae for the shear forces and bending moments consist of two parts. The first part represents the shear force/bending moment share which is due to the external load components q_x and q_y acting at the shear centre of the building, developing horizontal displacements in directions x and y. The second part represents the shear force/bending moment share which is due to the external torque m_z developing rotation around the shear centre and consequently additional displacements. The two parts are marked $\{q_x\}$, $\{q_y\}$ and $\{m_z\}$ in the Tables.

As a function of k ($k = 0.412$), Table 5.3 gives the location of $M_{t,\max}$ at $z = 0.948H = 22.75$ m and, according to formula (5.106), the maximum value is

$$M_{t,\max} = 193.05 \left\{ 24 - 22.75 + \sinh\frac{0.412 \times 22.75}{24} \left(\sinh(0.412) + \frac{1}{0.412} \right) \times \right.$$

$$\left. \times \frac{24}{\cosh(0.412)} - 24\cosh\frac{0.412 \times 22.75}{24} \right\} = 193.05 \times 0.631 = 121.8 \text{ kNm}.$$

The warping torsional moment assumes maximum at $z = 0$, which is obtained from formula (5.107) as

$$M_{\omega,\max} = m_z H = 193.05 \times 24 = 4633.2 \text{ kNm}.$$

The performance of the building is clearly not acceptable. The maximum horizontal displacement in direction x is 59 times greater than the recommended maximum displacement of

$$u = v = \frac{H}{500} = \frac{24}{500} = 0.048 \text{ m}.$$

Even in direction y, the maximum displacement is more than twice as big as the recommended maximum.

As these excessive displacements are mainly caused by the lack of bracing in direction x, the efficiency of the bracing system is improved for 'Layout 2' by turning around shear walls No. 3 and No. 4 by 90 degrees. They are also moved to the left-hand side of the building (Fig. 6.2).

6.2 CASE 2

Table 6.4 gives the geometrical and stiffness data for 'Layout 2'.

Table 6.4 Geometrical and stiffness characteristics for 'Layout 2'

Element	$\bar{x}_i$	$\bar{y}_i$	$I_{x,i}$	$I_{y,i}$	x_i	y_i	$I_{x,i}x_i^2$	$I_{y,i}y_i^2$	$I_{\omega,i}$	J_i
1	24	2.5	2.0833	0.0033	0.035	-5.0	0.00	0.08	0	0.0133
2	24	12.5	2.0833	0.0033	0.035	5.0	0.00	0.08	0	0.0133
3	2.5	0.0	0.0033	2.0833	-21.465	-7.5	1.54	117.19	0	0.0133
4	2.5	15.0	0.0033	2.0833	-21.465	7.5	1.54	117.19	0	0.0133
Σ			4.1733	4.1733			3.08	234.54	237.6	0.0533

164 Illustrative example

The coordinates of the shear centre are obtained from formulae (5.96):

$$\bar{x}_o = 23.965 \text{ m} \quad \text{and} \quad \bar{y}_o = 7.5 \text{ m}.$$

When the location of the shear centre is known, the warping constant can be obtained from formula (5.97):

$$I_\omega = \sum_1^n (I_{\omega,i} + I_{x,i} x_i^2 + I_{y,i} y_i^2) = 237.6 \text{ m}^6.$$

The coordinates of the centroid of the layout in the coordinate system whose origin is in the shear centre are obtained from formula (2.10):

$$x_c = \frac{L}{2} - \bar{x}_o = \frac{24}{2} - 23.965 = -11.965 \text{ m}, \quad y_c = \frac{B}{2} - \bar{y}_o = \frac{15}{2} - 7.5 = 0.$$

The distance between shear centre O and geometrical centre C is calculated from formula (2.9):

$$t = \sqrt{x_c^2 + y_c^2} = 11.965 \text{ m},$$

with which – in making use of formula (2.11) – the value of the radius of gyration can be calculated:

$$i_p = \sqrt{\frac{24^2 + 15^2}{12} + 11.965^2} = \sqrt{209.93} = 14.49 \text{ m}.$$

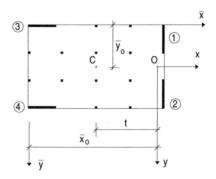

Fig. 6.2 8-storey building for the example: 'Layout 2'.

Torsion parameter k is obtained from formula (3.19):

$$k = H\sqrt{\frac{GJ}{EI_\omega}} = 24\sqrt{\frac{8.3333 \times 0.0533}{20 \times 237.6}} = 0.232.$$

6.2.1 Critical load

The basic critical loads are calculated from formulae (3.11) and (3.16):

$$N_{cr,X} = \frac{7.84 r_s EI_Y}{H^2} = \frac{7.84 \times 0.834 \times 2 \times 10^4 \times 4.1733}{24^2} = 947.5 \text{ MN},$$

$$N_{cr,Y} = \frac{7.84 r_s EI_X}{H^2} = \frac{7.84 \times 0.834 \times 2 \times 10^4 \times 4.1733}{24^2} = 947.5 \text{ MN},$$

$$N_{cr,\varphi} = \frac{\alpha r_s EI_\omega}{i_p^2 H^2} = \frac{8.04 \times 0.834 \times 2 \times 10^4 \times 237.6}{209.93 \times 24^2} = 263.4 \text{ MN},$$

where the value of critical load parameter α (8.04) is obtained from Table 3.2 in section 3.1 as a function of

$$k_s = \frac{k}{\sqrt{r_s}} = \frac{0.232}{\sqrt{0.834}} = 0.254.$$

Because of the monosymmetrical arrangement of the bracing system, the coupling of the basic modes can be calculated using formula (3.28). The value of the mode coupling parameter is obtained from Table 3.3 as

$$\varepsilon = 0.231$$

as a function of

$$r_2 = \frac{N_{cr,\varphi}}{N_{cr,Y}} = \frac{263.4}{947.5} = 0.280$$

and

$$\tau_X = \frac{x_c}{i_p} = 0.826.$$

The combined critical load is

166 Illustrative example

$$N_{combined} = \varepsilon N_{cr,Y} = 0.231 \times 947.5 = 218.8 \text{ MN}.$$

As the sway critical load in the perpendicular direction is greater than the combined critical load, the critical load of the building is

$$N_{cr} = 218.8 \text{ MN}.$$

6.2.2 Fundamental frequency

The first natural frequencies for the basic modes are calculated from formulae (3.52) and (3.55):

$$f_X = \frac{0.56 r_f}{H^2} \sqrt{\frac{EI_Y}{\rho A}} = \frac{0.56 \times 0.892}{24^2} \sqrt{\frac{2 \times 10^7 \times 4.1733 \times 9.81}{2.5 \times 24 \times 15}} = 0.827 \text{ Hz},$$

$$f_Y = \frac{0.56 r_f}{H^2} \sqrt{\frac{EI_X}{\rho A}} = \frac{0.56 \times 0.892}{24^2} \sqrt{\frac{2 \times 10^7 \times 4.1733 \times 9.81}{2.5 \times 24 \times 15}} = 0.827 \text{ Hz},$$

$$f_\varphi = \frac{\eta r_f}{i_p H^2} \sqrt{\frac{EI_\omega}{\rho A}} = \frac{0.565 \times 0.892}{14.49 \times 24^2} \sqrt{\frac{2 \times 10^7 \times 237.6 \times 9.81}{2.5 \times 24 \times 15}} = 0.435 \text{ Hz},$$

where the value of modifier r_f (0.892) is obtained from Table 3.8. The value of frequency parameter η in the formula of f_φ ($\eta = 0.565$) is obtained from Table 3.9 in section 3.2.1 as a function of torsion parameter k.

The coupling of the basic modes is calculated using formula (3.65). The value of the mode coupling parameter is obtained from Table 3.3 as

$$\varepsilon = 0.229$$

as a function of

$$r_2 = \frac{f_\varphi^2}{f_Y^2} = \frac{0.435^2}{0.827^2} = 0.277$$

and

$$\tau_X = \frac{x_c}{i_p} = 0.826.$$

The combined frequency is

$$f_{combined} = \sqrt{\varepsilon} f_Y = \sqrt{0.229} \times 0.827 = 0.396 \text{ Hz}.$$

As the lateral frequency in the perpendicular direction is greater than the combined frequency, the fundamental frequency of the building is

$$f = 0.396 \text{ Hz}.$$

6.2.3 Maximum stresses and deformations

The torque around the shear centre is obtained from formula (5.6) where μ is zero as the horizontal load is uniformly distributed:

$$m_z = q_x y_c + q_y x_c = 21.45 \times 11.965 = 256.65 \text{ kNm/m}.$$

The maximum top translations of the equivalent column are obtained from formulae (5.98):

$$u_{max} = \frac{-18 \times 24^4}{8 \times 2 \times 10^7 \times 4.173} = -0.009 \text{ m}, \quad v_{max} = \frac{-21.45 \times 24^4}{8 \times 2 \times 10^7 \times 4.173} = -0.011 \text{ m}.$$

The maximum top rotation is calculated from formula (5.99):

$$\varphi_{max} = \frac{256.65 \times 24^2}{8.3333 \times 10^3 \times 0.0533} \left(\frac{\cosh(0.232) - 1}{0.232^2 \cosh(0.232)} - \frac{\tanh(0.232)}{0.232} + \frac{1}{2} \right) =$$

$$= 0.00219 \text{ rad}$$

or, in making use of formula (5.33), in degrees:

$$\varphi_{max} = 57.3 \times 0.00219 = 0.126°.$$

This rotation also represents the rotation of the building. The maximum horizontal displacements of the building are obtained using formulae (5.34):

$$v_{max} = v_A = v + x_A \tan\varphi = -0.011 - 23.965 \times \tan(0.126) = -0.064 \text{ m},$$

$$u_{max} = u_B = u + y_B \tan\varphi = -0.009 - 7.5 \times \tan(0.126) = -0.026 \text{ m}.$$

168 Illustrative example

The maximum shear force factor [formula (5.100)] and the maximum bending moment factor [formula (5.101)] are needed for the calculation of the maximum shear forces and bending moments in the bracing shear walls:

$$\eta_{T_{max}} = 1 - \frac{1}{\cosh(0.232)} + \frac{\tanh(0.232)}{0.232} = 1.009,$$

$$\eta_{M_{max}} = \frac{2}{0.232^2 \cosh(0.232)} - \frac{2}{0.232^2} + \frac{4\tanh(0.232)}{0.232} - \frac{2}{\cosh(0.232)} = 1.004.$$

The maximum shear forces in the bracing elements are obtained from formulae (5.102) and (5.103). The values are summarized in Table 6.5. The maximum bending moments are obtained from formulae (5.104) and (5.105) and are tabulated in Table 6.6.

Table 6.5 Maximum shear forces for 'Layout 2'

Bracing element	$T_{x,i}$ $\{q_x\}$	$T_{x,i}$ $\{m_z\}$	$T_{x,i}$	$T_{y,i}$ $\{q_y\}$	$T_{y,i}$ $\{m_z\}$	$T_{y,i}$
1	-0.35	0.44	0.09	-257.0	1.9	-255.1
2	-0.35	-0.44	-0.79	-257.0	1.9	-255.1
3	-215.65	408.62	192.97	-0.4	-1.9	-2.3
4	-215.65	-408.62	-624.27	-0.4	-1.9	-2.3
Σ	-432.0	0.0	-432.0	-514.8	0.0	-514.8

Table 6.6 Maximum bending moments for 'Layout 2'

Bracing element	$M_{x,i}$ $\{q_x\}$	$M_{x,i}$ $\{m_z\}$	$M_{x,i}$	$M_{y,i}$ $\{q_y\}$	$M_{y,i}$ $\{m_z\}$	$M_{y,i}$
1	4.1	-5.2	-1.1	3083.9	-22.1	3061.8
2	4.1	5.2	9.3	3083.9	-22.1	3061.8
3	2587.8	-4881.6	-2293.8	4.9	22.1	27.0
4	2587.8	4881.6	7469.4	4.9	22.1	27.0

As a function of k ($k = 0.232$), Table 5.3 gives the location of $M_{t,\max}$ at $z = 0.982H = 23.57$ m and, according to formula (5.106), the maximum value is

$$M_{t,\max} = 256.65\left\{24 - 23.57 + \sinh\frac{0.232\times 23.57}{24}\left(\sinh(0.232) + \frac{1}{0.232}\right)\times\right.$$
$$\left.\times\frac{24}{\cosh(0.232)} - 24\cosh\frac{0.232\times 23.57}{24}\right\} = 256.65\times 0.21 = 53.9\text{ kNm}.$$

The warping torsional moment assumes maximum at $z = 0$, which is obtained from formula (5.107) as:

$$M_{\omega,\max} = m_z H = 256.65\times 24 = 6159.6\text{ kNm}.$$

The efficiency of the bracing system has improved considerably with 'Layout 2' and the maximum horizontal displacements in directions x and y decreased by 99% and 42%, respectively. The value of the global critical load increased by 7193% and the smallest coupled frequency by 742%. However, the maximum displacement in direction y still exceeds the recommended maximum:

$$v_{\max} = v_A = 0.064\text{ m} > 0.048\text{ m} = \frac{H}{500}.$$

The efficiency can still be improved by slightly modifying the position of the bracing walls for 'Layout 3'.

6.3 CASE 3

'Layout 3' in Fig. 6.3 only shows a small change compared to the previous case: the shear walls are at the same location but shear walls No. 2 and No. 4 are rotated by 90 degrees. Table 6.7 gives the geometrical and stiffness data for the new arrangement.

The coordinates of the shear centre – from formula (5.96) – are

$$\bar{x}_o = 12.0\text{ m} \quad\text{and}\quad \bar{y}_o = 7.5\text{ m}.$$

The warping constant is obtained from formula (5.97):

$$I_\omega = \sum_1^n (I_{\omega,i} + I_{x,i} x_i^2 + I_{y,i} y_i^2) = 835.1\text{ m}^6.$$

170   Illustrative example

The geometrical centre coincides with the shear centre so the coordinates of the geometrical centre in the coordinate system whose origin is in the shear centre are

$$x_c = 0.0 \text{ m} \quad \text{and} \quad y_c = 0.0 \text{ m}.$$

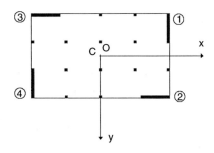

Fig. 6.3 8-storey building for the example: 'Layout 3'.

The radius of gyration – with $t = 0$ – is obtained from formula (2.11):

$$i_p = \sqrt{\frac{24^2 + 15^2}{12} + 0} = \sqrt{66.75} = 8.17 \text{ m}.$$

Table 6.7 Geometrical and stiffness characteristics for 'Layout 3'

Element	$\bar{x}_i$	$\bar{y}_i$	$I_{x,i}$	$I_{y,i}$	x_i	y_i	$I_{x,i}x_i^2$	$I_{y,i}y_i^2$	$I_{\omega,i}$	J_I
1	24.0	2.5	2.0833	0.0033	12.0	-5.0	300.0	0.08	0	0.0133
2	21.5	15.0	0.0033	2.0833	9.5	7.5	0.3	117.19	0	0.0133
3	2.5	0.0	0.0033	2.0833	-9.5	-7.5	0.3	117.19	0	0.0133
4	0.0	12.5	2.0833	0.0033	-12.0	5.0	300.0	0.08	0	0.0133
Σ			4.1733	4.1733			600.6	234.54	835.1	0.0533

Torsion parameter k is obtained from formula (3.19):

$$k = H\sqrt{\frac{GJ}{EI_\omega}} = 24\sqrt{\frac{8.3333 \times 0.0533}{20 \times 835.1}} = 0.124.$$

6.3.1 Critical load

The basic critical loads are calculated from formulae (3.11) and (3.16):

$$N_{cr,X} = \frac{7.84 r_s EI_Y}{H^2} = \frac{7.84 \times 0.834 \times 2 \times 10^4 \times 4.1733}{24^2} = 947.5 \text{ MN},$$

$$N_{cr,Y} = \frac{7.84 r_s EI_X}{H^2} = \frac{7.84 \times 0.834 \times 2 \times 10^4 \times 4.1733}{24^2} = 947.5 \text{ MN},$$

$$N_{cr,\varphi} = \frac{\alpha r_s EI_\omega}{i_p^2 H^2} = \frac{7.90 \times 0.834 \times 2 \times 10^4 \times 835.1}{66.75 \times 24^2} = 2862.1 \text{ MN},$$

where the value of critical load parameter α (7.90) is obtained from Table 3.2 in section 3.1 as a function of

$$k_s = \frac{k}{\sqrt{r_s}} = \frac{0.124}{\sqrt{0.834}} = 0.136.$$

As the shear centre of the bracing system and the geometrical centre of the plan of the building coincide, the basic critical loads do not couple and the smallest one is the global critical load:

$$N_{cr} = 947.5 \text{ MN}.$$

6.3.2 Fundamental frequency

The first natural frequencies for the basic modes are calculated from formulae (3.52) and (3.55):

$$f_X = \frac{0.56 r_f}{H^2} \sqrt{\frac{EI_Y}{\rho A}} = \frac{0.56 \times 0.892}{24^2} \sqrt{\frac{2 \times 10^7 \times 4.1733 \times 9.81}{2.5 \times 24 \times 15}} = 0.827 \text{ Hz},$$

$$f_Y = \frac{0.56 r_f}{H^2} \sqrt{\frac{EI_X}{\rho A}} = \frac{0.56 \times 0.892}{24^2} \sqrt{\frac{2 \times 10^7 \times 4.1733 \times 9.81}{2.5 \times 24 \times 15}} = 0.827 \text{ Hz},$$

172 Illustrative example

$$f_\varphi = \frac{\eta r_f}{i_p H^2}\sqrt{\frac{EI_\omega}{\rho A}} = \frac{0.561 \times 0.892}{8.17 \times 24^2}\sqrt{\frac{2 \times 10^7 \times 835.1 \times 9.81}{2.5 \times 24 \times 15}} = 1.435 \text{ Hz}.$$

The value of frequency parameter η ($\eta = 0.561$) in the third formula is obtained from Table 3.9 in section 3.2.1 as a function of torsion parameter k.

The basic modes do not combine, so the smallest natural frequency of the building is

$$f = 0.827 \text{ Hz}.$$

6.3.3 Maximum stresses and deformations

Because of $x_c = y_c = 0$, the value of the external torsional moment is zero:

$$m_z = 0.$$

The maximum top translations of the equivalent column are obtained from formulae (5.98):

$$u_{max} = \frac{-18 \times 24^4}{8 \times 2 \times 10^7 \times 4.173} = -0.009 \text{ m}, \quad v_{max} = \frac{-21.45 \times 24^4}{8 \times 2 \times 10^7 \times 4.173} = -0.011 \text{ m}.$$

As the external torsional moment is zero, the building does not rotate around the shear centre. The maximum translations of the building are equal to the maximum translations of the equivalent column:

$$u_{max} = u_B = u = -0.009 \text{ m}, \quad v_{max} = v_A = v = -0.011 \text{ m}.$$

Table 6.8 Maximum shear forces for 'Layout 3'

Bracing element	$T_{x,i}$ $\{q_x\}$	$T_{x,i}$ $\{m_z\}$	$T_{x,i}$	$T_{y,i}$ $\{q_y\}$	$T_{y,i}$ $\{m_z\}$	$T_{y,i}$
1	-0.35	0.0	-0.35	-257.0	0.0	-257.0
2	-215.65	0.0	-215.65	-0.4	0.0	-0.4
3	-215.65	0.0	-215.65	-0.4	0.0	-0.4
4	-0.35	0.0	-0.35	-257.0	0.0	-257.0
Σ	-432.0	0.0	-432.0	-514.8	0.0	-514.8

As the external torque equals zero, there is no need to calculate the maximum shear force factor and the maximum bending moment factor. The maximum shear forces and bending moments in the bracing elements are obtained from formulae (5.102), (5.103), (5.104) and (5.105) and their values are summarized in Tables 6.8 and 6.9.

The efficiency of the bracing system has further improved compared to 'Layout 2'. The value of the global critical load increased by 333% and the smallest natural frequency by 109%. The maximum horizontal displacements in directions x and y decreased by 65% and 83%, respectively, compared to the previous case. The maximum displacements in both directions are now smaller than the recommended maximum values:

$$u_{max} = 0.009 \text{ m} < 0.048 \text{ m} = \frac{H}{500}, \qquad v_{max} = 0.011 \text{ m} < 0.048 \text{ m} = \frac{H}{500}.$$

Table 6.9 Maximum bending moments for 'Layout 3'

Bracing element	$M_{x,i}$ $\{q_x\}$	$M_{x,i}$ $\{m_z\}$	$M_{x,i}$	$M_{y,i}$ $\{q_y\}$	$M_{y,i}$ $\{m_z\}$	$M_{y,i}$
1	4.1	0.0	4.1	3083.8	0.0	3083.8
2	2587.8	0.0	2587.8	4.9	0.0	4.9
3	2587.8	0.0	2587.8	4.9	0.0	4.9
4	4.1	0.0	4.1	3083.8	0.0	3083.8

The comparison of the values of the maximum shear forces and bending moments with those of the previous cases also shows significant improvement: the maximum values of the shear forces (in walls No. 3 and No. 4 for 'Layout 1' and in wall No. 4 for 'Layout 2') dropped by about 50–60%. There is also a similar reduction in the values of the maximum bending moments. These favourable changes are due to the fact that the elements of the bracing system with 'Layout 3' are arranged optimally: the bending *and* torsional stiffnesses are maximum *and* the external torque is minimum (=zero).

174 Illustrative example

6.4 EVALUATION

The results of this example and the evaluation of over one hundred bracing systems show that the structural performance of a building is greatly influenced by the arrangement of the bracing system [Zalka, 1997]. As the third arrangement (Case 3) for the example above demonstrated, even a small change in the arrangement can significantly alter the response of the building to the external load. An optimum (or near optimum) arrangement can be achieved by following the three guidelines below.

The first step is to produce a well balanced system with adequate bending stiffnesses in both directions x and y. When the adequacy of the bending stiffnesses is judged, the external load should always be considered: the bending stiffness should be 'proportional' to the load in the direction in question. – The system in 'Layout 1' has very small bending stiffness in direction x ($I_y = 0.0133$) and by orders of magnitude greater bending stiffness in direction y ($I_x = 8.3333$), while the external load in both directions is of the same order of magnitude ($q_x = 18$; $q_y = 21.45$).

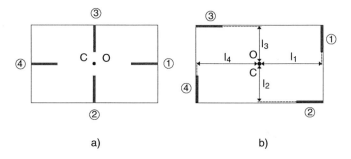

Fig. 6.4 Increasing the warping stiffness of the system. a) $I_\omega \approx 0$, b) $I_\omega = $ max.

When the bracing system has the necessary bending stiffness in both directions, the efficiency of the system can usually be further improved by maximizing its torsional stiffness, thus reducing the effect of the rotation of the building around the shear centre. The most effective way of doing this is to increase the warping stiffness of the system. In the typical case when the bracing system is dominated by shear walls, the warping stiffness is increased by increasing the 'torsion arm' of the shear walls. The 'torsion arm' is the perpendicular distance between the plane of the wall and the shear centre of the bracing system. Figure 6.4/a shows a system having shear walls with zero torsion arms where the warping torsional stiffness of the system is zero. The shear walls in Fig. 6.4/b have long torsion arms (l_1, l_2, l_3 and l_4): therefore the system has great warping stiffness.

Reducing the torsional effect of the external load is also very important. The resultant of the external horizontal load is usually assumed to pass through the centroid of the plan of the building and imposes a torque around the shear centre, unless the shear centre and the centroid coincide. This torque can have a significant and detrimental effect on the performance of the bracing system. This effect can be considerably reduced by shortening the distance between the shear centre and the centroid. – 'Layout 3' (Fig. 6.3) shows a good example for eliminating the effect of torsion: the shear centre and the centroid coincide and both the deformations of the building as well as the shear forces and bending moments in the bracing elements are much smaller than in the previous cases.

When all the three recommendations above are implemented, the resulting bracing system is well balanced and provides effective resistance against horizontal displacements and rotations usually in the form of a symmetrical (or nearly symmetrical) system.

7

Global critical load ratio

A number of publications and guidelines are available for the elastic stability analysis of the *elements* of complex systems. The choice is much more limited when the stability of the *whole* system is investigated. The practical significance of the elastic global critical load has been recognized but research in this direction has concentrated mainly on framed structures [Dulácska and Kollár, 1960; Kollár, 1972; Council..., 1978a and 1978b; Stevens, 1983]. Theoretical research and recent developments in computer modelling of structures show that it is essential to understand whole building behaviour as an analysis based on global approach can lead to safer and more economical structures. Dowrick [1976] drew attention to the importance of the overall stability of structures but national and international codes have been slow to incorporate global approaches. Halldorsson and Wang [1968] suggested that a 'general safety factor' should be used for building structures as its importance is comparable to that of the 'overturning safety factor' used in the design of dams. Chwalla [1959] recommended some forty years ago the introduction of the factor. The reason for not taking the 'general safety factor' on board was probably the lack of simple and clear methods for the determination of the global critical load at that time.

National [BS5950, 1990; DIN 18800, 1990; MSZ series, 1986] and European [Eurocode 1, 1995; Eurocode 2, 1992; Eurocode 3, 1992] codes do not use a consistent approach to sway instability and do not address torsional instability although the importance of these phenomena and their direct link with global safety are well known [Council..., 1978a and 1978b; CEB, 1983; ISE, 1988].

It is shown in this chapter that, based on the global critical load introduced in Chapter 3, the global critical load ratio can be easily produced and effectively used for assessing the safety and the performance of the bracing system.

7.1 GLOBAL CRITICAL LOAD RATIO – GLOBAL SAFETY FACTOR

The simple and expressive procedures presented in Chapter 3 for the determination of the global critical loads and Dunkerley's summation theory make it possible to introduce the elastic *global critical load ratio* as follows [MacLeod and Zalka, 1996]. The global critical load ratio is the ratio of the total vertical load (N) and the global critical load (N_{cr}):

$$v = \frac{N}{N_{cr}}. \tag{7.1}$$

When the structure is subjected to two load systems, e.g. concentrated top load and UDL on the floors, then the global critical load ratio assumes the form:

$$v = \frac{F}{F_{cr}} + \frac{N}{N_{cr}}, \tag{7.2}$$

where F is the total concentrated load on top floor level, F_{cr} is the corresponding critical concentrated load, N_{cr} is the critical load for buildings under uniformly distributed floor load and

$$N = QLBn \tag{7.3}$$

is the total uniformly distributed vertical load measured on ground floor level. In formula (7.3), Q is the intensity of the uniformly distributed floor load, L and B are the plan length and breadth of the building and n is the number of stories.

It is noted here that the reciprocal of the critical load ratio can be considered a *global safety factor* [Zalka and Armer, 1992]. Relationship (7.1) is used in this sense in DIN 18800 [1990]; Kollár [1977] also used this approach although he did not actually used the term 'global safety factor' when he investigated the stability of a building using different bracing system arrangements.

It follows from the definition of the global critical load ratio that theoretically any value smaller than one indicates stable equilibrium. However, it is advisable to limit its value. In accordance with code regulations [BS5950, 1990; DIN 18800, 1990; Eurocode 3, 1992], the recommended limit [MacLeod and Zalka, 1996] is

178   Global critical load ratio

$$\nu \leq 0.1. \tag{7.4}$$

If condition (7.4) is satisfied, then the vertical load bearing elements can be considered as braced (by the bracing system) and neglecting the second-order effects (due to sway and torsion) may result in a maximum 10% error. This approximation seems to be acceptable in view of the levels of approximation inherent in many other characteristics such as load, material qualities, stiffness, etc. – The behaviour and design of storey-height columns in a braced building is discussed briefly in section 3.1.6.b and in detail in [Kollár and Zalka, 1999].

If condition (7.4) is not satisfied, then the stability of the building can still be acceptable but stable equilibrium must be demonstrated by using some more accurate procedure which takes into account the second-order effects.

According to a widely accepted rule of thumb in the structural design of buildings, there is an 'absolute bottom line', as far as global safety is concerned: the value of the global safety factor must be at least four. When the global critical load ratio is used, this translates to

$$\nu_{max} \leq 0.25. \tag{7.5}$$

7.2 GLOBAL CRITICAL LOAD RATIO – PERFORMANCE INDICATOR

The most important feature of the global critical load ratio is, in accordance with the definition, that it represents the 'level' of stability of the building. In addition, the global critical load ratio can be conveniently used to monitor the structural performance of the bracing system. This will be demonstrated using the data and the results of the examples presented in Chapter 6.

Two horizontal load cases are considered for the stress analysis: a horizontal load of 28 kN/m with an angle of 50° with axis x (Table 7.1) as in Chapter 6, and a horizontal load of 28 kN/m but with an angle of 90° with axis x (Table 7.2). The characteristic results of the stress analysis are given in the second, third and fourth columns of the tables. The global critical load ratio is calculated assuming a uniformly distributed floor load of intensity $Q = 8.0$ kN/m² (representing a total vertical load of 23.04 MN) for each of the three layouts. The global critical load ratios are given in the last column in Tables 7.1 and 7.2. The weight per unit volume of the building is assumed to be $\gamma = 2.50$ kN/m^3 for the frequency analysis.

With the 8-storey building featured in the example in the previous chapter, the shear forces, the bending moments, the rotations and particularly the horizontal displacements could be considered as characteristics of the performance of the bracing system. As the efficiency of the bracing system was improved by rearranging the elements of the bracing system, the maximum deformations of the building decreased. A similar tendency was observed regarding the change in the value of the global critical loads – and the global critical load ratios which originate from them – and the fundamental frequency of the building: in parallel with the decrease of the maximum translations, the value of the global critical loads and the fundamental frequencies increased and the value of the global critical load ratios decreased (Table 7.1).

Table 7.1 Performance indicators. Angle of horizontal load and axis x: 50°

Layout	Maximum rotation [°]	Maximum translation [mm]	Maximum translation ratio [-]	Fundamental frequency [Hz]	Critical load ratio [-]
1	0.286	2843	59.22	0.047	7.680
2	0.126	64	1.33	0.396	0.105
3	0.000	11	0.23	0.827	0.024

The question emerges, which characteristic is best used for assessing the performance of the bracing system: the maximum horizontal displacement, the fundamental frequency or the global critical ratio? It can be said against the fundamental frequency that it plays a relatively minor role in structural engineering design: the designer of an ordinary building rarely carries out a dynamic analysis. The advantage of using the maximum displacement would be its practical importance: it is very useful to know the displacements of a building. However, there are two disadvantages here: the maximum displacement is not a nondimensional quantity and its value also depends on the direction of the horizontal load. The first disadvantage can easily be eliminated by introducing the maximum displacement ratio as the ratio of the maximum displacement to the recommended maximum displacement $H/500$ (the fourth column in Table 7.1). However, the second disadvantage is an inherent one and can lead to unfavourable situations as a fatal weakness of the bracing system may be underestimated. The building used for the numerical example reveals this shortcoming. If an external load in direction y is applied, then, although the performance of the bracing system is correctly characterized by the maximum displacements, the displacements do not reveal the *drastic* difference between 'Layout 1' and the other two layouts (Table 7.2), which is obvious when the horizontal load has a 50° angle with axis x (Table 7.1). The fatal weakness of the bracing system that it has practically no bending stiffness in direction x (Layout 1) does not come through as spectacularly as with monitoring the global critical load ratio.

Table 7.2 Performance indicators. Angle of horizontal load with axis y: 0°

Layout	Maximum rotation [°]	Maximum translation [mm]	Maximum translation ratio [-]	Fundamental frequency [Hz]	Critical load ratio [-]
1	0.373	153	3.19	0.047	7.680
2	0.164	84	1.75	0.396	0.105
3	0.000	14	0.29	0.827	0.024

On the other hand, the global critical load ratio is a nondimensional quantity and its value does not depend on the direction of the load, i.e. it assumes a uniformly distributed vertical load. The great advantage of the global critical load ratio approach is that it 'automatically' considers the most dangerous circumstances, i.e. it automatically takes into consideration whether the system tends to develop sway, torsional or combined sway-torsional deformations. In addition, the global critical load ratio is very sensitive to changes in the size and/or in the arrangement of the elements of the bracing system. Its importance is also underlined by the fact that it has a direct link with the safety of the building. The calculation of the global critical load ratio is simple: the examples in Chapter 6 show that it can be carried out in minutes and it can be repeated even faster when the necessary modifications in the bracing system are made.

It is therefore recommended that, in addition to the conventional element-based approach, the global critical load ratio approach also be used in the design process in assessing the performance and 'health' of the bracing system. The use of the global critical load ratio offers the following additional advantages:

- Construction costs can be reduced by finding the most effective bracing system (using the least material).
- Safety levels can be increased if necessary, without increasing construction costs.

7.3 FURTHER APPLICATIONS

As the magnitude of the vertical load on the building increases in relation to the critical load, the value of the stresses and deformations which are caused by the horizontal load increases and the value of the natural frequencies decreases. The critical load ratio can be used to allow for these second-order effects.

In making use of the magnification factor

$$\frac{1}{1-v}, \qquad (7.6)$$

the accuracy of the formulae given in Chapters 4 and 5 for the stresses and deformations can be improved. Factor v in formula (7.6) is the global critical load ratio. Values of the stresses and deformations multiplied by magnification factor (7.6) take into account the fact that the stresses and deformations of a building under horizontal loads increase due to the vertical load.

The accuracy of the formulae for the natural frequencies can also be improved in a similar way. In the dynamic analysis of beam-columns it is shown that the axial compressive forces reduce the value of the natural frequencies of vibration by a factor of $(1 - v)^{1/2}$ [Timoshenko and Young, 1955]. In allowing for this effect, the accuracy of the formulae given in section 3.2 can be improved by incorporating the factor into the formulae:

$$f_c = f\sqrt{1-v}. \qquad (7.7)$$

In formula (7.7), f_c is the frequency which includes the effect of the vertical load, f is the frequency which is calculated by ignoring this effect (using the formulae given in section 3.2) and v is the relevant critical load ratio. The individual formulae for the basic natural frequencies are given in section 3.2.5.a.

8

Use of frequency measurements for the global analysis

The application of the equivalent column concept in Chapters 3 and 5 resulted in simple procedures and closed-form solutions for the global structural analysis. A weakness of the model used for the analysis is that when the calculation model is established only the bracing walls, cores and frameworks (the primary structural elements of the 'building skeleton') are taken into account. Although this is a widely accepted assumption in structural analysis, tests on real buildings show that the behaviour changes considerably during the construction process and the final behaviour can be quite different. The effect of the floor slabs on the lateral stiffness is normally not significant [Ellis and Ji, 1996] but other structural and non-structural elements can play a more important role. The lateral stiffness of completed buildings can be much greater than that of the skeleton of the building. Tests on the 8-storey Cardington Steel Building after completion showed that a bare frame analysis can be as much as 500% conservative in predicting the lateral stiffness of the building [Daniels, 1994]. Torsion can be significantly more dangerous than predicted by some design models.

An efficient way of improving the accuracy of the theoretical procedures is to incorporate experimental data into the calculation. The aim of this chapter is to present such a method which takes into account all the contributing structural and non-structural elements. The experimental data needed for the procedure are the natural frequencies which are used for establishing the characteristic stiffnesses. The stiffnesses are then used to create more accurate formulae for the global stress and stability analyses of the building.

The simple procedure presented in this chapter can be used to calculate the critical loads and the maximum horizontal displacements and the rotation of buildings with a doubly symmetrical bracing system.

8.1 STIFFNESSES

The procedure to be presented is based on the characteristic stiffnesses which are obtained using experimental data. The lateral stiffnesses for multistorey buildings are obtained from formulae (3.52) by simple rearrangement:

$$EI_Y = \frac{\rho A H^4 f_X^2}{0.3136 r_f^2}, \qquad EI_X = \frac{\rho A H^4 f_Y^2}{0.3136 r_f^2}, \qquad (8.1)$$

where

ρ is the mass density per unit volume of the building defined by formula (3.45),
f_X, f_Y are the first natural frequencies of lateral vibration in the principal directions, measured on the building,
A is the area of the plan of the building,
H is the height of the building,
r_f is a modifier given in Table 3.8.

The torsional stiffness is obtained in a similar manner by rearranging formula (3.55):

$$EI_\omega = \frac{\rho A H^4 i_p^2 f_\varphi^2}{\eta^2 r_f^2}, \qquad (8.2)$$

where

f_φ is the first pure torsional frequency, measured on the building,
i_p is the radius of gyration of the area of the plan of the building,
η is the frequency parameter given in Table 3.9 in section 3.2.1.

The contribution of the Saint-Venant stiffness to the overall torsional resistance of the building is accounted for by frequency parameter η in formula (8.2).

When the warping stiffness of the bracing system is zero (e.g. a single bracing core of closed cross-section), the torsional stiffness is obtained from formula (3.56) where $i = 1$ is substituted for the fundamental torsional frequency:

$$GJ = 16 \rho A H^2 i_p^2 f_\varphi^2. \qquad (8.2a)$$

184 Use of frequency measurements

$$GJ = 16\rho AH^2 i_p^2 f_\varphi^2. \tag{8.2a}$$

The stiffnesses for single-storey buildings are obtained from formulae (3.67) and (3.72) as

$$EI_Y = \frac{4\pi^2 AH^3 f_X^2 \rho^*}{3}, \qquad EI_X = \frac{4\pi^2 AH^3 f_Y^2 \rho^*}{3} \tag{8.3}$$

and

$$EI_\omega = \frac{4\pi^2 AH^3 i_p^2 f_\varphi^2 \rho^*}{k+3}, \tag{8.4}$$

where

ρ^* is the mass density per unit area of the top floor, defined by formula (3.68).

When the warping stiffness of the bracing system is zero, the torsional stiffness is obtained from formula (3.70):

$$GJ = 4\pi^2 H i_p^2 A\rho^* f_\varphi^2. \tag{8.4a}$$

8.2 CRITICAL LOADS

In combining the stiffnesses given by formulae (8.1) to (8.4) and the theoretical basic critical loads presented in section 3.1.1, simple closed-form formulae are obtained for the basic critical loads which now contain the frequencies measured on the building.

8.2.1 Multistorey buildings under uniformly distributed floor load

In making use of formulae (3.11) and (8.1), the formulae for the basic sway critical loads are obtained as

$$N_{cr,X} = 25 r\rho AH^2 f_X^2, \qquad N_{cr,Y} = 25 r\rho AH^2 f_Y^2, \tag{8.5}$$

where, based on formulae (3.15) and (3.53a) and Tables 3.1 and 3.8

Critical loads 185

$$r = \frac{r_s}{r_f^2} = \frac{n+2.06}{n+1.588} \qquad (8.6)$$

is a modifier which takes into account the fact that the load and the mass are concentrated on floor levels. Values for r are given in Table 8.1.

Table 8.1 Values for modifier r

n	1	2	3	4	5	6	7	8	9	10
r	1.296	1.238	1.103	1.084	1.072	1.062	1.055	1.049	1.045	1.041
n	12	14	16	18	20	25	30	40	50	>50
r	1.035	1.030	1.027	1.024	1.022	1.018	1.015	1.011	1.009	1.000

The pure torsional critical load is obtained by combining formulae (3.16) and (8.2):

$$N_{cr,\varphi} = \frac{\alpha \rho r A f_\varphi^2 H^2}{\eta^2}. \qquad (8.7)$$

Critical load parameter α and frequency parameter η are to be found in Figs 3.3 and 3.11 and Tables 3.2 and 3.9 in Chapter 3 as a function of k_s and k, respectively. When values for k (and for k_s) are not available, e.g. the size and/or the arrangement of the elements of the bracing system are not known, or when the effect of warping stiffness EI_ω is by orders of magnitude greater than that of the Saint-Venant stiffness GJ (i.e. when $k \ll 1$ holds), then the approximation $k = 0$ (and $k_s = 0$) can be used. This approximation leads to

$$N_{cr,\varphi} = 25 r \rho A H^2 f_\varphi^2. \qquad (8.8)$$

Formula (8.8) is always conservative and in most practical cases (when $k < 0.5$ holds) results in a critical load smaller than the exact one by not more than 10%.

Formulae (8.7) and (8.8) cannot be used when the warping stiffness is zero. In such cases, the combination of formulae (3.17) and (8.2a) yields the critical load of pure torsional buckling:

$$N_{cr,\varphi} = 16 \rho A H^2 f_\varphi^2. \qquad (8.9)$$

8.2.2 Concentrated top load; single-storey buildings

With single-storey buildings when the load is concentrated on top floor level, the basic sway critical loads are obtained by combining formulae (3.30) and (8.3):

$$F_{cr,X} = 32.47 f_X^2 H A \rho^*, \qquad F_{cr,Y} = 32.47 f_Y^2 H A \rho^*. \qquad (8.10)$$

The combination of formulae (3.31) and (8.4) leads to the critical load of pure torsional buckling as

$$F_{cr,\varphi} = \frac{4\pi^2 k^2 + \pi^4}{k+3} f_\varphi^2 H A \rho^*. \qquad (8.11)$$

In formulae (8.10) and (8.11) ρ^* is the mass density per unit area of the top floor, defined by formula (3.68).

When the warping stiffness dominates over the Saint-Venant torsional stiffness (i.e. $k \to 0$), formula (8.11) simplifies to

$$F_{cr,\varphi} = 32.47 f_\varphi^2 H A \rho^*. \qquad (8.12)$$

Formulae (8.11) and (8.12) cannot be used in the special case when the warping stiffness is zero. In such a case, the combination of (formulae (3.31a) and (8.4a) leads to

$$F_{cr,\varphi} = 4\pi^2 f_\varphi^2 H A \rho^*. \qquad (8.13)$$

8.3 DEFORMATIONS

By making use of the stiffnesses defined by formulae (8.1) to (8.4), the deformations of buildings subjected to horizontal load can be easily calculated. When the formulae for the horizontal displacements are derived, however, an important aspect has to be borne in mind. The theoretical formulae in Chapter 5 yield the displacements in the 'arbitrary' directions x and y (normally parallel with the sides of the building for rectangular layouts), but the formulae of the lateral stiffnesses defined by formulae (8.1) and (8.3) are related to principal directions X and Y.

8.3.1 Multistorey buildings subjected to horizontal load of trapezoidal distribution

After combining formulae (5.16), (5.17) and (5.19), setting $I_{xy} = 0$ and substituting formula (8.1) for the lateral stiffnesses, the horizontal displacements of the equivalent column in the principal directions are obtained as

$$u(z) = \frac{0.3136 q_{0X} r_f^2}{\rho A f_X^2 H^4}(Z_1 + \mu Z_2) \tag{8.14}$$

and

$$v(z) = \frac{0.3136 q_{0Y} r_f^2}{\rho A f_Y^2 H^4}(Z_1 + \mu Z_2), \tag{8.15}$$

where

q_{0X}, q_{0Y} are the uniform part of the horizontal load in the principal directions,
r_f is a modifier, according to Table 3.8 in section 3.2.1,
ρ is the mass per unit volume of the building,
A is the area of the plan of the building,
H is the height of the building,
μ is the slope of the load function according to formula (5.2),
Z_1, Z_2 are auxiliary functions defined by formulae (5.18),
f_X, f_Y are fundamental frequencies in the principal directions.

The equivalent column develops maximum displacements in the principal directions at $z = 0$:

$$u_{max} = u(H) = \frac{q_{0X} r_f^2}{\rho A f_X^2}(0.0392 + 0.0287\mu) \tag{8.16}$$

and

$$v_{max} = v(H) = \frac{q_{0Y} r_f^2}{\rho A f_Y^2}(0.0392 + 0.0287\mu). \tag{8.17}$$

188 Use of frequency measurements

To calculate the rotations of the equivalent column, the Saint-Venant stiffness in formula (5.27) should first be replaced by the warping stiffness by making use of formula (3.19). The resulting expression can then be combined with formula (8.2) of the warping stiffness which contains the fundamental torsional frequency measured on the building. Some rearrangement then results in the rotation as

$$\varphi(z) = \frac{\eta^2 r_f^2 m_{z0}}{k^4 \rho A f_\varphi^2 i_p^2 \cosh k} \left\{ (1+\mu)\left(\cosh\frac{kz}{H} - 1\right) + (1 + \frac{\mu}{2} - \frac{\mu}{k^2})k \times \right.$$
$$\left. \times \left(\sinh(k - \frac{kz}{H}) - \sinh k\right) + \frac{zk^2}{H^2}\cosh k\left(H - \frac{z}{2} + \mu(\frac{H}{2} - \frac{H}{k^2} - \frac{z^2}{6H})\right)\right\}. \quad (8.18)$$

Maximum rotation develops at $z = 0$:

$$\varphi_{max} = \frac{\eta^2 r_f^2 m_{z0}}{k^2 \rho A f_\varphi^2 i_p^2}\left(\frac{(1+\mu)(\cosh k - 1)}{k^2 \cosh k} - (1 + \frac{\mu}{2} - \frac{\mu}{k^2})\frac{\tanh k}{k} + \right.$$
$$\left. + \frac{1}{2} + \frac{\mu}{3} - \frac{\mu}{k^2}\right). \quad (8.19)$$

The rotation of the equivalent column can be calculated in a much simpler manner when the torsional resistance of the bracing system is dominated by the warping stiffness and the effect of the Saint-Venant torsional stiffness is negligible. If this condition is satisfied, as is the case in most practical structural engineering applications, then formula (5.31) can be used. After carrying out the necessary modifications of formula (8.2) of the torsional stiffness ($k \to 0$), and making some rearrangement, the formula for the rotations of the equivalent column is obtained as

$$\varphi(z) = \frac{0.3136 m_{z0} r_f^2}{f_\varphi^2 i_p^2 H^4 \rho A}(Z_1 + \mu Z_2). \quad (8.20)$$

Maximum rotation develops at the top of the equivalent column:

$$\varphi_{max} = \varphi(H) = \frac{m_{z0} r_f^2}{f_\varphi^2 i_p^2 \rho A}(0.0392 + 0.0287\mu). \quad (8.21)$$

Formulae (8.18) to (8.21) cannot be used when the warping stiffness of the equivalent column is zero. In such cases the rotation can be calculated by combining formulae (5.29) and (8.4a):

$$\varphi(z) = \frac{m_{z0}z}{16H^2 i_p^2 \rho A f_\varphi^2}\left(H - \frac{z}{2} + \mu(\frac{H}{2} - \frac{z^2}{6H})\right). \qquad (8.22)$$

Maximum rotation develops at $z = 0$:

$$\varphi_{max} = \varphi(H) = \frac{m_{z0}}{32 i_p^2 \rho A f_\varphi^2}\left(1 + \frac{2\mu}{3}\right). \qquad (8.23)$$

8.3.2 Concentrated force at top level; single-storey buildings

The horizontal displacements of the equivalent column under a concentrated force are given by formulae (5.75a) and (5.75b). After substituting expressions (8.3) for the lateral stiffnesses, the horizontal displacements are obtained as

$$u(z) = \frac{3F_X}{4\pi^2 f_X^2 A \rho^*}\left(\frac{z^2}{2H^2} - \frac{z^3}{6H^3}\right) \qquad (8.24)$$

and

$$v(z) = \frac{3F_Y}{4\pi^2 f_Y^2 A \rho^*}\left(\frac{z^2}{2H^2} - \frac{z^3}{6H^3}\right), \qquad (8.25)$$

where

F_X, F_Y are the horizontal forces in the principal directions,
ρ^* is the mass density per unit area defined by formula (3.68).

Maximum displacements develop at the top of the equivalent column:

$$u_{max} = u(H) = \frac{F_X}{4\pi^2 f_X^2 A \rho^*} \qquad (8.26)$$

and

$$v_{max} = v(H) = \frac{F_Y}{4\pi^2 f_Y^2 A\rho^*}. \tag{8.27}$$

The rotations of the equivalent column are obtained by combining formulae (5.77), (3.19) and (8.4):

$$\varphi(z) = \frac{M_z(k+3)}{4\pi^2 k^3 A\rho^* i_p^2 f_\varphi^2}\left(\frac{kz}{H} - \sinh\frac{kz}{H} + (\cosh\frac{kz}{H} - 1)\tanh k\right). \tag{8.28}$$

Maximum rotation develops at $z = H$:

$$\varphi_{max} = \varphi(H) = \frac{M_z(k+3)}{4\pi^2 k^2 A\rho^* i_p^2 f_\varphi^2}\left(1 - \frac{\tanh k}{k}\right). \tag{8.29}$$

The above formulae for the rotations simplify when the Saint-Venant stiffness is negligibly small. The combination of formulae (5.79a) and (8.4) (with $k = 0$) results in

$$\varphi(z) = \frac{3M_z}{4\pi^2 i_p^2 f_\varphi^2 A\rho^*}\left(\frac{z^2}{2H^2} - \frac{z^3}{6H^3}\right). \tag{8.30}$$

Maximum rotation develops at the top of the equivalent column:

$$\varphi_{max} = \varphi(H) = \frac{M_z}{4\pi^2 i_p^2 f_\varphi^2 A\rho^*}. \tag{8.31}$$

Formulae (8.29) to (8.31) cannot be used when the warping stiffness of the equivalent column is zero. In such cases, the rotation is obtained by substituting formula (8.4a) for GJ in formula (5.79c):

$$\varphi(z) = \frac{M_z z}{4\pi^2 H i_p^2 f_\varphi^2 A\rho^*}. \tag{8.32}$$

Maximum rotation develops at the top of the equivalent column:

$$\varphi_{max} = \varphi(H) = \frac{M_z}{4\pi^2 i_p^2 f_\varphi^2 A\rho^*}. \tag{8.33}$$

8.3.3 Deformations of the building

The rotation of the building is identical to the rotation of the equivalent column. When the displacements of the building are calculated, however, in addition to the displacements of the equivalent column, the additional displacements due to the rotation of the building also have to be taken into account – as described in section 5.3.

9

Equivalent wall for frameworks; Buckling analysis of planar structures

The structural types covered in this chapter include frameworks on pinned and fixed supports, with or without cross-bracing, in-filled frameworks, shear walls and coupled shear walls. The simple procedures and closed-form solutions produced here make it possible for the structural designer to carry out planar stability checks on multistorey, multibay structures in minutes. In addition, a simple procedure is given for the establishment of an equivalent wall for frameworks and coupled shear walls. In using the equivalent wall, the frameworks and coupled shear walls can be included in the bracing system, the equivalent column introduced in Chapter 2 can be set up and the 3-dimensional global analyses presented in the previous chapters can be carried out.

9.1 INTRODUCTION

There are basically two ways to produce the critical load of frameworks and coupled shear walls. Structural designers can rely on commercially available computer programs based on the 'exact' method or they can use approximate methods. Both approaches have their advantages and disadvantages.

Well-established methods have been developed for the calculation of stresses, deformations and frequencies in frameworks and coupled shear walls and a number of computer packages are available for the exact analysis. These computer procedures are now standard items in design offices. The routine application of these programs has also proved their reliability and accuracy.

Partly because of the large size of the structures and partly because of the sometimes ill-conditioned stiffness matrixes, the situation regarding the stability analysis is somewhat different. The stability analysis of

Introduction

multistorey, multibay frameworks and coupled shear walls is a formidable task and the larger the structure, the more complicated the solution becomes. The complexity of the problem and the great number of input and output data needed for the analysis of large structures often make it difficult to get a clear picture of the behaviour and to achieve optimum structural solution. The reliability of some computer programs is also questionable in certain stiffness regions. Even sophisticated FE procedures may have difficulties in determining the critical load. The LUSAS 'User Manual' [1995] warns that the solution is not without its problems and convergence problems might emerge in the iterative procedure. Indeed, the warning was found to be justified when the accuracy analysis of the methods presented in this chapter was carried out.

Some excellent approximate procedures have been developed. However, they are either still too complicated for hand calculations or their range of applicability is limited [Stevens, 1967; Wood, 1974 and 1975; Horne, 1975; Council, 1978a and 1978b; MacLeod and Marshall, 1983; Lightfoot, McPharlin and Le Messurier, 1979; Wood and Marshall 1983; MacLeod, 1990]. It is a common 'feature' of these approximate methods that their limitations have not been fully investigated and very little information has been published about their accuracy.

The main objective of this chapter is to produce simple, descriptive, albeit approximate, closed-form solutions for the stability analysis of frameworks and coupled shear walls and to introduce an equivalent wall which can be used for the 3-dimensional global stability analysis. It is also the intention of this chapter to establish simple models for the stability analysis, which offer a clear picture of the behaviour and show clearly the effects of the most representative geometrical and stiffness characteristics.

The formulae and procedure presented here are based on the application of the continuum method and the summation theorems of civil engineering. By generalizing Asztalos' [1972] method and using the continuum model, a simple procedure was developed and closed-form solutions were given for the approximate stability analysis of frameworks and coupled shear walls of rectangular network [Zalka and Armer, 1992]. The method was further developed by Kollár [1986] who also demonstrated that different planer bracing elements could be treated in a unified manner by using a model that is based on a sandwich column with thin or thick faces. He also indicated that the accuracy of the continuum model could be further improved. The results of the latest developments [Zalka, 1998b] with the improved formulae are presented in this chapter.

9.2 CHARACTERISTIC DEFORMATIONS, STIFFNESSES AND PART CRITICAL LOADS

A great number of material, geometrical and stiffness characteristics should be considered when the critical load is calculated. As it is not possible to take everything into consideration, the question emerges what to include and what to neglect when the mathematical and physical models of the structure are established.

All the important characteristics must be taken into consideration; otherwise the results are not accurate enough. To decide which characteristics are important and which are not, it is necessary to know their contribution to the resistance of the structure. It is also important to know the contribution of the different characteristics if the performance of the structure is to be improved. One way to achieve this is the establishment of the characteristic deformations and, as deformations and stiffnesses are in close relationship, the related stiffnesses. A part critical load can then be attached to each type of stiffness. Finally, using these part critical loads, the overall critical load of the structure can be easily produced.

The establishment of the continuum model, the derivation of the governing differential equations and comprehensive numerical analyses show that the deformation of a framework or a system of coupled shear walls can be superimposed using three different types of deformation: the full-height *bending* deformation of the individual columns/walls (Fig. 9.1/a), the full-height *bending* deformation of the structure as whole unit (Fig. 9.1/b) and the *shear* deformation of the structure (Fig. 9.1/c) [Zalka, 1992]. A characteristic stiffness can be attached to each type of deformation, which then leads to the corresponding part critical load. The overall critical load is finally obtained by combining these part critical loads.

The structure can utilize its bending stiffness in two ways. The columns may develop full-height buckling and the role of the (relatively flexible) beams is restricted to connecting the columns (Fig. 9.1/a). The structure behaves as if it was a set of columns and overall buckling is characterized by the buckling of the columns of height H, having a bending stiffness of

$$E_c I_c = \sum_{1}^{n} E_{c,i} I_{c,i}, \tag{9.1}$$

where $E_{c,i}$ and $I_{c,i}$ are the modulus of elasticity and the second moment of area of the *i*th column and n stands for the number of columns. (The full-height bending of the individual columns is sometimes called 'local' bending as $I_{c,i}$ is calculated with respect to the local centroidal axis of the ith column, marked with l in Fig. 9.6/a.)

Assuming fixed support for the columns, the bending critical load which is associated with the local bending stiffness, is obtained using Timoshenko's classical formula [Timoshenko and Gere, 1961] for columns under uniformly distributed vertical load as

$$N_l = \frac{7.837 r_s E_c I_c}{H^2} \qquad (9.2)$$

for structures under uniformly distributed floor load and as

$$F_l = \frac{\pi^2 E_c I_c}{4H^2} \qquad (9.3)$$

for structures under concentrated top load, where H is the height of the structure and r_s is a reduction factor whose role is discussed later on and whose values are given in Table 3.1 in section 3.1.1.

The effect of the full-height bending buckling of the individual columns on the overall critical load is normally not significant. However, this effect does contribute to the overall behaviour and in certain cases, for example when frameworks have relatively stiff columns or when coupled shear walls have wide walls and relatively weak lintels, the contribution can be considerable.

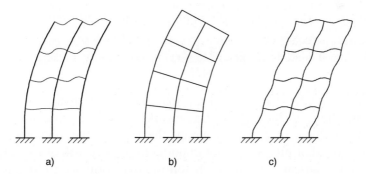

Fig. 9.1 Characteristic deformations. a) Full-height bending of the individual columns, b) full-height bending of the framework as a whole, c) shear deformation.

A structure may develop full-height bending in a different manner. The bending deformation of the framework may be characterized by the bending of the framework as a whole structural unit when the columns act as longitudinal fibres (in tension and compression) and the role of the (relatively stiff) beams is to transfer shear so as to make the columns work together in this fashion (Fig. 9.1/b). The bending stiffness associated with this bending deformation is defined by

$$E_c I_g = \sum_1^n E_{c,i} A_{c,i} t_i^2 . \tag{9.4}$$

The term 'global' bending is also used to describe the full-height bending of the framework as a whole unit as I_g, the 'global' second moment of area of the cross-sections of the columns, is calculated with respect to their 'global' centroidal axis (marked with g in Fig. 9.6/a):

$$I_g = \sum_1^n A_{c,i} t_i^2 . \tag{9.5}$$

In the above formulae, t_i is the distance of the ith column from the centroid of the cross-sections and $A_{c,i}$ is the cross-sectional area of the ith column.

The bending critical load which is associated with the full-height bending of the structure as a whole unit is obtained from

$$N_g = \frac{7.837 r_s E_c I_g}{H^2} \tag{9.6}$$

for structures under uniformly distributed floor load and from

$$F_g = \frac{\pi^2 E_c I_g}{4H^2} \tag{9.7}$$

for structures under concentrated top load. Due to the nature of the full-height bending of the whole framework when the columns act as longitudinal fibres of a solid column, the type of the supports of the columns – fixed or pinned – is not important; the only task of the supports is to prevent vertical movements. It follows that formulae (9.6) and (9.7) are equally valid for frameworks on fixed and also on pinned supports.

Reduction factor r_s in formulae (9.2) and (9.6) allows for the fact that the actual load of the structure consists of concentrated forces on floor levels and is not uniformly distributed over the height as is assumed for the derivation of the original formulae for buckling. The continuous load

Characteristic deformations 197

is obtained by distributing the concentrated forces *downwards* (cf. Figs 3.1 and 9.4) resulting in a more favourable load distribution. The effect of this unconservative manoeuvre can be accounted for by applying Dunkerley's summation theorem (cf. section 3.1.1) which leads to the introduction of the reduction factor

$$r_s = \frac{n}{n+1.588}. \qquad (9.8)$$

This phenomenon is similar to the one discussed in section 3.1.1 and values for the reduction factor r_s are given there, in Table 3.1, as a function of the number of storeys n.

The full-height bending of the whole structure normally plays an important role in the behaviour and it has a significant effect on the value of the overall critical load.

The shear stiffness which, by definition, also represents the shear critical load, is linked to the shear deformation of the structure. The shear deformation is characterized by lateral sway which is mainly resisted by the stiffening effect of the beams (Fig. 9.1/c; Fig. 9.2/a). The shear deformation and shear stiffness are associated by two phenomena: full-height sway (due to the stiffening effect of the beams over the height of the framework) and storey-height sway (because the stiffening effect is only concentrated on storey levels). Consequently, the corresponding shear critical load originates from two sources and its value is obtained in two steps.

First, in assuming that the stiffening effect of the beams is continuously distributed over the height and there is no additional sway between the beams (Fig. 9.2/b), the part critical load which is associated with this full-height shear deformation is obtained as

$$K_g = 2\sum_{1}^{n-1} \frac{6E_{b,i}I_{b,i}}{l_i h}, \qquad (9.9)$$

where $E_{b,i}$, $I_{b,i}$ and l_i are the modulus of elasticity, the second moment of area and the length of the ith beam and h is the storey height.

Second, in assuming that the structure only develops sway between the storey levels, a storey-height section of the structure is investigated. Assuming frameworks on fixed supports, each storey behaves in the same way (Fig. 9.2/c) and the part critical load which characterizes this storey-height shear deformation is obtained as

198 Equivalent wall

$$K_l = \sum_1^n \frac{\pi^2 E_{c,i} I_{c,i}}{h^2}. \tag{9.10}$$

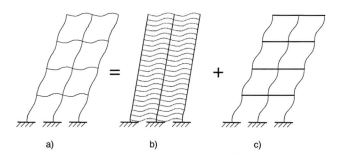

Fig. 9.2 Shear deformations. a) Total shear deformation, b) full-height (continuous) shear, c) storey-height shear.

According to the Föppl–Papkovich theorem [Tarnai, 1999], the two part critical loads can be combined and the shear critical load of the structure is obtained as

$$K = \left(\frac{1}{K_l} + \frac{1}{K_g}\right)^{-1} = K_g \frac{K_l}{K_g + K_l}. \tag{9.11}$$

Formula (9.11) is of general validity. In the special case when all the columns, beams and bays are identical, respectively, i.e. with $l_i = l$, $I_{c,i} = I_c$, $I_{b,i} = I_b$ and $E_{b,i} = E_{c,i} = E$, formula (9.11) assumes the form

$$K = \frac{12 E n I_c}{h^2 \left(\dfrac{12}{\pi^2} + \dfrac{n}{n-1}\lambda\right)}, \tag{9.11a}$$

where

$$\lambda = \frac{\dfrac{I_c}{h}}{\dfrac{I_b}{l}}. \tag{9.11b}$$

For single-bay frameworks ($n = 2$), formula (9.11a) simplifies to

$$K = \frac{24EI_c}{h^2\left(\dfrac{12}{\pi^2} + 2\lambda\right)}. \tag{9.11c}$$

It is important to note that the value of the shear critical load does not depend on the distribution of the vertical load, and neither on the height of the structure. The type of support of the structure may only indirectly affect the value of the shear stiffness (through the storey-height shear stiffness).

It is noted here that the terms 'global' shear stiffness and 'local' shear stiffness are also used for the full-height and storey-height shear stiffnesses as the associated shear deformations affect the whole structure and the storey-height part of the whole structure, respectively.

Theoretical investigations [Hegedűs and Kollár, 1987] and accuracy analyses [Zalka and Armer, 1992] show that the three types of deformation do not always influence the value of the critical load by the same 'weight' and cannot even develop at the same time in certain stiffness regions. This can be taken into account by using combination factor r which is defined as the ratio of the storey-height shear stiffness to the sum of the storey-height and full-height shear stiffnesses:

$$r = \frac{K_l}{K_g + K_l}. \tag{9.12}$$

Using the part critical loads associated with the full-height bending of the individual columns, the full-height bending of the framework as a whole and the shear deformation and some modifiers, it is possible to produce simple closed-form solutions for the critical load of regular frameworks and coupled shear walls on different types of support, with different structural arrangements and under different vertical loads. These formulae are presented in the following sections.

9.3 FRAMEWORKS ON FIXED SUPPORTS

During buckling, frameworks on fixed supports normally develop a deformation which is the combination of the three characteristic types of deformation (Fig. 9.3).

Three simple methods will be shown for the analysis in the following sections. They represent three totally different approaches but they have

200 Equivalent wall

one thing in common: each method is based on the use of the three part critical loads N_l, N_g and K and the critical load is obtained by the combination of the part critical loads. The methods are described in detail in this section in relation to frameworks on fixed supports, and then in the following sections they will be applied to frameworks on pinned supports, frameworks with cross-bracing, infilled frameworks and coupled shear walls.

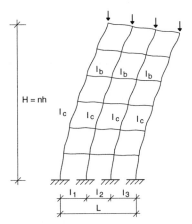

Fig. 9.3 Frameworks on fixed supports.

9.3.1 The application of summation theorems

Once the part critical loads are available, for example for frameworks under uniformly distributed load, the repeated application of the Föppl–Papkovich theorem and the Southwell theorem [Tarnai, 1999] results in a very simple formula for the critical load. In considering the full-height bending of the structure as a whole unit and its shear deformation, it is first assumed that the structure is stiffened against full-height bending as a whole, then against developing shear deformation. The reciprocal summation of the corresponding critical loads leads to the critical load of a framework with both full-height bending and shear stiffnesses. However, the individual columns of the framework may also develop full-height bending. The effect of this phenomenon can be taken into account by applying the Southwell theorem: the critical load which belongs to the full-height buckling of the individual columns (N_l) and the combined critical load of the full-height buckling as a whole (N_g) and shear critical load (K) have to be added up:

$$N_{cr} = N_l + \left(\frac{1}{N_g} + \frac{1}{K}\right)^{-1}. \tag{9.13}$$

To make formula (9.13) more easily comparable to other formulae to be introduced later, it is rearranged to

$$N_{cr} = \frac{N_l + K + N_l K/N_g}{1 + K/N_g}. \tag{9.13a}$$

This formula can also be used for the calculation of the critical load of frameworks under concentrated top load: F_l and F_g have to be substituted for N_l and N_g.

9.3.2 The continuum model

Although the continuum model was originally developed for the stress analysis of proportional frameworks [Csonka, 1956; 1965a; 1965b] (cf. section 4.5), it has been shown that its application to the stability analysis of regular frameworks results in reliable elastic critical loads [Kollár, 1986; Zalka, 1998b].

The continuum model is obtained by cutting the beams at the contraflexure points and then combining the individual columns of the framework to form an equivalent column. The stiffening effect of the beams is taken into account by applying concentrated bending moments to the columns at floor levels, which are then uniformly distributed, leading to the shear stiffness of the model. In a similar manner, the uniformly distributed load on the beams is first replaced by concentrated forces at floor levels, then these forces are uniformly distributed over the height of the column (Fig. 9.4).

The equilibrium of an elementary section of the equivalent column leads to the governing differential equation

$$y'''' + \left(\frac{N(z) - K}{r_s E_c I_c} y'\right)' = 0, \tag{9.14}$$

202 Equivalent wall

where $N(z) = qz$ is the vertical load of the structure at z and q is the intensity of the uniformly distributed vertical load. Reduction factor r_s is given by formula (9.8) and in Table 3.1. The relationship between the intensity of the uniformly distributed load on the beams (p) and the intensity of the uniformly distributed vertical load (q) is

$$q = \frac{pLn}{H}, \qquad (9.15)$$

where L is the total width of the framework and n is the number of storeys.

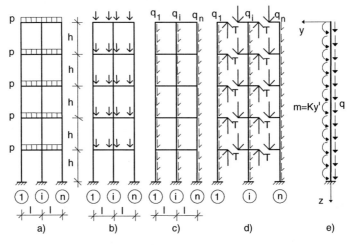

Fig. 9.4 Origination of the continuum model for frameworks on fixed supports.
a) Framework, b) replacement of beam-load, c) distribution of load over the height, d) cutting through the beams, e) equivalent column.

The governing differential equation is accompanied by the boundary conditions

$$y(0) = 0, \qquad (9.16a)$$

$$y'(H) = 0, \qquad (9.16b)$$

$$y''(0) = 0, \qquad (9.16c)$$

$$y'''(H) = 0 \qquad (9.16d)$$

in the coordinate system whose origin is fixed to the top of the equivalent column (Fig. 9.4/e).

The model directly takes into consideration the full-height bending of the individual columns and the shear deformation of the framework while the full-height bending of the structure as a whole is incorporated into the model and into the formulae afterwards. The solution of the eigenvalue problem defined by equation (9.14) leads to

$$N_{cr} = \frac{r(\alpha - \beta)N_l + K + rN_l K/N_g}{1 + K/N_g}, \qquad (9.17)$$

where N_l, N_g, K and r are the part critical loads and the combination factor, as introduced in section 9.2 and defined by formulae (9.2), (9.6), (9.11) and (9.12).

Table 9.1 Critical load parameter α for frameworks on fixed supports

β	α	β	α	β	α	β	α
0.0000	1.0000	0.05	1.1487	2	5.624	80	106.44
0.0005	1.0015	0.06	1.1782	3	7.427	90	118.38
0.001	1.0030	0.07	1.2075	4	9.100	100	130.25
0.002	1.0060	0.08	1.2367	5	10.697	200	246.24
0.003	1.0090	0.09	1.2659	6	12.241	300	359.51
0.004	1.0120	0.10	1.2949	7	13.749	400	471.29
0.005	1.0150	0.20	1.5798	8	15.227	500	582.06
0.006	1.0180	0.30	1.8556	9	16.682	1000	1127.5
0.007	1.0210	0.40	2.1226	10	18.118	2000	2199.1
0.008	1.0240	0.50	2.3817	20	31.820	5000	5360.5
0.009	1.0270	0.60	2.6333	30	44.862	10000	10567
0.010	1.0300	0.70	2.8780	40	57.545	100000	102579
0.020	1.0598	0.80	3.1163	50	69.991	1000000	1011864
0.030	1.0896	0.90	3.3488	60	82.265	2000000	2018802
0.040	1.1192	1.00	3.5758	70	94.405	>2000000	$\beta+1$

The critical load parameter α (the eigenvalue of the problem) in formula (9.17) is given in Fig. 9.5 and in Table 9.1 as a function of stiffness parameter

$$\beta = \frac{K}{N_l}. \qquad (9.18)$$

The solution of eigenvalue problem (9.14) is demonstrated in Appendix B in detail.

204 Equivalent wall

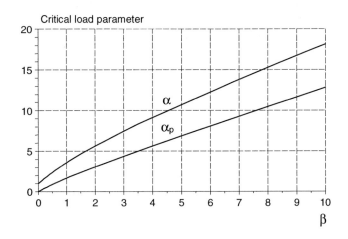

Fig. 9.5 Critical load parameters α and α_p for fixed and pinned frameworks.

The solution is obtained in a similar – but much simpler – way when the framework is subjected to concentrated load at the top of the structure:

$$F_{cr} = \frac{rF_l + K + rF_l K/F_g}{1 + K/F_g}, \qquad (9.19)$$

where K, F_l and F_g are the shear critical load, the full-height bending buckling load of the individual columns and the full-height bending buckling load of the structure as a whole, respectively, and r is the combination factor, given in section 9.2.

9.3.3 The sandwich model

The sandwich model [Plantema, 1961; Allen, 1969] offers a different approach to the problem. The faces of the sandwich column represent the columns of the framework (with its 'local' and 'global' bending stiffnesses) and the connecting media represents the beams (with their shear stiffness) (Fig. 9.6/a/b).

According to this approach, the framework is analysed using an equivalent sandwich column with either thick or thin faces. The model with thin faces will be used here. This model takes into consideration the 'global' bending and shear stiffnesses but neglects the 'local' bending stiffness of the framework. This approximation is in line with frameworks on pinned supports as, because of the pinned supports, the individual columns of the framework cannot 'utilize' their own bending stiffness. As a rule, the approximation is also justified with frameworks on fixed supports as the bending stiffness of the individual columns is normally insignificant compared to the 'global' bending stiffness in structural engineering practice and therefore can be ignored as a conservative approximation. However, if for some reason it is necessary to take into consideration the 'local' bending stiffness of the columns or/and if the distribution of the load is different from the uniformly distributed case, a more sophisticated (and more complicated) method can be used which was developed by Hegedűs and Kollár [1984] for the buckling analysis of sandwich columns with thick faces subjected to axial load of arbitrary distribution.

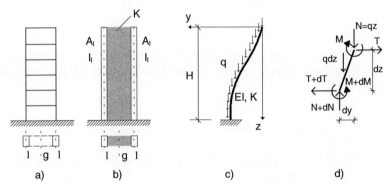

Fig. 9.6 Origination and analysis of the sandwich model. a) Frame, b) sandwich, c) buckled shape, d) elementary section.

The governing differential equation of the problem is obtained by considering the equilibrium of an elementary section of the sandwich column (Fig. 9.6/c/d) which is characterized by the 'global' bending stiffness and the shear stiffness of the system [Zalka and Armer, 1992]:

$$y'''' - \frac{q}{K}(3y''' + zy'''') + \frac{q}{r_s E_c I_g}(y' + zy'') = 0. \qquad (9.20)$$

In the governing differential equation q represents the intensity of the uniformly distributed vertical load, $E_c I_g$ is the 'global' bending stiffness of the columns, r_s is the modifier as defined by formula (9.8) and given in Table 3.1 and K is the shear stiffness of the system.

The governing differential equation is supplemented by the boundary conditions

$$y(0) = 0, \qquad (9.21a)$$

$$y'(H)(1 - \frac{qH}{K}) = 0, \qquad (9.21b)$$

$$y''(0) - \frac{q}{K} y'(0) = 0, \qquad (9.21c)$$

$$y'''(0) - \frac{2q}{K} y''(0) = 0, \qquad (9.21d)$$

which express that the lateral translation equals zero at the top of the column, the tangent to the equivalent column at the bottom of the column is parallel to axis z, the bending moments vanish at the top of the column and the shear forces are also zero at the top of the column, in the coordinate system whose origin is fixed at the top of the equivalent column (Fig. 9.6/c).

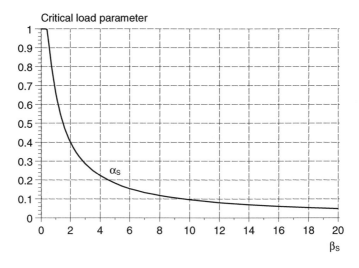

Fig. 9.7 Critical load parameter α_S.

The eigenvalue problem defined by the governing differential equation (9.20) and boundary conditions (9.21a) to (9.21d) is unique, inasmuch as the eigenvalue appears both in the differential equation and in the boundary conditions. As a rule, problems of this kind are difficult to handle and the situation is made worse by the fact that in certain stiffness regions the eigenvalues are calculated from ill-conditioned matrixes. Accidentally, this is also the case with some of the eigenvalues originating from the continuum model (and from discrete models).

Table 9.2 Critical load parameter α_S for sandwich columns

β_S	α_S	β_S	α_S	β_S	α_S	β_S	α_S
0.0	1.0000	2.0	0.4005	4.0	0.2230	20	0.04884
0.1	1.0000	2.1	0.3852	4.1	0.2181	25	0.03926
0.2	1.0000	2.2	0.3711	4.2	0.2135	30	0.03282
0.3	1.0000	2.3	0.3579	4.3	0.2090	35	0.02819
0.4	0.9972	2.4	0.3457	4.4	0.2047	40	0.02471
0.5	0.9325	2.5	0.3342	4.5	0.2006	45	0.02199
0.6	0.8663	2.6	0.3235	5.0	0.1824	50	0.01981
0.7	0.8051	2.7	0.3134	5.5	0.1672	55	0.01803
0.8	0.7501	2.8	0.3039	6.0	0.1543	60	0.01654
0.9	0.7011	2.9	0.2950	6.5	0.1433	65	0.01527
1.0	0.6575	3.0	0.2866	7.0	0.1337	70	0.01419
1.1	0.6186	3.1	0.2787	7.5	0.1253	80	0.01243
1.2	0.5838	3.2	0.2711	8.0	0.1179	90	0.01105
1.3	0.5526	3.3	0.2640	8.5	0.1114	100	0.00995
1.4	0.5243	3.4	0.2572	9.0	0.1055	200	0.00499
1.5	0.4988	3.5	0.2508	10	0.09544	300	0.00333
1.6	0.4755	3.6	0.2447	12	0.08015	400	0.00250
1.7	0.4543	3.7	0.2389	14	0.06908	500	0.00200
1.8	0.4349	3.8	0.2333	16	0.06069	1000	0.00100
1.9	0.4170	3.9	0.2280	18	0.05413	>1000	$1/(1+\beta_S)$

The generalized power series procedure (demonstrated in Appendix B), however, also used for the analysis of the continuum model, can easily handle eigenvalues in the boundary conditions. The flexibility of the procedure also makes it possible to deal with convergence problems in a relatively simple and effective way. The procedure can be manually controlled and closely monitored. When the convergence process becomes slow or/and unstable in a certain stiffness region, the characteristics of the iteration procedure can be adjusted and the accuracy and reliability of the

208 Equivalent wall

procedure improved. This may slow the procedure, but once an ill-conditioned stiffness region is covered in such a manner, the results can be tabulated and made available for future use.

The solution of eigenvalue problem (9.20) using the technique outlined above leads to a very simple formula and the critical load of frameworks subjected to a uniformly distributed floor load is obtained from

$$N_{cr} = \alpha_S K, \qquad (9.22)$$

where K is the shear stiffness defined by formula (9.11). Critical load parameter α_S is given in Fig. 9.7 and in Table 9.2 as a function of stiffness ratio β_S, defined by

$$\beta_S = \frac{K}{N_g}, \qquad (9.23)$$

where N_g is the part critical load characterizing the bending buckling of the framework as a whole, as defined by formula (9.6).

When the framework is subjected to concentrated forces at the top of the structure, the sandwich model leads to

$$F_{cr} = \frac{K}{1 + K/F_g}. \qquad (9.24)$$

In this formula F_g is the part critical load characterizing the bending buckling of the framework as a whole, as given by formula (9.7).

9.3.4 Design formulae

The importance of the elastic critical load in design practice has been recognized and means by which elastic instability are considered have been presented [Stevens, 1983]. As any investigation dealing with instability in practice needs the elastic critical load (among other characteristics), simple formulae based on the models introduced above are given in this section for design application.

Theoretical and accuracy analyses show that both the continuum and the sandwich models can be used for practical structural design. For frameworks subjected to uniformly distributed load on the beams, each method has stiffness regions where its accuracy is better than that of the other method. As both the continuum and sandwich models approach the same problem from a different direction, it seems to be sensible to combine the two relevant formulae. Indeed, a detailed accuracy analysis [Zalka, 1998b] shows that by combining the two formulae the error range can be narrowed, resulting in more accurate critical loads. The combination is carried out in two steps.

First, formula (9.22) is supplemented by the term rN_l. This step, based on Southwell's summation theorem, makes it possible to take into consideration approximately the 'local' bending deformations of the columns in the sandwich model. Second, the modified formula is combined with formula (9.17) resulting in the formula

$$N_{cr} = \frac{rN_l(1+\alpha-\beta+2\beta_s)+K(1+\alpha_s+\alpha_s\beta_s)}{2(1+\beta_s)} \qquad (9.25)$$

for the calculation of the critical load.

Critical load parameters α and α_S are given in Table 9.1 and Table 9.2 as a function of β and β_S, defined by formulae (9.18) and (9.23), respectively. The part critical loads N_l and K are given by formulae (9.2) and (9.11). Combination factor r is calculated using formula (9.12).

As for concentrate forces on top floor level, the continuum solution is slightly more accurate than the sandwich solution and therefore the continuum formula

$$F_{cr} = \frac{rF_l(1+K/F_g)+K}{1+K/F_g} \qquad (9.26)$$

is advocated for the calculation of the critical load.

Theoretically, the structure is in stable equilibrium if the value of the critical load is greater than the total vertical load, i.e. if, for uniformly distributed load on the beams, the inequality

$$\frac{N}{N_{cr}} < 1 \qquad (9.27)$$

holds, where N is the total vertical load measured at ground floor level.

It is sometimes necessary to consider the simultaneous effect of the two load systems: concentrated load on top of the structure and uniformly distributed load on floor levels. Dunkerley's reciprocal theorem can be applied in such cases and formula (9.27) can be extended:

210 Equivalent wall

$$\frac{F}{F_{cr}} + \frac{N}{N_{cr}} < 1. \tag{9.27a}$$

To achieve adequate safety in practice, however, it is necessary for the critical load to exceed the total vertical load by a certain margin. The required level of safety in practice is discussed in detail in [Stevens, 1967; MacLeod and Zalka, 1996] and it is only mentioned here that if the criterion

$$\frac{F}{F_{cr}} + \frac{N}{N_{cr}} < 0.1 \tag{9.28}$$

is fulfilled, then the structure is stable enough with the necessary lateral stiffness to be considered as a bracing element in a building.

9.4 FRAMEWORKS ON PINNED SUPPORTS

There are cases when fixed supports on ground floor level cannot be constructed or the rigid connection between the superstructure and the substructure is not favourable for some reason. In such cases, frameworks on pinned supports are used, with or without ground floor beams (Fig. 9.8).

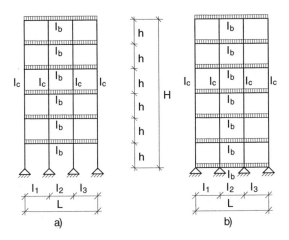

Fig. 9.8 Frameworks on pinned supports. a) Without ground floor beams, b) with ground floor beams.

Both the continuum model and the sandwich model introduced in section 9.3 for frameworks on fixed supports can be used for the analysis. When the continuum model is used, differential equation (9.14) still holds but with the boundary conditions

$$y(0) = 0, \tag{9.29a}$$

$$y''(0) = 0, \tag{9.29b}$$

$$y''(H) = 0, \tag{9.29c}$$

$$y'''(0) - y'(0)\frac{K}{E_c I_c} = 0. \tag{9.29d}$$

The solution of the eigenvalue problem results in critical load parameter α_p, whose values are given in Fig. 9.5 and in Table 9.3, as a function of stiffness ratio β:

$$\beta = \frac{N_l}{K}. \tag{9.30}$$

Table 9.3 Critical load parameter α_p for frameworks on pinned supports

β	α_p	β	α_p	β	α_p	β	α_p
0.0000	0.0000	0.05	0.0987	2	3.060	80	90.33
0.0005	0.0010	0.06	0.1182	3	4.359	90	101.13
0.001	0.0020	0.07	0.1375	4	5.616	100	111.91
0.002	0.0040	0.08	0.1568	5	6.847	200	218.61
0.003	0.0060	0.09	0.1760	6	8.061	300	324.21
0.004	0.0080	0.10	0.1950	7	9.261	400	429.18
0.005	0.0100	0.20	0.3811	8	10.451	500	533.75
0.006	0.0120	0.30	0.5596	9	11.633	600	638.01
0.007	0.0140	0.40	0.7316	10	12.808	800	845.88
0.008	0.0160	0.50	0.8981	20	24.303	1000	1053.1
0.009	0.0180	0.60	1.0598	30	35.545	5000	5153.0
0.010	0.0200	0.70	1.2174	40	46.646	10000	10242
0.020	0.0398	0.80	1.3715	50	57.655	100000	101113
0.030	0.0595	0.90	1.5226	60	68.595	200000	201764
0.040	0.0792	1.00	1.6709	70	79.482	>200000	β

212 Equivalent wall

9.4.1 Frameworks without ground floor beams

The behaviour of pinned frameworks without ground floor beams (Fig. 9.8/a) is somewhat different from that of fixed frameworks. First, the individual columns cannot develop full-height buckling because, due to the pinned supports, they are not stable in themselves. Second, because of the pinned supports and the lack of ground floor beams, these structures are more sensitive to storey-height shear at ground floor level as the columns of the first storey region tend to develop sway buckling. If these differences are taken into account when the part critical loads are established, the procedure presented for frameworks on fixed supports can be applied and similar formulae can be created for the critical load.

When the storey-height shear deformation was investigated with frameworks on fixed supports, it turned out that any storey could be the object of the investigation as each storey behaved in the same way (Fig. 9.2/c). The situation with frameworks on pinned supports is different as the first storey section is the most vulnerable to storey-height shear (Fig. 9.9). Due to the pinned lower support on the ground floor level, the critical load which characterizes storey-height shear is defined by

$$K_l = \sum_1^n \frac{\pi^2 E_{c,i} I_{c,i}}{4h^2}. \tag{9.31}$$

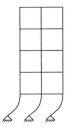

Fig. 9.9 Storey-height shear deformation of frameworks on pinned supports without ground floor beams.

The formula for the critical load associated with full-height shear [formula (9.9)] is unchanged and the shear critical load is computed combining formulae (9.9) and (9.31) as

$$K = K_g \frac{K_l}{K_g + K_l}. \qquad (9.31a)$$

When the framework is under UDL on floor levels (Fig. 9.8/a), the formula for the critical load is obtained in a way similar to that of frameworks on fixed supports and the combination of the continuum solution and the sandwich solution leads to

$$N_{cr} = \frac{rN_l(\alpha_p - \beta) + K(1 + \alpha_s + \alpha_s \beta_s)}{2(1 + \beta_s)}, \qquad (9.32)$$

where K, N_l and r are defined by formulae (9.31a), (9.2) and (9.12), respectively. Critical load parameters α_p and α_s are given in Tables 9.3 and 9.2 as a function of stiffness ratios β and β_S, defined by formulae (9.18) and (9.23).

Both the continuum model and the sandwich model lead to the same formula for the critical loads of frameworks subjected to concentrated load on top of the structure:

$$F_{cr} = \frac{K}{1 + K/F_g}. \qquad (9.33)$$

9.4.2 Frameworks with ground floor beams

When the framework is subjected to UDL on the beams (Fig. 9.8/b), the procedure described in section 9.3.4 is applied and the critical load is obtained as a combination of the results which belong to the two models. To this end, the term

$$rN_l(1-r), \qquad (9.34)$$

which approximately takes into consideration the stiffening effect of the ground floor beams, is added to the sandwich solution. The combination of the two formulae then results in the critical load for the uniformly distributed load as

$$N_{cr} = \frac{rN_l(1 + \alpha_p - \beta + \beta_S - r\beta_S - r) + K(1 + \alpha_s + \alpha_s \beta_S)}{2(1 + \beta_S)}, \qquad (9.35)$$

214 Equivalent wall

where K, N_l and r are defined by formulae (9.11), (9.2) and (9.12), respectively. Critical load parameters α_p and α_s are given in Tables 9.3 and 9.3 as a function of stiffness ratios β and β_s, defined by formulae (9.18) and (9.23).

It should be borne in mind when the full-height bending critical loads N_l and N_g are calculated that with frameworks with ground floor beams the uniformly distributed axial load on the equivalent column does not represent an approximation and therefore there is no need to reduce the part critical loads, i.e. $r_s = 1.0$ should be used in formulae (9.2) and (9.6).

For frameworks under concentrated top load, both the continuum model and the sandwich model result in the same formula for the critical load:

$$F_{cr} = \frac{K}{1 + K/F_g}. \tag{9.36}$$

In formula (9.36) F_g is the full-height bending critical load defined by formula (9.7).

9.5 FRAMEWORKS WITH GROUND FLOOR COLUMNS OF DIFFERENT HEIGHT

It is a practical case that the height of the columns on ground floor level is different from those above the first floor level. In most cases, they are higher than the ones on the other storey levels (Fig. 9.10/a/b/c) and therefore they create a more unfavourable situation, as far as stability is concerned.

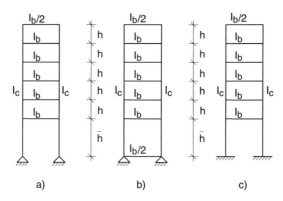

Fig. 9.10 Frameworks with ground floor columns of different height.

Such frameworks are non-regular but their behaviour only differs from the corresponding regular ones in storey-height shear buckling. Because of the 'softer' first storey region, local failure would occur through the buckling of the ground floor columns. It follows that the formulae given for the critical load can still be used if the formula for the storey-height shear critical load is modified. In assuming frameworks with higher columns on ground floor level, the storey-height shear critical load should be computed from

$$K_l = \sum_1^n \frac{\pi^2 E_{c,i} I_{c,i}}{h^{*2}}, \qquad (9.37)$$

with

$$h^* = 2\bar{h} \qquad (9.38)$$

for the framework in Fig. 9.10/a and with

$$h^* = \bar{h} \qquad (9.39)$$

for the frameworks in Figs 9.10/b and 9.10/c. In the above formulae $\bar{h}$ marks the height of the first storey columns.

The critical load is then obtained by applying the corresponding formulae given for the critical load in the previous sections.

It should be borne in mind that, because of the higher first storey columns, an approximation is made on the nature of the applied load, which may result in a slightly overestimated critical load for frameworks with higher first storey columns. This unconservative effect can be approximately compensated for by using a reduction factor r_s which belongs to a framework one storey lower than the actual one.

This case underlines again the warning that one should be very cautious with the application of the continuum method to non-regular structures [Hegedűs, 1987]. Both the nature and the extent of the deviation from the regular case should be carefully examined. Even after establishing the differences, the structural engineer may have to face a situation when neither the magnitude nor the sign of the error is known and the reliability of the method may be questionable.

9.6 ANALYSIS OF COUPLED SHEAR WALLS BY THE FRAME MODEL

Coupled shear walls can be regarded as special frameworks where walls (instead of columns) are connected with beams or lintels (Fig. 9.11). However, there are two basic differences between coupled shear walls and frameworks.

1) The two end-sections of the beams connecting the walls cannot develop bending because the walls, whose stiffness is practically infinitely great, do not let them. The centroidal axes of these sections are characterized by straight lines (Fig. 9.11/b). (This is also the case with frameworks but because of the relatively short sections in the case of frameworks, this effect is negligible and is normally not taken into account.)

2) As a rule, the depth of the cross-section of the beams is relatively great and the beams are relatively short and therefore their shear deformation should be taken into account.

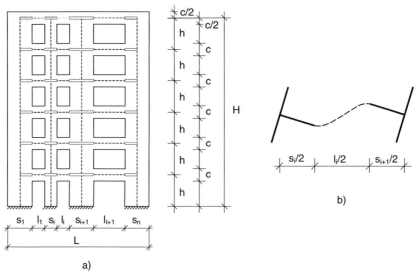

Fig. 9.11 Coupled shear walls. a) Typical arrangement, b) deformation of a beam with stiff end-sections.

Bearing in mind the above differences, the equations and formulae derived for the stability analysis of frameworks can be used for the stability analysis of coupled shear walls if certain modifications are made. Both differences only affect the full-height shear stiffness and the necessary modifications can be built into formula (9.9), which then assumes the form:

$$K_g^* = \sum_1^{n-1} \frac{6E_{b,i}I_{b,i}\left((l_i+s_i)^2 + (l_i+s_{i+1})^2\right)}{l_i^3 h\left(1+12\dfrac{\rho E_{b,i}I_{b,i}}{l_i^2 G_{b,i}A_{b,i}}\right)}, \qquad (9.40)$$

where

$G_{b,i}$ is the modulus of elasticity in shear of the beams,
$A_{b,i}$ is the cross-sectional area of the beams,
l_i is the distance between the ith and $(i+1)$th walls,
s_i is the width of the ith wall,
ρ is a constant whose value depends on the shape of the cross-section of the beams ($\rho = 1.2$ for rectangular cross-sections).

With the modified full-height shear critical load, the formula for the modified shear critical load is obtained as with frameworks:

$$K^* = K_g^* \frac{K_l}{K_g^* + K_l}, \qquad (9.41)$$

where K_l is the storey-height shear critical load:

$$K_l = \sum_1^n \frac{\pi^2 E_{c,i} I_{c,i}}{h^2}. \qquad (9.42)$$

The formulae derived for the critical loads of frameworks on fixed supports can now be used. When the coupled shear walls are subjected to UDL on the floor levels, the critical load is obtained from

$$N_{cr} = \frac{rN_l(1+\alpha-\beta+2\beta_s) + K^*(1+\alpha_s+\alpha_s\beta_s)}{2(1+\beta_s)}. \qquad (9.43)$$

Combination factor r in formula (9.43) is now given by

$$r = \frac{K_l}{K_g^* + K_l}, \qquad (9.44)$$

where K_g^* is the modified full-height shear critical load [formula (9.40)].

218 Equivalent wall

The part critical load N_l characterizing the full-height buckling of the individual walls in formula (9.43) is defined by formula (9.2).

Parameter α is given in Fig. 9.5 and in Table 9.1 as a function of stiffness ratio β:

$$\beta = \frac{K^*}{N_l}. \qquad (9.45)$$

Values for parameter α_S are given in Table 9.2 and in Fig. 9.7 as a function of stiffness ratio

$$\beta_S = \frac{K^*}{N_g}. \qquad (9.45a)$$

When the coupled shear walls are subjected to concentrated load on the top floor level, the critical load is obtained from

$$F_{cr} = \frac{rF_l(1+K^*/F_g)+K^*}{1+K^*/F_g}. \qquad (9.46)$$

The full-height bending part critical loads (F_l) and (F_g) in formula (9.46) are defined by formulae (9.3) and (9.7), respectively.

It is noted here that slightly different terminology is used in design practice for frameworks and for coupled shear walls. The distance between the axes of the columns (l_i) is normally used with frameworks; on the other hand, the distance between the wall sections (l_i) is used in the case of coupled shear walls. In spite of the similarity between the two cases, this is why both sets of formulae are given, using the corresponding terminology.

9.7 FRAMEWORKS WITH CROSS-BRACING

Frameworks with cross-bracing have a long history in civil engineering practice in different areas: offshore structures, transmission towers, building frames, roofs, grandstands, etc. Laced and other types of built-up columns, mainly used in steel structures, represent another area of application. When frameworks with cross-bracing are used in multistorey industrial and commercial building structures, their main task usually is to provide the structure with sufficient stiffness to resist lateral loads. Cross-bracing systems in structural engineering are becoming popular because of their economic use of material and the speed of construction.

In parallel with their widespread use for providing multistorey buildings with adequate lateral stiffness, considerable attention has been paid to developing methods for the structural analysis of cross-bracing systems. As the stress analysis can be performed using conventional procedures and commercially available computer packages, new research has centred on the stability analysis [DeWolf and Pelliccione, 1979; Wang and Boresi, 1992; Vafai, Estekanchi and Mofid, 1995]. Two characteristic ways of instability have been identified: the storey-height buckling of members in compression and the full-height buckling of the whole structure. Experimental results have been documented and theoretical solutions have been given for evaluating the effective length of compression braces, used for storey-height buckling analysis [Stoman, 1988 and 1989; Thevendran and Wang, 1993]. However, although most frameworks with cross-bracing become unstable through full-height buckling, only few papers have been devoted to the full-height buckling analysis and they have only dealt with concentrated end forces [Gjelsvik, 1990 and 1991], which is mainly of theoretical interest.

The aim of this section is to present simple closed-form solutions for the full-height buckling analysis of multistorey, multibay frameworks under both concentrated top load and UDL on floor levels.

The behaviour of frameworks with cross-bracing is similar to those without cross-bracing, inasmuch as it can be characterized by the same basic types of deformation (and the corresponding stiffnesses). It follows that the procedures introduced for sway frameworks of rectangular network in the previous sections can be used with some modifications. The deformation of frameworks with cross-bracing is normally dominated by the full-height bending of the structure as a whole, which is modified by shear deformation. As the effect of the full-height bending deformation of the individual columns is insignificant in practical cases, the sandwich model with thin faces will be applied for the analysis. The sandwich approach results in a very simple solution with good accuracy.

The full-height bending behaviour of frameworks as a whole with and without cross-bracing is the same and therefore formulae (9.6) and (9.7) for the full-height bending critical load can be readily used. It is the shear deformation and the shear stiffness (and the corresponding shear critical load) which are different due to the cross-bracing, and these characteristics will be determined in the next section.

9.7.1 Shear stiffness and shear critical load

The shear stiffness characterizing the shear deformation of the framework represents the effect of the beams and the cross-bracing and its value depends on their geometrical arrangement and stiffness characteristics. Values of the shear stiffness can be obtained by analysing the shear deformation of a typical one-storey section of the structure.

Based on the analysis of a one-storey section of a one-bay framework (Fig. 9.12), the angular displacement is given by

$$\gamma = \frac{\delta}{h} \tag{9.47}$$

or by

$$\gamma = \frac{Q}{K} \tag{9.48}$$

where

- δ is the lateral displacement (storey drift),
- h is the storey height,
- Q is the shear force,
- K is the shear stiffness.

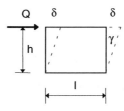

Fig. 9.12 Shear deformation of a one-storey section.

Combining formulae (9.47) and (9.48), the shear stiffness can be obtained from

$$K = \frac{hQ}{\delta}. \tag{9.49}$$

Knowing the lateral displacement of different types of cross-bracing, the shear stiffness can be determined using formula (9.49).

The lateral displacement of the single cross-bracing shown in Fig. 9.13/a, for example, consists of two parts [Timoshenko and Gere, 1961]. Due to the lengthening of the diagonal bar, a lateral displacement of

$$\delta_d = \frac{Qd^3}{A_d E_d l^2} \tag{9.50}$$

develops, where

- d is the length of the diagonal bar,
- A_d is the cross-sectional area of the diagonal bar,
- E_d is the modulus of elasticity of the diagonal bar,
- l is the bay of the framework.

The corresponding shear stiffness is obtained by using formula (9.49):

$$K_d = A_d E_d \frac{hl^2}{d^3}. \tag{9.51}$$

The shortening of the horizontal bar results in a lateral displacement of

$$\delta_h = \frac{Ql}{A_h E_h}, \tag{9.52}$$

where

- A_h is the cross-sectional area of the horizontal bar,
- E_h is the modulus of elasticity of the horizontal bar.

Formula (9.49) gives the corresponding shear stiffness as

$$K_h = A_h E_h \frac{h}{l}. \tag{9.53}$$

The two part stiffnesses can be combined by using the Föppl–Papkovich summation theorem [Tarnai, 1999], resulting in the exact value of the shear stiffness:

$$K = \left(\frac{1}{K_d} + \frac{1}{K_h}\right)^{-1} = \left(\frac{d^3}{A_d E_d hl^2} + \frac{l}{A_h E_h h}\right)^{-1}. \tag{9.54}$$

With double cross-bracing (Fig. 9.13/b), one diagonal is in tension and the other is in compression while the horizontal bar does not take part in the transmission of the shear force. The system is equivalent to the one which has no horizontal bar (Fig. 9.13/c) and the lateral displacement is given as

$$\delta_d = \frac{Qd^3}{2A_d E_d l^2}. \tag{9.55}$$

Formula (9.49) yields the shear stiffness as

$$K = K_d = 2A_d E_d \frac{hl^2}{d^3}. \tag{9.56}$$

The situation is similar when there is only one set of continuous diagonals with no horizontal bars (Fig. 9.13/d) and the shear stiffness is obtained from

$$K = A_d E_d \frac{hl^2}{d^3}. \tag{9.57}$$

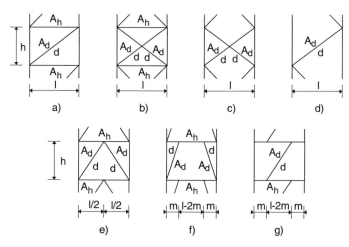

Fig. 9.13 Different types of cross-bracing. a) Single, b)–c) double, d) continuous, e)–f) K-bracing, g) knee-bracing.

Similar formulae can be derived for other cross-bracing arrangements with known storey drift. Simple formulae for the storey drift of cross-bracings with different geometrical arrangements have been presented, for

Frameworks with cross-bracing

example, by Stafford Smith and Coull [1991]. Based on the storey drift, the shear stiffness for the K-bracing system shown in Fig. 9.13/e can be expressed as

$$K = \left(\frac{2d^3}{A_d E_d h l^2} + \frac{l}{4 A_h E_h h} \right)^{-1}. \tag{9.58}$$

The situation with the cross-bracing in Fig. 9.13/f is somewhat different as the bending of the horizontal bar also affects the shear stiffness:

$$K = \left(\frac{d^3}{2 A_d E_d m^2 h} + \frac{m}{2 A_h E_h h} + \frac{h(l-2m)^2}{12 I_h E_h l} \right)^{-1}, \tag{9.59}$$

where I_h is the second moment of area of the horizontal bar.

Finally, the shear stiffness of the knee-braced frame in Fig. 9.13/g is obtained from

$$K = \left(\frac{d^3}{A_d E_d h(l-2m)^2} + \frac{l-2m}{A_h E_h h} + \frac{hm^2}{3 I_h E_h l} \right)^{-1}. \tag{9.60}$$

The above formulae for the shear stiffness assume one-bay frameworks. For multibay frameworks, the shear stiffness is obtained by adding up the shear stiffnesses of each bay:

$$K = \sum_{1}^{n-1} K_i, \tag{9.61}$$

where n is the number of columns and K_i refers to the shear stiffness of the ith bay.

By definition, the shear stiffness of the framework also represents the shear critical load of the structure. As with frameworks without cross-bracing, the formulae for the shear critical load show that the value of the shear critical load does not depend on the distribution of the vertical load nor on the height and type of the support of the structure.

9.7.2 Critical loads

Once the shear critical load is known, the critical load for frameworks with cross-bracing can be calculated using either the continuum model or the sandwich model. According to the results of a detailed accuracy analysis, the formulae resulting from the application of the sandwich model are of sufficient accuracy and they are also very simple.

When the framework is subjected to uniformly distributed load on floor levels, the critical load is obtained from

$$N_{cr} = \alpha_S K, \qquad (9.62)$$

where K represents the shear critical load, as given in section 9.7.1 for different types of cross-bracing. Values for critical load parameter α_S are given in Fig. 9.7 and in Table 9.2 as a function of stiffness ratio

$$\beta_S = \frac{K}{N_g}. \qquad (9.62a)$$

The full-height bending critical load N_g is to be calculated using formula (9.6).

When the structure is subjected to concentrated load on top, the formula for the critical load is

$$F_{cr} = \frac{K}{1 + K/F_g}, \qquad (9.63)$$

where F_g is the full-height bending critical load defined by formula (9.7).

It should be borne in mind that the procedure presented in this section yields the *full-height* critical load of the framework and the possibility of the storey-height buckling of a compression member should also be considered in the analysis.

Although it is difficult to determine which structure develops storey-height and which full-height buckling, based on the investigation of hundreds of structures, a simple criterion has been established which can be used to indicate the nature of buckling.

The investigation showed that, assuming geometrical and stiffness characteristics normally used in practice, the 'slender' frameworks (i.e. those much higher than wider) tended to develop full-height buckling. All structures that fulfilled the criterion

$$\eta = \frac{H}{W} > 4 \tag{9.64}$$

developed full-height buckling. In formula (9.64), H and W are the total height and width of the framework, respectively.

As the height/width or 'global slenderness' ratio of the framework decreases, the structure is more susceptible to storey-height buckling. Indeed, most structures with $\eta < 4$ became unstable because one element in compression (normally on the ground floor level) failed.

Criterion (9.64) is met by the majority of framed structures in practice as the general height/width ratio for plane frame structures seems to be in the range of 5 to 9 [Schueller, 1977]. It should be noted that criterion (9.64) was established empirically using data of an accuracy analysis where the structures had a bay/storey-height ratio of 1 and therefore can only be used as an indicator.

9.7.3 Structures with global regularity

The above formulae for the critical loads of frameworks with cross-bracing assume regular structures whose geometrical and stiffness characteristics are identical at each floor level. However, numerical investigations have shown that regularity can be interpreted in a broader sense.

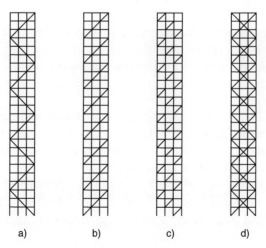

Fig. 9.14 24-storey frameworks with 'global' regularity.

226 Equivalent wall

Strictly speaking, none of the 24-storey structures in Fig. 9.14 is regular. Still, the approximate formulae give satisfactory results for the critical loads: the deviation from the 'exact' critical loads is less than 7% for frameworks on pinned and fixed supports, subjected to concentrated top load or uniformly distributed load on the floors. This follows from the fact that these bracing types have 'global' regularity, i.e. they are regular in 3-storey units.

Worked examples and a detailed accuracy analysis regarding frameworks with cross-bracing are available in [Zalka, 1998a and 1999].

9.8 INFILLED FRAMEWORKS

As pointed out in Chapter 8, the lateral and torsional stiffnesses of real buildings can be far greater than those predicted by theoretical models. The reason behind this discrepancy may be that most theoretical models rely on a skeleton of the building, involving the primary load bearing elements, and neglect the effects of the secondary structural elements. The most important contributors in this area are frameworks filled with masonry walls.

The theory of the behaviour of infilled frameworks is well documented and a number of publications are available dealing with different structural aspects [Polyakov, 1956; Mainstone and Weeks, 1972; Stafford Smith and Coull, 1991; Madan *et al.*, 1997]. The objective of this section is to introduce a simple procedure for the calculation of the critical load of infilled frameworks.

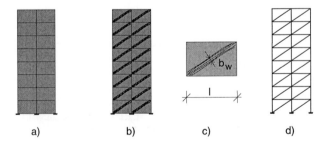

Fig. 9.15 Model for in-filled frameworks. a) Two-bay in-filled frame, b) equivalent diagonal 'bars' in compression, c) size of diagonal bar, d) model for the analysis.

Research shows that when an infilled framework is subjected to external loads, the most important contribution of the masonry structure in a storey-high panel can be represented by a diagonal strip (Fig. 9.15/b) in compression. This section can be modelled by a diagonal 'bar' whose cross-sectional area is

$$A_d = tb_w, \qquad (9.65)$$

where the width of the diagonal strip (Fig. 9.15/c) can be approximated by

$$b_w = 0.15d, \qquad (9.66)$$

where

- d is the length of the diagonal bar,
- t is the thickness of the masonry wall.

This leads to a framework with single bracing as a possible model for the analysis (Fig. 15/d). Frameworks with different types of cross-bracing are investigated in section 9.7 and the procedure presented there can be readily applied to infilled frameworks.

Accordingly, the critical load of infilled frameworks can be calculated from

$$N_{cr} = \alpha_S K, \qquad (9.67)$$

where K is the shear critical load:

$$K = \sum_{1}^{n-1} \left(\frac{d_i^3}{A_{d,i} E_d h l_i^2} + \frac{l_i}{A_{h,i} E_h h} \right)^{-1}. \qquad (9.68)$$

In formula (9.68)

- d_i is the length of the diagonal bar in the ith bay of the framework,
- l_i is the width of the ith bay,
- $A_{d,i}$ is the cross-sectional area of the diagonal bar in the ith bay [formula 9.65)],
- $A_{h,i}$ is the cross-sectional area of the beam of the frame in the ith bay,
- E_d is the modulus of elasticity of the diagonal bar,
- E_h is the modulus of elasticity of the horizontal bar,
- h is the storey height,
- n is the number of columns of the framework.

228 Equivalent wall

The modulus of elasticity of the diagonal bar (E_d) refers to the material of the masonry structure; for masonry, its value normally varies between 1000 N/mm² and 4000 N/mm².

Values for critical load parameter α_s are given in Fig. 9.7 and in Table 9.2 as a function of stiffness ratio

$$\beta_s = \frac{K}{N_g}. \tag{9.69}$$

In formula (9.69)

$$N_g = \frac{7.837 r_s E_c I_g}{H^2} \tag{9.70}$$

is the full-height bending critical load, where

- H is the height of the infilled framework,
- r_s is the reduction factor whose values are given in Table 3.1,
- E_c is the modulus of elasticity of the columns of the framework,
- I_g is global second moment of area of the framework [formula (9.5)].

9.9 EQUIVALENT WALL FOR 3-DIMENSIONAL ANALYSIS

The stability, frequency and global stress analyses of bracing systems consisting of bracing cores, shear walls *and* frameworks/coupled shear walls can be considerably simplified, if the frameworks/coupled shear walls are replaced by equivalent solid walls. The cores and (real and equivalent) walls can be combined to form a single cantilever whose 3-dimensional analysis leads to the simple closed-form solutions presented in Chapters 3 and 5.

The replacement of the frames/coupled shear walls by equivalent walls can be based on making the critical load of the framework/coupled shear walls equal to that of the equivalent wall. In making use of the critical load of a framework/coupled shear walls (N_{cr}) – and assuming UDL on the floor levels – the thickness of the equivalent wall (t^*) can be derived from

$$N_{cr} = N_{cr,wall} = \frac{7.837 r_s E t^* L^3}{12 H^2}, \tag{9.71}$$

which leads to

$$t^* = \frac{1.53 N_{cr} H^2}{r_s E L^3}, \qquad (9.72)$$

where

- N_{cr} is the critical load of the frame/coupled shear walls,
- H, L, E are the height, width and the modulus of elasticity of the equivalent wall (identical to the height, width and the modulus of elasticity of the framework/coupled shear walls),
- r_s reduction factor (whose values are given in Table 3.1).

When an equivalent wall is incorporated into the bracing system of shear walls and cores developing predominantly bending type deformation, the accuracy and reliability of the 3-dimensional analysis basically depends on how well the equivalent wall 'fits in' the bracing system, as far as deformations are concerned, i.e. to what extent the characteristic deformation of the equivalent wall conforms to the deformation of the bracing system.

As far as the overall deformation of a bracing system consisting of shear walls and cores *and* frameworks and coupled shear walls is concerned, the bracing system falls into two categories. Consequently, two possibilities should be considered.

1) The characteristic deformation of the equivalent wall (representing a framework or coupled shear walls) is of bending type. In this case, conformity is ensured as both the shear walls and cores and the equivalent wall develop bending-type deformation. – The 3-dimensional analysis can be safely used as there is no interaction among the members of the bracing system which would change the basic 3-dimensional behaviour (based on sway in bending and torsion in warping as well as their combination) assumed for the derivation of the governing differential equations of the 3-dimensional analysis.

2) The characteristic deformation of the equivalent wall is of shear type. In this case, the bracing system develops a combination of bending and shear type deformation. – The 3-dimensional analysis results in approximate solutions. The degree of approximation depends on to what extent the overall deformation of the bracing system defers from bending type deformation.

When a decision is made which of the above categories a bracing system belongs to, it is important to know what type of deformation the elements of the system develop. The following guidelines (and the

230 Equivalent wall

comparative analysis given in section 9.10) help to categorize the bracing elements.

- Bracing elements which normally develop predominantly bending type deformation are:
  shear walls, cores, coupled shear walls with relatively slender lintels, frameworks with cross-bracing (with H/W>4), frameworks on fixed supports (with H/W>4), infilled frameworks (with H/W>4).
- Bracing elements which tend to develop predominantly shear type deformation are:
  frameworks on pinned supports, coupled shear walls with relatively strong lintels, wide frameworks (with H/W<4).

As the most effective bracing elements are shear walls, cores and frameworks with cross-bracing, most bracing systems fall in Category 1 and fulfil the basic conditions for the 3-dimensional analysis.

When shear walls and cores are supplemented by frameworks developing shear deformations (Category 2), the contribution of the frameworks is normally very small compared to that of the shear walls and cores. Because of their relatively small contribution, they can either be ignored in the analysis (as a conservative approximation) or if they are incorporated into the bracing system as equivalent walls, they do not normally alter the bending type deformation of the shear walls and cores significantly and consequently the level of accuracy of the analysis is only slightly reduced.

In certain cases all (or most of) the elements of the bracing system develop shear deformation. The 3-dimensional analysis can still be applied but the following considerations have to be taken into account. The analysis can be safely used only for doubly symmetrical cases when the basic buckling modes do not couple. The results for unsymmetrical cases are only approximate and have to be treated with caution.

9.10 SHEAR WALLS

Of all the possible bracing elements, shear walls represent the simplest problem, as far as the calculation of the critical load is concerned. When the load is a concentrated force at top floor level, the formula for the critical load is

$$F_{cr} = \frac{\pi^2 EI}{4H^2}, \qquad (9.73)$$

and when the shear wall is subjected to uniformly distributed normal load at every floor level, the critical load is obtained from

$$N_{cr} = \frac{7.837 r_s EI}{H^2}, \qquad (9.74)$$

where

E	is the modulus of elasticity,
I	is the second moment of area of the shear wall,
H	is the height of the shear wall (Fig. 9.16),
r_s	is a reduction factor whose values are given in Table 3.1 in section 3.1.1.

In Fig. 9.16, t and L mark the thickness and the width of the shear wall.

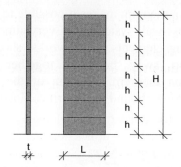

Fig. 9.16 Shear wall.

9.11 SYMMETRICAL CROSS-WALL SYSTEM BUILDINGS

A set of parallel shear walls, coupled shear walls and frameworks represents a typical building system. In the symmetrical case when the centre of the bracing system of a building and that of the applied vertical load coincide, which is often the case with cross-wall system buildings, stability failure may occur in three different ways. The system can develop sway buckling in both principal planes and pure torsional buckling. Sway buckling is investigated in this section and torsional buckling is covered in

232 Equivalent wall

section 3.1. Figure 9.17 shows the layout of such a cross-wall system building, where

- L, B are the plan length and breadth of the building,
- s is the distance between the bracing elements,
- m is the number of bracing elements in the building.

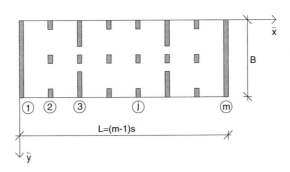

Fig. 9.17 Cross-wall system building.

Southwell's summation theorem makes it possible to determine the critical load of symmetric cross-wall systems in a very simple manner. The floor slabs make the bracing elements (frameworks, shear walls and coupled shear walls) develop sway buckling together and, according to Southwell's summation theorem, the sum of the critical loads of the bracing elements is a lower bound to the critical load of the whole system. Formulae

$$F_{cr}^{bui} = \sum_{1}^{m} F_{cr,j} \tag{9.75}$$

and

$$N_{cr}^{bui} = \sum_{1}^{m} N_{cr,j} \tag{9.76}$$

give the total critical load of cross-wall system buildings subjected to concentrated forces at top floor level and uniformly distributed load at floor levels. Parameter m in formulae (9.75) and (9.76) is the number of the frameworks/shear walls/coupled shear walls bracing the building and $F_{cr,j}$ and $N_{cr,j}$ are the critical loads of the individual elements. Simple formulae for their computation are given in sections 9.3 to 9.8.

The building shown in Fig. 9.17 may not be stable in direction x. Stability in this direction should also be ensured. This can be done by perpendicular (coupled) shear walls, frameworks or by cores. In the first case, and if the system is symmetrical, a stability analysis similar to the one just presented, can be carried out. In the general case, when the system is not symmetrical, combined sway-torsional buckling has to be investigated, as described in section 3.1. Even if the system is doubly symmetrical, the possibility of pure torsional buckling has to be considered. This investigation can also be carried out according to the procedure presented in section 3.1.

9.12 PLANAR BRACING ELEMENTS: A COMPARISON

Different structural elements (shear walls, coupled shear walls, frameworks with single or double bracing, in-filled frames, frameworks on fixed supports, framework on pinned supports with or without ground floor beams) can be used for bracing purposes with different efficiency. The best way to compare them, as far as their efficiency is concerned, is to compare their critical loads.

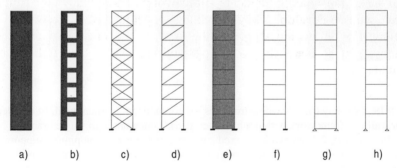

Fig. 9.18 One-bay (3.5 m wide) structures for comparison. a) Shear wall, b) coupled shear walls, c) framework with double bracing, d) framework with single bracing, e) in-filled frame, f) framework on fixed supports, g) framework on pinned supports with ground floor beam, h) framework on pinned supports without ground floor beam.

This section presents the results of such an exercise involving two sets of structures. The one-bay and two-bay structures have the same overall size (total height and width), respectively. The total height of the structures in both groups varies from 4 storeys to 99 storeys. The storey height is 3 m. The modulus of elasticity is 3×10^4 N/mm^2 for the reinforced concrete

234 Equivalent wall

shear walls and the columns and beams of the frameworks, 3×10^3 N/mm^2 for the masonry in the infilled frames and 2×10^5 N/mm^2 for the steel cross-bracing. The reinforced concrete elements and the masonry in the infilled framework have a thickness of 0.3 m. The cross-section of the cross-bracing in all cases is $0.25\times0.015 = 0.00375$ m^2.

The total width (external size) of the structures in the one-bay group (Fig. 9.18) is 3.5 m. The width of the wall sections in the coupled shear walls is 1.2 m, the depth and length of the beams are 1.0 m and 1.1 m, respectively. The depths of the columns and the beams of the frameworks are 0.5 m and 0.3 m, respectively.

The total width of the structures in the two-bay group (Fig. 9.19) is 6.3 m. The width of the wall sections in the coupled shear walls is 1.0 m, the depth and length of the beams are 1.5 m and 1.65 m, respectively. The depths of the columns and the beams of the frameworks are 0.3 m and 0.5 m, respectively.

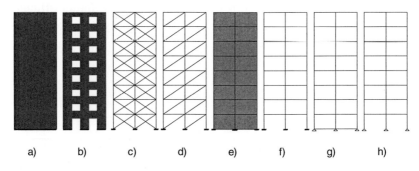

Fig. 9.19 Two-bay (6.3 m wide) structures for comparison. a) Shear wall, b) coupled shear walls, c) framework with double bracing, d) framework with single bracing, e) in-filled frame, f) framework on fixed supports, g) framework on pinned supports with ground floor beams, h) framework on pinned supports without ground floor beams.

The results of the analysis (the critical loads in MN, calculated using the formulae given in this chapter for structures subjected to uniformly distributed vertical load) are given in Tables 9.4 and 9.5, where the abbreviations used in the first column are:

    SW: Shear wall,
    CSW: Coupled shear walls,
    FDX: Framework with double cross-bracing,
    FSX: Framework with single cross-bracing,

IFF: Infilled framework,
FF: Framework on fixed supports,
PFG: Framework on pinned supports with ground floor beams,
PF: Framework on pinned supports without ground floor beams.

Table 9.4 Critical loads for one-bay structures

Storeys	4	8	12	16	20	24	28	40	60	99
SW	1253	365	172	99.6	64.9	45.6	33.8	16.8	7.58	2.81
CSW	864	317	157	92.5	60.7	42.9	31.8	15.8	7.10	2.63
FDX	433	189	98.5	59.3	39.4	28.2	20.9	10.6	4.76	1.77
FSX	242	154	88.5	55.7	37.8	27.2	20.4	10.4	4.73	1.77
IFF	188	139	83.9	53.9	37.0	26.8	20.2	10.3	4.73	1.76
FF	41.2	31.1	27.0	24.2	20.6	17.3	14.4	8.79	4.48	1.77
PFG	29.3	25.8	23.8	22.2	19.6	16.6	13.9	8.55	4.38	1.74
PF	20.3	18.7	17.6	16.6	15.5	13.5	11.7	7.60	4.08	1.68

The thick line in the tables separates shear and bending type behaviour. The structures on the left-hand side of the thick separator develop predominantly shear and those on the right-hand side predominantly bending type deformation.

Table 9.5 Critical loads for two-bay structures

Storeys	4	8	12	16	20	24	28	40	60	99
SW	7306	2129	1002	580	378	266	197	98.2	44.2	16.4
CSW	1779	972	556	350	238	171	129	65.5	29.9	11.1
FDX	942	438	232	141	94.0	66.9	50.0	25.2	11.4	4.24
FSX	501	349	206	131	89.6	64.7	48.7	24.9	11.4	4.23
IFF	386	311	194	126	87.3	63.4	48.0	24.7	11.3	4.23
FF	55.2	52.1	49.4	46.6	41.6	35.7	30.3	19.0	9.97	4.04
PFG	53.6	51.5	49.1	46.6	42.4	36.4	31.0	19.4	10.1	4.09
PF	15.7	15.5	15.2	14.9	14.6	14.2	13.8	11.6	7.49	3.55

It is the low-rise – medium-rise region where the performance of the different types of structure really differs. A shear wall developing bending deformation has by orders of magnitude greater critical load and is by orders of magnitude more effective a bracing element than a framework of the same height and width developing shear deformation. As the height –

and the height/width ratio – of the structures increase, the difference in performance becomes smaller and smaller. Eventually all structures develop bending deformation and although a shear wall always remains the most effective bracing structure, its advantage over the frameworks is significantly reduced while the performance of the different types of framework – regardless of the type – is practically identical.

9.13 SUPPLEMENTARY REMARKS

The results of detailed accuracy analyses involving hundreds of frameworks and coupled shear walls demonstrate that the accuracy of methods is acceptable for checking the stability of practical structural engineering structures [Zalka, 1998a and 1998b]. A short summary of the result of the accuracy analyses is given in this section.

The approximate formulae tend to result in conservative critical loads. The most accurate critical loads were obtained for frameworks with cross-bracing, which are also the most effective and economic structures favoured for bracing purposes. The error range was between -2% and 8% (for the maximum unconservative and conservative critical load).

Good, conservative approximations were obtained for the critical loads of frameworks on fixed supports. The error of most critical loads was between 0% and 10% and the maximum error was 19%.

The method for coupled shear walls resulted in acceptable estimates for the critical loads which were normally up to 15% greater than those obtained from the finite element analysis. The error range widened to -6% and 23% (for the maximum unconservative and conservative critical load) in some cases.

The least accurate critical loads were obtained for frameworks on pinned supports. The error of the approximate formulae was between 0% and 20% in most practical cases. In some cases, however, the error range widened to -3% and 39% (for the maximum unconservative and conservative critical load). As frameworks on pinned supports are rarely used for bracing purposes, critical loads of this level of accuracy still offer good indication as to the overall suitability of the framework in a multistorey building.

It has to be noted that the lateral and torsional stiffness of building structures are normally provided by shear walls and cores and the contribution of the frameworks is small compared to that of the shear walls and cores and therefore the accuracy of their critical loads is of secondary importance.

The finite element methods which are considered 'exact' occasionally produce unreliable results. In some cases, some FE results indicated greater critical loads for higher frameworks than for lower frameworks, when all the other characteristics of the frameworks were the same. In other cases, unconservative errors in the region of 150–250% were experienced. However, it should be mentioned that the stiffness characteristics in such cases assumed extreme values. The problems were probably caused by the ill-conditioned stiffness matrix. These cases underline the importance of having alternative procedures available to check suspicious looking results.

It is worth mentioning that some well-publicized commercially available FE procedures had difficulties with the stability analysis in certain stiffness regions when the system was too large. Incorrect load factors were obtained for different 12-storey shear walls with two rows of opening. The situation is made more difficult (and more dangerous) by the fact that with complex structures like coupled shear walls it is sometimes difficult even to realize that the results are not reliable, especially, when the results are only 'slightly' incorrect. With the 12-storey structures investigated, the incorrect buckling shape was a warning sign. However, when the buckling shape looks reasonable, it is very difficult indeed to find out that the critical load may not be correct. This also shows how important it is to know the characteristic types of deformation and the buckled shape to be expected from the structure.

The results obtained by using the methods given in this chapter also have to be carefully evaluated, especially when the method is applied to non-regular structures. Both the nature and the extent of the deviation from the regular case should be carefully examined. Even after establishing the differences, the structural designer may have to face a situation when neither the magnitude nor the sign of the error is known.

Finally, it is mentioned here that detailed stability analysis of frameworks is given by Appeltauer and Kollár [1999] and the theoretical background to the application of the sandwich theory to the stability analysis of building structures (including frameworks) is given by Hegedűs and Kollár [1999)].

10

Test results and accuracy analysis

To demonstrate the accuracy of the methods developed for the stress, stability and frequency analyses – and to learn more about 3-dimensional behaviour – a series of tests on small-scale models of 10-storey buildings (Fig. 10.1) was carried out at the Building Research Establishment[*]. It was also the aim of the tests to examine the effect of storey-high columns on the global lateral and torsional stiffness of multistorey structures.

Comparisons were also made with the results of similar approximate methods. This chapter gives a short summary of the results of the tests and the comparisons. The details of the tests are available in separate publications [Zalka and White, 1992 and 1993]. The accuracy of planar procedures is discussed in section 9.11.

10.1 DESCRIPTION OF THE MODELS

The floor slabs of the models were made of fibre boards. The vertical load bearing elements were represented by 96 copolymer columns at each storey, arranged in a rectangular network with a space of 200 mm in both directions. A relatively great number of columns was chosen to make their contribution to the lateral and torsional stiffnesses of the model noticeable. The size of the layout was 2200 mm in direction x and 1400 mm in direction y. Each storey height was 150 mm and the total height of the model was 1500 mm. The lateral and torsional stiffnesses were provided by a combination of U-shaped and L-shaped perspex cores and shear walls.

A circular bending test involving 50 specimens was carried out to establish the modulus of elasticity of the cores and shear walls. The modulus of elasticity was found to be $E = 3150$ N/mm^2. The shear modulus assumed the value $G = 1312.5$ N/mm^2.

Two bracing system arrangements were constructed to enable the analysis of planar (Model 'M1') and torsional (Model 'M2') behaviour.

[*] Figures 10.1, 10.2, 10.9, 10.10 and 10.13 are reproduced by permission of BRE.

Description of the models 239

Fig. 10.1 Ten-storey building model.

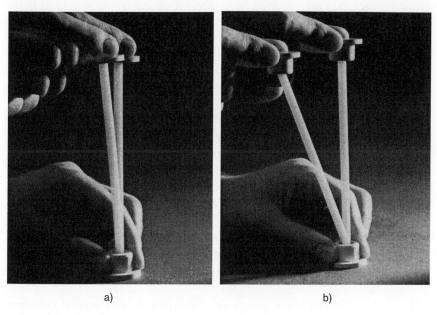

a) b)

Fig. 10.2 Columns of storey height. a) Fixed, b) pinned.

240   Test results

Both arrangements were constructed using fixed (Fig. 10.2/a) and 'pinned' (Fig. 10.2/b) columns. It should be pointed out that the columns described as 'pinned' did not represent columns with perfect pinned ends. Due to limitations in manufacturing the joints, there was some friction between the end sections of the columns and the rubber 'O' ring housed in the floor adapters. This friction prevented the end of the columns from developing totally free rotation and caused some increase in stiffness.

The models were first subjected to concentrated horizontal loading at different locations. The load was applied by using a simple pulley mechanism attached to the top floor of the model (Fig. 10.1). Load increments of approximately 5 N were chosen. Instruments were mounted onto a rigid dexion reference framework which spanned two sides over the full height of the model. The lateral displacements of the model were measured in directions x and y (parallel with the sides of the models). When the experimental values were compared to the theoretical values, the formulae given in section 5.6 for concentrated top loading were used.

After the horizontal loading exercise, the models were tested for stability and finally their dynamic performance was investigated.

The models with both bracing system arrangements and with both sets of columns were subjected to a series of loads but only the most characteristic cases are reported here. The following notation is used in the text and in the figures:

- *Fm*   Model with *F*ixed columns; results from *m*easurements
- *Ft*   Model with *F*ixed columns; *t*heoretical results
- *Pm*   Model with *P*inned columns; results from *m*easurements
- *Pt*   Model with *P*inned columns; *t*heoretical results
- F_y Concentrated force in direction *y*, at location shown
- e_y Translation in direction *y*, at location shown
- F_x Concentrated force in direction *x*, at location shown
- e_x Translation in direction *x*, at location shown
- *C*   Centroid: geometrical centre of ground plan
- O_{Fm} Location of shear centre *O* of the model with *f*ixed columns, according to *m*easurements
- O_{Ft} Location of shear centre *O* of the model with *f*ixed columns, according to *t*heoretical formulae
- O_{Pm} Location of shear centre *O* of the model with *p*inned columns, according to *m*easurements
- O_{Pt} Location of shear centre *O* of the model with *p*inned columns, according to *t*heoretical formulae
- •   Shear centre

10.2 HORIZONTAL LOAD ON MODEL 'M1'

Model 'M1' was nearly symmetrical, as far as the stiffnesses were concerned (Fig. 10.3); the shear centre of the bracing system and the geometrical centre of the plan of the building nearly coincided.

The model was first subjected to a series of concentrated horizontal forces at top floor level and the location of the shear centre was established. The theoretical formulae (2.1) and (2.2) in Chapter 2 gave very good predictions for the location of the shear centre. The distance between the theoretical and measured shear centres was 80 mm (6% related to the width of the model) with the model with pinned columns (O_{Pt} and O_{Pm}), and 40 mm (3%) with the model with fixed columns (O_{Ft} and O_{Fm} in Fig. 10.3).

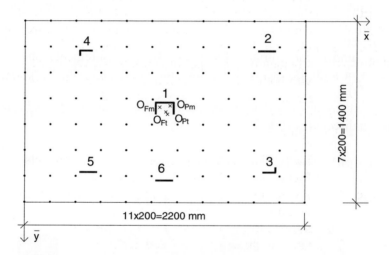

Fig. 10.3 Bracing system arrangement for model 'M1'.

It is interesting to note that the location of the shear centre hardly changed when the pinned columns were replaced by fixed columns. This follows from the fact that the bracing system (without the columns) had a shear centre which situated very near the centroid and the columns in their doubly symmetrical arrangement did not change the situation.

242   Test results

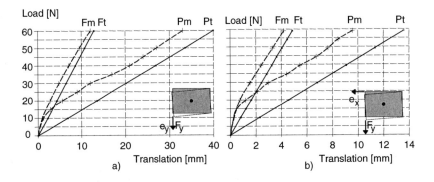

Fig. 10.4 Translations of model 'M1' under horizontal force F_y. a) Translations e_y, b) translations e_x.

The models were then subjected to concentrated horizontal forces at different corners. Figure 10.4 shows the results of the theoretical and experimental values of the translations on top floor level in directions x and y, under concentrated force F_y. The measurements show good agreement with the theoretical values for the model with fixed columns ($Fm - Ft$).

The situation was similar when the model was subjected to a horizontal force in direction x (Fig. 10.5).

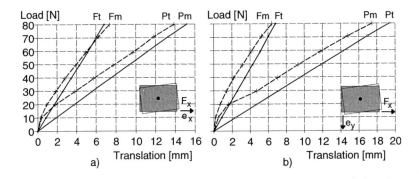

Fig. 10.5 Translations of model 'M1' under horizontal force F_x. a) Translations e_x, b) translations e_y.

For the model with pinned columns, the measurements show consistently smaller translations than the theoretical values. As mentioned in section 10.1, the reason for this deviation is attributed to the fact that the rotation of the pinned joints was restricted by some friction due to manufacturing limitations. This phenomenon with the same tendency was observed with all the models with pinned columns.

10.3 HORIZONTAL LOAD ON MODEL 'M2'

With the second arrangement ('M2' in Fig. 10.6), the distribution of the stiffnesses of the bracing system was unsymmetrical and the centroid and the shear centre were well apart. When the model was constructed using pinned columns, the shear centre situated near the U-core. The distance between the theoretical and measured shear centres (O_{Pt} and O_{Pm}) was 150 mm (7% related to the corresponding width of the model).

Due to the 'stabilizing' effect of the fixed columns, i.e. a doubly symmetrical system of columns was added to the unsymmetrical system of bracing elements, the shear centre moved closer to the centroid of the layout but the model still remained fairly unsymmetrical which was the requirement regarding this arrangement.

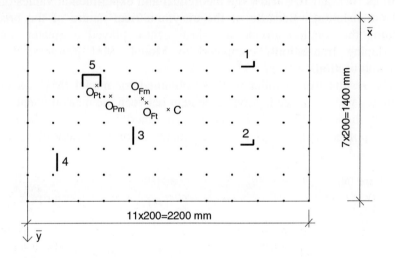

Fig. 10.6 Bracing system arrangement for model 'M2'.

Again, the theoretical predictions for the location of the shear centre were very good; the difference between the experimental and theoretical values (O_{Fm} and O_{Ft}) was only 65 mm (4%). Once again, the greater, but still acceptable, difference with the pinned model can be attributed to the friction in the 'pinned' joints. Following the loading sequence, aimed at the determination of the location of the shear centre, the models were subjected to horizontal forces at the corner points in both direction x and direction y. The horizontal translations were measured at several locations over the height of the model in direction y and in direction x.

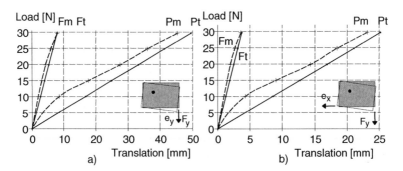

Fig. 10.7 Translations of model 'M2' under horizontal force F_y. a) Translations e_y, b) translations e_x.

Figures 10.7 and 10.8 show the theoretical and experimental values of top level translations. Because of the unsymmetrical nature of the bracing system, the rotation around the shear centre played a greater role in developing translations, compared to Model 'M1' presented in the previous section.

The results were similar to those obtained for model 'M1', as far as accuracy was concerned: good agreement between the measurements (Fm) and the theoretical values (Ft) for the fixed model and, due to the friction in the 'pinned' joints, slightly overestimated theoretical values (Pt) for the model with pinned columns.

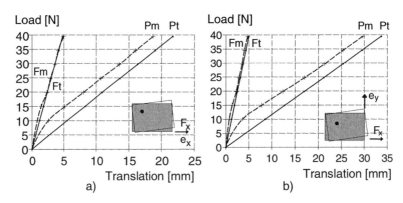

Fig. 10.8 Translations of model 'M2' under horizontal force F_x. a) Translations e_x, b) translations e_y.

10.4 COMPARATIVE ANALYSIS OF THE FORMULAE FOR HORIZONTAL LOAD

All known approximate methods developed for the stress analysis of bracing systems developing unsymmetrical bending and torsion use one or two or all three of the following assumptions. They either assume uniformly distributed horizontal load and/or neglect the effect of the Saint-Venant torsional stiffness on the load share on the individual bracing elements and/or neglect the product of inertia of the bracing system. As the method presented in Chapter 5 does not use any of these assumptions and therefore can be considered as the generalization of such simpler methods, the analysis of some special cases offers excellent opportunities for comparison. Some of these special cases are: uniformly distributed horizontal load ($\mu = 0$); the case when the effect of the Saint-Venant stiffness is neglected, i.e. when $\eta_T = 1.0$ and $\eta_M = 1.0$ hold; and the case when $I_{xy} = 0$ holds (zero product of inertia).

Two more special cases can be considered for comparison. When the external load passes through the shear centre and the bracing system only develops unsymmetrical bending, the formulae presented in the thesis simplify and become identical to Tarnai's [1996] formulae given in matrix form for unsymmetrical bending. Another special case, when the bracing system under uniformly distributed horizontal load ($\mu = 0$) only develops mixed torsion, leads to Vlasov's formula [1940] for the rotation of beam-columns under uniformly distributed torque.

Szmodits [1975], using a technique similar to the one applied here, but assuming $\mu = 0$, $\eta_T = 1.0$ and $\eta_M = 1.0$ as mentioned above, and Kollár and Póth [1994], relying on static considerations, derived closed-form solutions for the uniformly distributed load case. Both sources were also used for the numerical analysis. The relevant formulae showed good agreement.

The formulae for load distribution were also checked and found in line with those presented by Beck and Schäfer [1969], König and Liphardt [1990] and MacLeod [1990], assuming $\mu = 0$, $I_{\omega,i} = 0$ and $I_{xy,i} = 0$.

Worked examples [Stiller, 1965; Beck and Schäfer, 1969; Rosman, 1967] were also used for comparing the load share on the bracing elements under concentrated top load and uniformly distributed horizontal load. The deviation of the results was less than 5%.

246 Test results

10.5 DYNAMIC TESTS

Dynamic tests were also carried out on the models. Only a summary of the results is given here; the detailed description of the tests and the results are available in separate publications [Boughton, 1994; Zalka, 1994b].

The formulae and tables presented in section 3.2 gave

$$f = 0.79 \text{ Hz}$$

for the fundamental frequency of Model 'M2' with pinned columns.

A series of tests was carried out using a small vibrator on top of the model (Fig. 10.9). The vibrator was linked to a computer which monitored the behaviour and evaluated the measurements. The fundamental frequency of the model with pinned columns was measured to be

$$f = 0.78 \text{ Hz}$$

which is 1% smaller than the theoretical value.

When the pinned columns were replaced by fixed ones, the formulae in section 3.2 predicted

$$f = 1.96 \text{ Hz}$$

for the fundamental frequency. According to the experimental measurements, the fundamental frequency was

$$f = 1.61 \text{ Hz}$$

which was 18% smaller than the theoretical value.

Lateral vibrations were coupled with pure torsional vibrations in both cases. When the behaviour of the model with fixed columns was investigated, the effect of the columns on the global stiffness of the model was taken into account using approximate formula (2.17) derived in section 2.2.

The effect of the compressive forces on the frequencies was neglected in the theoretical analysis.

Fig. 10.9 Dynamic tests: model 'M2' and the vibrator.

10.6 STABILITY TESTS

Theoretical research into the stability of structures has resulted in new design methods and a considerable amount of information but the calibration and verification of the methods have been progressing in a much slower pace. Due to the formidable task, even when tests *are* carried out, they concentrate on individual structural elements [Croll and Walker, 1972; Barbero and Tomblin, 1993] or they use a small-scale part of a complex structure (a framework/coupled shear walls) [Szittner, 1979] as opposed to the whole structure. To fill this gap, a series of tests was carried out on the small-scale models shown in the previous sections. A summary of the results is given below; details of the tests are available in a research report [Zalka and White, 1992].

The models were subjected to two types of vertical load. In all cases they had to carry their own weight, which was a uniformly distributed load on each floor with an intensity of $q = 0.08$ kN/m^2. In addition to this dead weight, the models were subjected to a vertical load on the top level. This load was placed on the top floor in increments and was distributed across the whole floor area by a small steel frame and two wooden slabs (Fig. 10.10). The term *critical concentrated load* in this section refers to this load.

Characteristic deformations were measured during the loading process and the load-deflection and load-rotation diagrams were produced. Based on the diagrams, the Southwell Plot [Croll and Walker, 1972] was then produced, with deformations along the vertical axis and the deformation-load ratio along the horizontal axis. The critical load of the structure was finally established as the slope of the linearized relationship (Fig. 10.12). The simultaneous presence of the uniformly distributed dead weight and the concentrated top load was taken into account by the Dunkerley summation theorem.

Stability tests 249

Fig. 10.10 Stability tests: loading and failure mode.

10.6.1 Model 'M1'

The procedure presented in section 3.1 yielded

$$V_{cr} = 3.13 \text{ kN}$$

for the critical concentrated load of Model 'M1' with pinned columns.

Both the theoretical investigations and the behaviour of the model during the tests showed that the buckling mode of this arrangement was that of sway buckling in plane yz. Accordingly, the translations in direction y on top floor level were used for the Southwell Plot. The vertical concentrated top load was increased in ten increments from 0.2 kN to 1.5 kN. The Southwell Plot resulted in the value of the critical concentrated load as

$$V_{cr} = 3.03 \text{ kN}.$$

The model proved to be much stronger when the pinned columns were replaced by fixed ones but the characteristic buckling mode did not change: the model developed sway buckling in plane yz, which was practically not coupled with the other two basic modes. In this case, the vertical concentrated load was increased to 2.5 kN. The theoretical prediction and the Southwell Plot – which showed some uncertainty in evaluation – put the value of the critical concentrated load in the region of 13 kN – 14 kN. This corresponds to a fourfold increase in the value of the critical load – and in the value of the lateral stiffness of the model – compared to the corresponding values with the model with pinned columns. Figure 10.11 shows the buckled shapes where *PC* refers to the model with pinned and *FC* to the model with fixed columns.

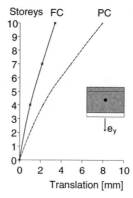

Fig. 10.11 Buckled shape of model 'M1'.

10.6.2 Model 'M2'

In accordance with the theoretical predictions, this model was very sensitive to torsion. This was also reflected in the value of the critical concentrated load which, using the procedure presented in section 3.1, was obtained as

$$V_{cr} = 1.32 \text{ kN}$$

for the model with pinned columns. This value was much smaller than the critical concentrated load of Model 'M1'.

The tests demonstrated that the characteristic deformation was rotation around the shear centre, which was slightly coupled with sway in direction y. The value of the vertical load distributed on the top floor level was increased in nine increments from 0.2 kN to 1.3 kN. The translations of the corner points of the models were monitored and the measurements were also used to produce the rotations of the model (Fig. 10.12/a). Based on the rotations and the ratio of the rotations to the vertical load, the Southwell Plot was produced (Fig. 10.12/b). The critical concentrated load was obtained as the slope of the experimental line:

$$V_{cr} = 2.69 \text{ kN.}$$

The failure mode of the model was torsional buckling as shown in Fig. 10.10.

Two reasons may explain the significant difference between the theoretical and experimental results. As a stabilizing factor, the friction in the joints of the pinned columns played a greater role in contributing to the resistance of the model when the model developed torsion. This friction not only increased the lateral and torsional stiffnesses (and the corresponding critical loads) but also changed the location of the shear centre in such a way that the distance between the shear centre and the geometrical centre became shorter. This phenomenon reduced the effect of torsion which also led to a greater global critical load. The evaluation of the loading procedure later revealed that with increasing rotations the top floor slab developed warping and slightly detached itself from the load distributing wooden slab. In this way, the load which was assumed to be uniformly distributed over the top floor lost its uniform nature and some load concentration occurred near the shear centre. This also increased the critical load.

252 Test results

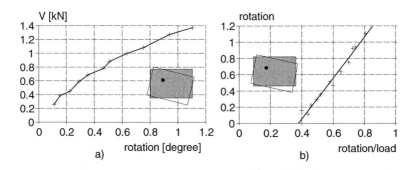

Fig. 10.12 Stability of model 'M2'. a) Load-rotation relationship, b) Southwell Plot.

When the pinned columns were replaced by fixed ones, the behaviour of the model was similar but, due to the 'healthy' contribution of the columns (greater stiffness and less eccentric load), the value of the critical load increased significantly. It was not possible to load this model up to failure, but the test results, the Southwell Plot and the theoretical investigations indicated that the critical concentrated load of the model was 12.9 kN. This represented a 2.0-fold and a 1.6-fold increase in the lateral stiffnesses in directions y and x, respectively, a 1.8-fold increase in the torsional stiffness and a 5.3-fold increase in the warping stiffness. Because of the more favourable mode coupling, due to the reduced eccentricity, the global critical load increased 6.2-fold, compared to the pinned case.

10.6.3 Deformation of the bracing elements

Both models 'M1' and 'M2' with both pinned and fixed columns behaved as the theoretical formulae predicted. The nature of the behaviour was not affected by the end conditions of the columns: The system with fixed columns and with pinned columns developed the same type of deformations. The nearly symmetrical system 'M1' developed lateral buckling in direction y (Fig. 10.11). Model 'M2' with pinned columns was loaded up to failure. The characteristic mode was torsion (Fig. 10.10). The structure lost stability when the two L-shaped cores marked with '1' and '2' in Fig. 10.6 broke between the ground floor and the first floor levels. These two bracing elements were the furthest from the shear centre and were subjected to the greatest deformations. The U-shaped core near the shear centre (marked with '5' in Fig. 10.6) developed considerable warping deformation at the bottom of the model (Fig. 10.13). The warping spread to the next two storeys then died away. The two shear walls marked with '3' and '4' in Fig. 10.6 developed slight warping in the first and second storey region. After dismantling the model, these shear walls fully recovered from the deformations.

Fig. 10.13 Deformation of U-core.

11

Evaluation; design guidelines

The techniques and methods presented in the book make it possible to carry out the global stress, stability and frequency (seismic) analyses of the bracing system of building structures. They are potentially useful both at the concept design stage and for final analysis. They can be used for

- checking of structural adequacy,
- assessing the suitability of structural layouts,
- verifying the results of other methods,
- evaluating computer packages,
- facilitating theoretical research,
- developing new techniques and procedures.

In addition to global analysis, the closed-form formulae and design diagrams can also be used for the sizing of the elements of the bracing system and for the structural design of individual beam-columns. The formulae presented for the stability analysis are directly applicable to the analysis of columns with thin-walled cross-sections, subjected to unsymmetrical bending and torsion (cf. section 3.1.6). The formulae derived for the stress and frequency analyses need some modifications before they can be applied to individual elements.

The techniques are simple and facilitate 'on the back of the envelope' calculations; the results are reliable, albeit approximate.

Based on the qualitative and quantitative analyses of the methods, the objective of this chapter is to group the most important information together by types of behaviour and to provide the structural engineer with design guidelines.

11.1 SPATIAL BEHAVIOUR

The single most important conclusion is that torsion plays a crucial role in 3-dimensional structural behaviour and by reducing torsion the overall performance and the effectiveness of the bracing system can be greatly improved. Both the nature (Saint-Venant or warping) and the extent of torsion are important. This is equally true when the effects of horizontal loading are analysed (when the deformation of the building and the stresses in the bracing elements are calculated) and also when the effects of the vertical load are considered (when the main objectives of the analysis are to determine the global critical load and the fundamental frequency).

The stability and the dynamic behaviour are characterized by very similar phenomena and it is not surprising that very similar guidelines have to be followed in the design procedure. These guidelines can also be applied to the stress analysis with taking into account the different nature of the load.

11.2 STABILITY ANALYSIS

Global stability and the value of the sway-torsional critical load basically depend on

- the values of the basic critical loads which belong to the basic (sway and pure torsional) modes and
- the coupling of the basic modes.

The sway critical loads are controlled by very simple relationships: their value is in direct proportion to the bending stiffness of the bracing system and in inverse proportion to the square of the height of the building. The situation with the pure torsional critical load is somewhat different. Its value is in direct proportion to the warping stiffness of the bracing system and in inverse proportion to the height of the building. (The effect of the Saint-Venant torsional stiffness is normally small.) However, in addition to these relationships, the torsional critical load also depends on horizontal geometrical characteristics of the building. The greater the size of the building and the greater the distance between the shear centre and the centre of the vertical load, the smaller the critical load. The latter characteristic also affects the coupling of the basic modes: the greater the distance between the shear centre and the centre of the vertical load, the more 'dangerous' the coupling is. Coupling always reduces the value of the critical load: the reduction can be as much as 141%.

In summarizing, the most important characteristics controlling the behaviour of the bracing system for stability are

- the lateral stiffnesses characterized by the second moments of area I_X and I_Y,
- the torsional stiffness, whose value is normally dominated by warping constant I_ω,
- the distance between the shear centre of the bracing system and the centre of the vertical load. (The centre of the vertical load coincides with the geometrical centre of the plan of the building for uniformly distributed load.)

The most effective ways to increase the value of the global critical load (and consequently to improve the performance of the bracing system for stability) are

- to increase the lateral stiffnesses (preferably in a balanced way in both principal directions),
- to increase the value of the warping constant. This can be most efficiently achieved by placing the bracing elements as far from the shear centre as possible in such a way that the perpendicular distance between the plane in which the bending stiffness of the element acts and the shear centre is maximum,
- to reduce the distance between the shear centre of the bracing system and the centre of the vertical load (the geometrical centre of the plan of the building for uniformly distributed load), optimally to zero.

11.3 FREQUENCY ANALYSIS

The value of the fundamental frequency basically depends on

- the values of the basic natural frequencies which belong to the basic (lateral and pure torsional) modes and
- the coupling of the basic modes.

The value of the frequencies for all three basic types of vibration (lateral and pure torsional) vibrations is in inverse proportion to the square root of the weight of the structure.

In addition to weight, lateral vibrations are defined by the lateral stiffnesses and the height of the building. The value of the lateral frequencies increases proportionally to the square root of the lateral stiffness and decreases in inverse proportion to the square of the height.

The value of the pure torsional (uncoupled) frequency is in proportion to the root of the warping stiffness of the bracing system and in inverse proportion to the square of the height of the building. The effect of the Saint-Venant torsional stiffness is normally small. The value of the pure torsional frequency is also influenced by horizontal characteristics of the building. Its value decreases as the size of the ground plan and the eccentricity of the mass of the building (the distance between the shear centre and the centre of the mass) increase. The coupling of the basic mode always reduces the value of the fundamental frequency. The extent of coupling depends on the 'eccentricity' of the mass: the more eccentric the mass, the greater the coupling of the basic modes.

The most important characteristics controlling the natural frequencies of the bracing system are

- the lateral stiffnesses characterized by the second moments of area I_X and I_Y,
- the torsional stiffness, whose value is normally dominated by warping constant I_ω,
- the distance between the shear centre of the bracing system and the centre of the mass of the building. (The centre of the mass coincides with the geometrical centre of the plan of the building for uniformly distributed mass),
- the weight of the building.

The most effective ways to increase the value of the fundamental frequency (and consequently to improve the dynamic performance of the bracing system) are

- to increase the lateral stiffnesses stiffnesses (preferably in a balanced way in both principal directions),
- to increase the value of the warping constant. This can be most efficiently achieved by placing the bracing elements as far from the shear centre as possible in such a way that the perpendicular distance between the plane in which the bending stiffness of the element acts and the shear centre is maximum,
- to reduce the distance between the shear centre of the bracing system and the centre of the mass (the geometrical centre of the plan of the building for uniformly distributed mass), optimally to zero,
- to reduce the weight of the building.

258 Evaluation; design guidelines

11.4 STRESSES AND DEFORMATIONS

The guidelines given above for achieving optimum stability and dynamic performance can also be used in the stress analysis: by increasing the overall performance of the bracing system the global deformations of the building and the maximum stresses in the bracing elements can be reduced.

The evaluation of the equations and formulae leads to the following observations. The most important characteristics of the behaviour of the bracing system are

- the lateral stiffnesses characterized by the second moments of area I_x and I_y,
- the torsional stiffness (normally dominated by warping constant I_ω),
- the eccentricity of the load, i.e. the perpendicular distance between the shear centre of the bracing system and the line of action of the external load.

The most effective ways to reduce the global deformations and to improve the performance of the bracing system under horizontal load are

- to increase the lateral stiffnesses (preferably in a balanced way when I_x and I_y are proportional to the external load in the relevant direction),
- to increase the value of the warping constant (I_ω). This can be most efficiently achieved by placing the bracing elements as far from the shear centre as possible in such a way that the perpendicular distance between the plane in which the bending stiffness of the element acts and the shear centre is maximum,
- to reduce the eccentricity of the load, i.e. to reduce the distance between the shear centre of the bracing system and the geometrical centre of the plan of the building (through which the external load passes).

When designing for torsion, the following points should be considered.

- In addition to the bending and torsional stiffnesses, the load share on the bracing elements, due to torsion, also depends on the ratio of the Saint-Venant torsional stiffness and the warping stiffness of the bracing system. This ratio can be best characterized using the nondimensional torsion parameter k.

- In the range $0 < k < \infty$, the value of the maximum shear force in the bracing elements, due to torsion, is always greater than the value which belongs to $k = 0$ (the case when the Saint-Venant stiffness is zero or is neglected). The maximum difference is 23% at $k = 2.8$. It follows that the assumption $k = 0$ (openly or tacitly made by most analytical methods) always represents an unconservative approximation.
- For dominant Saint-Venant stiffness (GJ), due to the rotation of the bracing system under horizontal load, the elements of the bracing system only develop shear forces near their support and the value of the bending moments decreases with increasing GJ.
- With increasing the value of k (from zero), the value of the maximum bending moments in the bracing elements, due to torsion, is first greater then smaller than the value which belongs to $k = 0$. The maximum increase is 5.4% at $k = 1.35$, then the value of the maximum bending moment tends to zero. Consequently, the assumption $k = 0$ represents a slightly unconservative approximation in the region $0 < k < 2.35$, then the approximation is increasingly conservative for $2.35 < k < \infty$.

It is interesting to note that with increasing Saint-Venant stiffness, 'activities' (load, shear force, rotation) over the height of the bracing elements tend to limit to the vicinity of the bottom. This shows striking similarities to the planar buckling of cantilevers developing bending and shear deformations as well as to the torsional buckling of cantilevers.

11.5 STRUCTURAL PERFORMANCE OF THE BRACING SYSTEM

The performance of the bracing system is best monitored by using the global critical load ratio (the ratio of the applied vertical load and the global critical load). The critical load ratio is a sensitive indicator which can be calculated in minutes.

The global critical load ratio – by definition – is in a direct relationship with the (global) safety of the structure: it indicates the 'level' of safety. In addition, the comparison of different global critical load ratios belonging to different bracing system arrangements also gives useful information regarding the change in the value of the fundamental frequency and the maximum global deformations of the building. Thus monitoring the most important characteristics of the bracing system, an optimum structural layout can be achieved in a fast and simple manner, simultaneously leading to maximum safety and economy.

260 Evaluation; design guidelines

The single most important conclusion is that torsion plays a crucial role in structural behaviour and by reducing torsion the overall performance and the level of global safety can be greatly improved.

It is interesting to note that the application of the *same* guidelines (cf. sections 11.2, 11.3 and 11.4) leads to optimum performance regarding stability, global deformations, stresses and dynamic behaviour.

11.6 STABILITY OF PLANAR STRUCTURES

The analysis of the planar structures considered for the investigation in Chapter 9 (frameworks on fixed and pinned support and with or without single or double cross bracing, infilled frames, shear walls and coupled shear walls) shows that they all can be characterized by four types of deformation: full-height 'local' bending of the individual columns, full-height 'global' bending of the framework as a whole, full-height (continuous) shear deformation and storey-height shear deformation (Figs 9.1 and 9.2). These deformations can be characterized by the corresponding stiffnesses. The critical load of these structures is obtained by combining the part critical loads which are associated with the four types of deformations. It is therefore important to analyse the buckled shape. The form of the buckled shape depends on a number of characteristics and also varies as the height of the structure increases.

11.6.1 Low-rise to medium-rise (4–25-storey) structures

The global shear stiffness of the framework, which is basically determined by the bending stiffness of the beams, normally has a major contribution to the overall stability of the framework.

With frameworks on fixed supports and frameworks on pinned supports with ground floor beams, subjected to concentrated forces on top of the columns, the buckled shape is normally dominated by the global bending stiffness, the effect of the shear stiffness is important and the local bending stiffness is negligible (Fig. 11.1).

When frameworks on fixed supports and frameworks on pinned supports with ground floor beams are subjected to UDL on the beams, the effect of the shear stiffness is dominant (Fig. 11.2).

Stability of planar structures 261

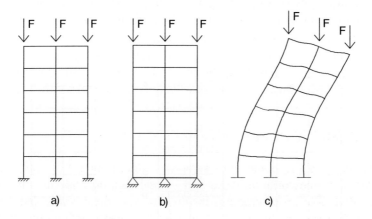

Fig. 11.1 Frameworks under concentrated top load.

When the frameworks have pinned supports which are not connected with beams, buckling failure develops through the horizontal sway of the ground floor columns in most cases (Fig. 11.3).

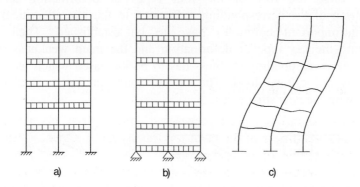

Fig. 11.2 Frameworks under UDL on the beams.

This, from the point of view of the framework as a whole, is considered to be local shear deformation. This local shear deformation is closely associated with the lateral stiffness of the ground floor columns, which basically determines the critical load of the framework.

Frameworks with (single or double) cross-bracing on fixed or pinned supports, under both concentrated top load and UDL on the beams, develop predominantly full-height bending deformation (as shown in Fig. 11.4 where the double curvature bending of the bars between the

262 *Evaluation; design guidelines*

nodes is not shown). The nature of the support (fixed of pinned) does not effect the behaviour (and the value of the critical load).

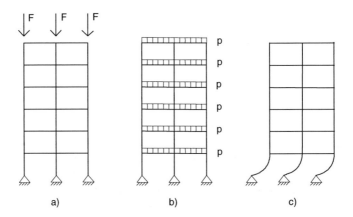

Fig. 11.3 Frameworks on pinned supports without ground floor beams.

The critical load – and the safety of a framework – can be increased as follows. First, the role of the four types of deformation and the contribution of the corresponding stiffnesses to the overall critical load (dominant/important/negligible) have to be established. Then, after identifying the key type of deformation and the main weakness of the system, the corresponding stiffness has to be increased which leads to a greater overall critical load. The critical load can be recalculated in minutes with the new stiffnesses.

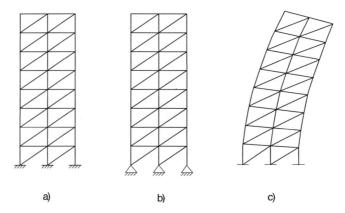

Fig. 11.4 Frameworks with single cross-bracing.

With coupled shear walls, of the four characteristic deformations, global bending and global shear deformations – and the corresponding stiffnesses/part critical loads – play an important role. Local bending deformation – and the corresponding stiffness/part critical load – normally have a minor effect on the overall critical load. (This contribution, however, can become significant with coupled shear walls with relatively wide walls.) The effect of local shear deformation, i.e. sway between the floor levels is normally negligible.

11.6.2 Tall (over 25-storey) structures

As the height of the frameworks and coupled shear walls increases, the effect of the full-height 'global' bending deformation of the structure becomes more and more important. Sooner or later the buckled shape of every planer structure is dominated by the full-height 'global' bending deformation. This point is illustrated in section 9.10 where the behaviour of 4 to 99 storey structures is investigated. According to Table 9.4 in section 9.10, each of the one-bay structures develops predominantly full-height 'global' bending deformation over the height of 24 storeys.

11.6.3 Structural performance of planar bracing elements

It goes without saying that solid shear walls represent the most effective type of bracing. Apart from some very rare occasions, they develop full-height bending deformation.

At the other end of the scale are frameworks on fixed and pinned supports, particularly on pinned supports without ground floor beams. Their buckled shape is normally dominated by local shear type deformation and the 'contribution' of the shear deformation reduces the critical load significantly for low and medium-rise frameworks. The performance of frameworks improves dramatically when the bays are 'filled in', by either (single or double) cross-bracing or by masonry panels.

It is interesting to note that the nature of the support structure (fixed/pinned) hardly influences the performance of infilled frameworks and frameworks with cross bracing as they normally develop full-height bending deformation.

Coupled shear walls represent a transition between frameworks and solid shear walls and their effectiveness largely depends on the relative stiffnesses of the wall-sections and connecting beams.

264 Evaluation; design guidelines

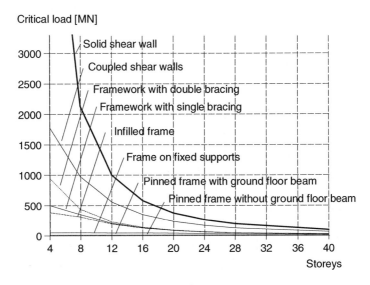

Fig. 11.5 One-bay bracing elements.

Figures 11.5 and 11.6 shows the critical loads of the one-bay and two-bay structures that were investigated in section 9.10. Details are given in Tables 9.4 and 9.5. The evaluation of the data in the tables and the figure leads to the following observations.

The structures can be divided into two categories. The structures in the first category are best described as structures with 'wall-like' behaviour. Shear walls, couples shear walls, frameworks with cross-bracing and infilled frameworks belong to this group. The structures here develop basically bending type deformation and they are fairly effective as bracing elements in a bracing system.

The behaviour of the structures in the second category can be characterized as 'frame-like' behaviour. Frameworks on fixed and on pinned supports, with or without ground floor beams, are in this group. They develop predominantly shear type deformation – until they reach a certain height – and their critical loads are much smaller than those of the 'wall-like' structures of the same height and width.

The difference between the two groups becomes smaller as the height of the structures increases. The behaviour of the structures in the 'frame-like' group is less and less dominated by shear type deformation and eventually all structures in the second group develop predominantly bending type deformation.

The values of the critical loads show that, especially for low-rise and medium-rise structures, there are huge differences between the structures in the two categories. The structures in the 'wall-like' category are much more efficient as bracing elements.

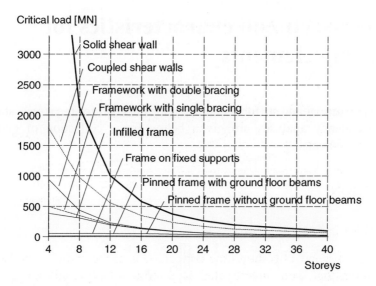

Fig. 11.6 Two-bay bracing elements.

Appendix A

Cross-sectional characteristics for bracing elements

The geometrical and stiffness characteristics needed for the stress, stability and frequency analyses are given below for a collection of common bracing elements.

I-SECTION

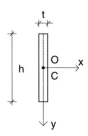

Fig. A1 I-section.

$$A = th, \quad I_x = \frac{th^3}{12}, \quad I_y = \frac{ht^3}{12}, \quad I_{xy} = 0,$$

$$J = \frac{ht^3}{3}, \quad I_\omega \approx 0 \quad \text{or}$$

$$I_\omega = \frac{t^3 h^3}{144} \quad \text{if the wall thickness is taken into account.}$$

TT-SECTION

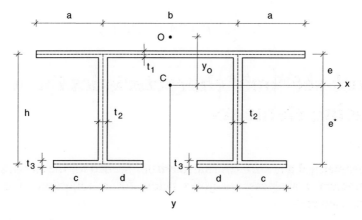

Fig. A2 TT-section.

$$A = A_f + 2A_g + 2A_a,$$

where

$$A_f = t_1(2a+b), \qquad A_g = t_2(h - \frac{t_1}{2} - \frac{t_3}{2}), \qquad A_a = t_3(c+d).$$

$$e = \frac{1}{A}\left(2A_a h + 2A_g(\frac{h}{2} - \frac{t_3}{4} + \frac{t_1}{4})\right), \qquad e^* = h - e.$$

$$y_o = -e - \frac{eAb^2}{4I_y} + \frac{2hA_{ag}}{I_y}, \qquad \text{where} \qquad A_{ag} = \frac{t_3}{3}(c^3 + d^3).$$

$$I_x = \frac{1}{12}\left(A_f t_1^2 + 2A_a t_3^2 + 2A_g(h - \frac{t_1}{2} - \frac{t_3}{2})^2\right) + A_f e^2 + 2A_a e^{*2} +$$

$$+ 2A_g\left(\frac{h}{2} - \frac{t_3}{4} + \frac{t_1}{4} - e\right)^2,$$

$$I_y = \frac{1}{12}\left(A_f(2a+b)^2 + t_3(b+2c)^3 - t_3(b-2d)^3 + 2A_g t_2^2\right) + \frac{b^2}{2}A_g,$$

$$I_{xy} = 0,$$

$$J = \frac{1}{3}\left(A_f t_1^2 + 2A_g t_2^2 + 2A_a t_3^2\right),$$

$$I_\omega = \frac{b^2 I_x}{4} + \frac{b^2 e^2 A}{4}\left(1 - \frac{b^2 A}{4I_y}\right) + 2h^2 I_{ag} - 2bfh^2 A_a + b^2 ehA\frac{I_{ag}}{I_y} - 4h^2 \frac{I_{ag}^2}{I_y},$$

where

$$f = \frac{c-d}{2}.$$

T-SECTION

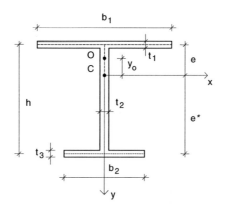

Fig. A3 T-section.

$$A = A_f + A_g + A_a,$$

where

$$A_f = t_1 b_1, \qquad A_a = t_3 b_2, \qquad A_g = t_2\left(h - \frac{t_1}{2} - \frac{t_3}{2}\right).$$

$$e = \frac{A_a}{A}h + \frac{A_g}{A}(\frac{h}{2} - \frac{t_3}{4} + \frac{t_1}{4}), \qquad e^* = h - e.$$

$$y_o = \frac{A_a b_2^2 h}{12 I_y} - e,$$

$$I_x = \frac{1}{12}\left(A_f t_1^2 + A_a t_3^2 + A_g (h - \frac{t_1}{2} - \frac{t_3}{2})^2 \right) + A_f e^2 + A_a e^{*2} +$$
$$+ A_g \left(\frac{h}{2} - \frac{t_3}{4} + \frac{t_1}{4} - e \right)^2,$$

$$I_y = \frac{1}{12}\left(A_f b_1^2 + A_a b_2^2 + A_g t_2^2 \right), \qquad I_{xy} = 0,$$

$$J = \frac{1}{3}\left(A_f t_1^2 + A_a t_3^2 + A_g t_2^2 \right), \qquad I_\omega = \frac{A_f b_1^2 A_a b_2^2 h^2}{144 I_y}.$$

L-SECTION

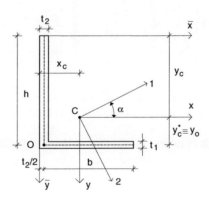

Fig. A4 L-section.

$$A = t_1 \left(b - \frac{t_2}{2} \right) + t_2 \left(h + \frac{t_1}{2} \right),$$

$$x_c = \frac{1}{2A}\left(t_2^2 (h - \frac{t_1}{2}) + t_1 (b + \frac{t_2}{2})^2 \right),$$

$$y_c = \frac{1}{2A}\left(t_2(h-\frac{t_1}{2})^2 + 2ht_1(b+\frac{t_2}{2})\right),$$

$$x_o = x_c - \frac{t_2}{2}, \qquad y_o = y_c^* = h - y_c.$$

$$I_x = \frac{t_2}{12}\left(h-\frac{t_1}{2}\right)^3 + \frac{t_1^3}{12}\left(b+\frac{t_2}{2}\right) + t_2\left(h-\frac{t_1}{2}\right)\left(y_c - \frac{h}{2} + \frac{t_1}{4}\right)^2 + \left(b+\frac{t_2}{2}\right)t_1 y_c^{*2},$$

$$I_y = \frac{t_1}{12}\left(b+\frac{t_2}{2}\right)^3 + \frac{t_2^3}{12}\left(h-\frac{t_1}{2}\right) + t_2\left(h-\frac{t_1}{2}\right)\left(x_c - \frac{t_2}{2}\right)^2 + t_1\left(b+\frac{t_2}{2}\right)\left(\frac{b}{2}+\frac{t_2}{4}-x_c\right)^2,$$

$$I_{xy} = t_2\left(h-\frac{t_1}{2}\right)\left(\frac{h}{2}-\frac{t_1}{4}-y_c\right)\left(\frac{t_2}{2}-x_c\right) + \left(b+\frac{t_2}{2}\right)\left(\frac{b}{2}+\frac{t_2}{4}-x_c\right)t_1 y_c^*.$$

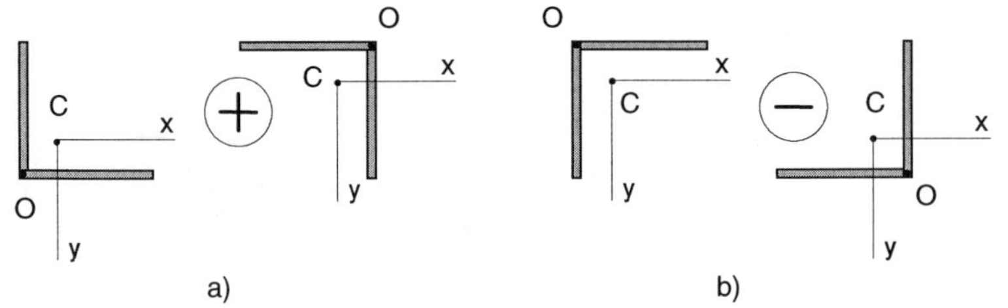

Fig. A5 Sign convention for the product of inertia. a) Positive, b) negative.

The sign of the product of inertia I_{xy} depends on the relative position of the section in relation to the coordinate axes. The sign convention is given in Fig. A5.

$$J = \frac{1}{3}\left(bt_1^3 + ht_2^3\right), \qquad I_\omega \approx 0 \qquad \text{or}$$

$$I_\omega = \frac{1}{36}\left(b^3 t_1^3 + h^3 t_2^3\right), \qquad \text{if the wall thickness is taken into account.}$$

$$\alpha = \frac{1}{2}\arctan\frac{2I_{xy}}{I_y - I_x}.$$

$$I_{1,2} = I_x \cos^2\alpha + I_y \sin^2\alpha \mp I_{xy} \sin 2\alpha.$$

□-SECTION

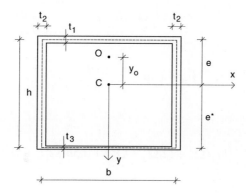

Fig. A6 □-section.

$$A = A_f + 2A_g + A_a,$$

where

$$A_f = t_1(b+t_2), \qquad A_a = t_3(b+t_2), \qquad A_g = t_2\left(h - \frac{t_1}{2} - \frac{t_3}{2}\right).$$

$$e = \frac{1}{A}\left(A_a h + A_g (h - \frac{t_3}{2} + \frac{t_1}{2})\right), \qquad e^* = h - e.$$

$$I_x = \frac{1}{12}\left(A_f t_1^2 + A_a t_3^2 + 2A_g(h - \frac{t_1}{2} - \frac{t_3}{2})^2\right) + A_f e^2 + A_a e^{*2} +$$

$$+ 2A_g\left(\frac{h}{2} - \frac{t_3}{4} + \frac{t_1}{4} - e\right)^2,$$

$$I_y = \frac{1}{12}\left((A_f + A_a)(b+t_2)^2 + 2A_g t_2^2\right) + A_g \frac{b^2}{2}, \qquad I_{xy} = 0.$$

$$y_o = -\frac{I_{\omega x}}{I_y} - e,$$

where

$$I_{\omega x} = \frac{b^2}{6}(\omega_1 t_1 + \omega_2 t_3) + \frac{bht_2}{2}(\omega_1 + \omega_2),$$

$$\omega_1 = -\Psi \frac{b}{2t_1}, \quad \omega_2 = \frac{bh}{2} - \Psi\left(\frac{b}{2t_1} + \frac{h}{t_2}\right), \quad \Psi = \frac{2A_o}{\int \frac{ds}{t}} = \frac{2bh}{\frac{b}{t_1} + \frac{b}{t_3} + \frac{2h}{t_2}}.$$

$$J = 2A_o \Psi = \frac{4h^2 b^2}{\frac{b}{t_1} + \frac{b}{t_3} + \frac{2h}{t_2}}.$$

$$I_\omega = \frac{2}{3}\left(ht_2(\Omega_1^2 + \Omega_1 \Omega_2 + \Omega_2^2) + 0.5b(\Omega_1^2 t_1 + \Omega_2^2 t_3)\right),$$

where

$$\Omega_1 = -\frac{bI_{\omega x}}{2I_y} + \omega_1, \quad \Omega_2 = -\frac{bI_{\omega x}}{2I_y} + \omega_2.$$

$+$-SECTION

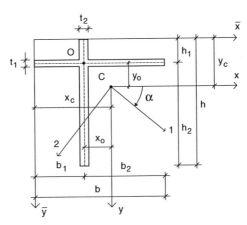

Fig. A7 $+$-section.

$$A = t_1 b + t_2(h - t_1),$$

where

$$b = b_1 + b_2, \qquad h = h_1 + h_2.$$

$$x_c = \frac{t_1 b^2}{2A} + \frac{t_2 b_1 (h - t_1)}{A}, \qquad y_c = \frac{t_2 h^2}{2A} + \frac{t_1 h_1 (b - t_2)}{A}.$$

$$x_o = b_1 - x_c, \qquad y_o = h_1 - y_c.$$

$$I_x = \frac{t_2 h^3 + (b - t_2) t_1^3}{12} + h t_2 \left(y_c - \frac{h}{2} \right)^2 + (b - t_2) t_1 (y_c - h_1)^2.$$

$$I_y = \frac{t_1 b^3 + (h - t_1) t_2^3}{12} + b t_1 \left(\frac{b}{2} - x_c \right)^2 + (h - t_1) t_2 (b_1 - x_c)^2.$$

$$I_{xy} = t_1 b (h_1 - y_c) \left(\frac{b}{2} - x_c \right) + t_2 (b_1 - x_c) \left(h_1 - \frac{t_1}{2} \right) \left(\frac{h_1}{2} - \frac{t_1}{4} - y_c \right) +$$
$$+ t_2 (b_1 - x_c) \left(h_2 - \frac{t_1}{2} \right) \left(h - \frac{h_2}{2} + \frac{t_1}{4} - y_c \right).$$

$$\alpha = \frac{1}{2} \arctan \frac{2 I_{xy}}{I_y - I_x}, \qquad J = \frac{1}{3} \left(b t_1^3 + (h - t_1) t_2^3 \right).$$

$$I_{1,2} = I_x \cos^2 \alpha + I_y \sin^2 \alpha \mp I_{xy} \sin 2\alpha.$$

$$I_\omega = \frac{1}{36} \left(t_2^3 (h_1 - \frac{t_1}{4})^3 + t_2^3 (h_2 - \frac{t_1}{4})^3 + t_1^3 (b_1 - \frac{t_2}{4})^3 + t_1^3 (b_2 - \frac{t_2}{4})^3 \right),$$

or

$$I_\omega \approx 0, \qquad \text{if the wall thickness is not taken into account.}$$

Z-SECTION

$$A = t_1(2b - t_2) + t_2(h + t_1).$$

$$I_x = \frac{t_2(h+t_1)^3 + (2b-t_2)t_1^3}{12} + (2b-t_2)t_1\frac{h^2}{4}.$$

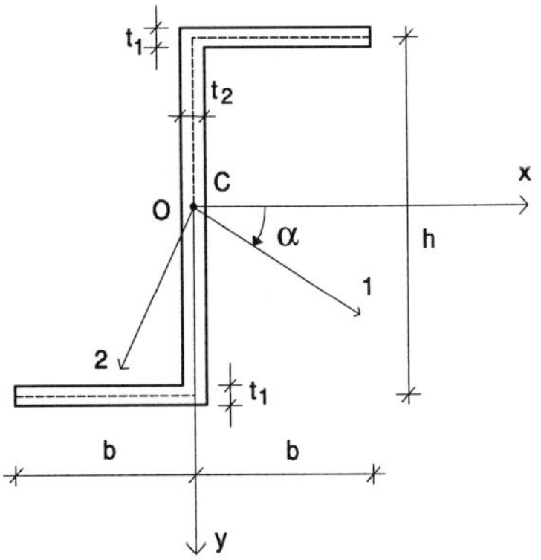

Fig. A8 Z-section.

$$I_y = \frac{(h+t_1)t_2^3}{12} + \frac{t_1}{6}\left(b - \frac{t_2}{2}\right)^3 + t_1(2b-t_2)\left(\frac{b}{2} + \frac{t_2}{4}\right)^2.$$

$$I_{xy} = -\frac{ht_1}{8}(4b^2 - t_2^2), \qquad \alpha = \frac{1}{2}\arctan\frac{2I_{xy}}{I_y - I_x}.$$

$$I_{1,2} = I_x \cos^2\alpha + I_y \sin^2\alpha \mp I_{xy}\sin 2\alpha.$$

$$J = \frac{1}{3}\left(2bt_1^3 + t_2^3 h\right), \qquad I_\omega = \frac{b^3 h^2 t_1}{12}\frac{bt_1 + 2ht_2}{2bt_1 + ht_2}.$$

CIRCULAR SECTIONS

Solid circular section

Fig. A9 Solid circular section.

$$A = R^2\pi, \qquad I_x = I_y = \frac{\pi R^4}{4}, \qquad I_{xy} = 0.$$

$$J = \frac{\pi R^4}{2}, \qquad I_\omega = 0.$$

Tube

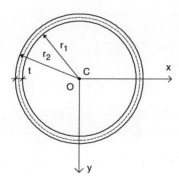

Fig. A10 Tube.

$$A = \pi(r_2^2 - r_1^2), \qquad I_x = I_y = \frac{\pi}{4}(r_2^4 - r_1^4), \qquad I_{xy} = 0.$$

$$J = 2\pi t \left(\frac{r_1 + r_2}{2}\right)^3, \qquad I_\omega = 0.$$

Open circular section

$$A = (r_2^2 - r_1^2)\beta.$$

$$x_c = \frac{2\sin\beta(r_2^3 - r_1^3)}{3\beta(r_2^2 - r_1^2)}, \qquad x_o = \frac{(r_1 + r_2)(\sin\beta - \beta\cos\beta)}{\beta - \sin\beta\cos\beta} - x_c.$$

$$I_x = \frac{r_2^4 - r_1^4}{4}(\beta - \sin\beta\cos\beta), \qquad I_{xy} = 0.$$

$$I_y = \frac{r_2^4 - r_1^4}{4}(\beta + \sin\beta\cos\beta) - \frac{1}{A}\left(\frac{2\sin\beta(r_2^3 - r_1^3)}{3}\right)^2.$$

$$J = \frac{r_1 + r_2}{3}\beta t^3, \qquad I_\omega = t\left(\frac{2}{3}\beta^3 - \frac{4(\sin\beta - \beta\cos\beta)^2}{\beta - \sin\beta\cos\beta}\right)\left(\frac{r_1 + r_2}{2}\right)^5.$$

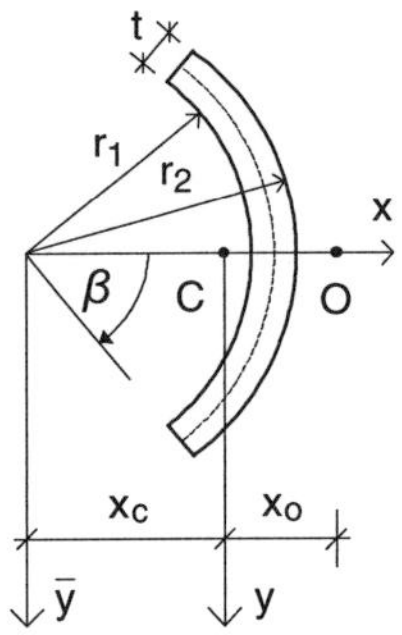

Fig. A11 Open circular section.

PERFORATED CORE

The formulae given for □-sections above can be applied to perforated cores. In using a uniform continuous medium, an equivalent thickness can be introduced for the wall section with the openings [Steinle and Hahn, 1988]. Parameters t^* and t_w^* in the formulae below stand for the equivalent thickness when the medium replaces the whole wall section of width b and the section of lintels of width w, respectively.

$$t_w^* = \frac{w}{\dfrac{b-t}{t^*} - \dfrac{b-t-w}{t}}, \qquad t^* = \frac{h}{b-t} \cdot \frac{1}{\dfrac{h_1}{2}\left(\dfrac{Gh_1^2}{12EI_s} + \dfrac{1}{A_s}\right) + l_1\left(\dfrac{h}{l}\right)^2\left(\dfrac{Gl_1^2}{12EI_r} + \dfrac{1}{A_r}\right)},$$

where

if $\quad a > c, \quad$ then $\quad l_1 = w + c \quad$ and $\quad h_1 = h,$

if $\quad a < c, \quad$ then $\quad l_1 = l \quad$ and $\quad h_1 = h - c + a,$

furthermore

$$A_s = ta, \qquad I_s = \frac{ta^3}{12}, \qquad A_r = tc, \qquad I_r = \frac{tc^3}{12}.$$

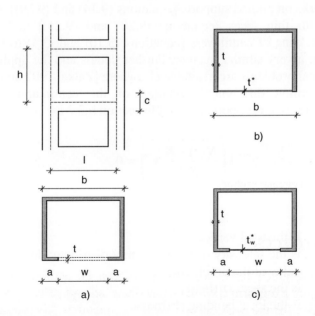

Fig. A12 Perforated core. a) Cross-section and elevation of perforated core, b) cross-section with full-width equivalent medium, c) cross-section with equivalent medium of width w.

Appendix B

The generalized power series method for eigenvalue problems

The solution of most of the problems discussed in this book is straightforward and details of the solution are not given. However, some of the stability problems are difficult to solve and even when a solution is found, there may be numerical difficulties due to the ill-conditioned eigenvalue problem. The power series method, generalized for such problems, is presented here and, as an example, is applied to produce the solution of the governing differential equation of the stability problem of frameworks on fixed supports. Other eigenvalue problems, including those of frameworks on pinned supports [equations (9.14) and (9.29)], sandwich columns with thin faces [equations (9.20) and (9.21)] and the pure torsional buckling of cantilevers [equations (3.10) and (3.4) to (3.7)] can be solved in a very similar manner. Further details and the application to 3-dimensional problems are available elsewhere [Zalka, 1992 and 1993].

The governing differential equation of the continuum model of frameworks on fixed supports is given in section 9.3.2 as

$$y'''' + \left(\frac{N(z) - K}{r_s E_c I_c} y'\right)' = 0, \qquad (B1)$$

where

- $N(z) = qz$ is the total vertical load,
- q is the intensity of the uniformly distributed load (the eigenvalue of the problem),
- K is the shear stiffness,
- $E_c I_c$ is the local bending stiffness,
- r_s is a modifier whose values are given in Table 3.1.

The origin of the coordinate system is fixed at the top of the equivalent column (Fig. 9.4/e) and the boundary conditions for frameworks on fixed supports are given in this coordinate system as follows.

The lateral translation is zero at the top of the column:

$$y(0) = 0. \tag{B2a}$$

The tangent to the equivalent column at the bottom of the column is parallel to axis z:

$$y'(H) = 0. \tag{B2b}$$

No bending moment develops at the top of the column:

$$y''(0) = 0. \tag{B2c}$$

The sum of the shear forces at the bottom of the column assumes zero value:

$$y'''(H) = 0. \tag{B2d}$$

The order of differential equation (B1) can be reduced by one. After integrating the differential equation once, making use of boundary condition (B2d) and introducing the non-dimensional quantities

$$\alpha = \frac{qH^3}{7.837 r_s E_c I_c} = \frac{N_{cr}}{N_l} \tag{B3}$$

and

$$\beta = \frac{KH^2}{7.837 r_s E_c I_c} = \frac{K}{N_l}, \tag{B4}$$

the equation can be rewritten as

$$y''' + \frac{7.837}{H^3}\alpha z y' - \frac{7.837}{H^2}\beta y' = 0. \tag{B5}$$

Boundary conditions (B2a), (B2b) and (B2c) complement the above differential equation.

The solution is found in the power series

280 Appendix B

$$y = c_0 + c_1 z + c_2 z^2 + c_3 z^3 + \ldots = \sum_{k=0}^{\infty} c_k z^k, \tag{B6}$$

where c_k denotes coefficients yet unknown. The derivatives of the solution function are

$$y' = c_1 + 2c_2 z + 3c_3 z^2 + 4c_4 z^3 + 5c_5 z^4 + \ldots = \sum_{k=1}^{\infty} k c_k z^{k-1}, \tag{B7a}$$

$$y'' = 2c_2 + 3 \cdot 2c_3 z + 4 \cdot 3c_4 z^2 + 5 \cdot 4c_5 z^3 + \ldots = \sum_{k=2}^{\infty} k(k-1) c_k z^{k-2}, \tag{B7b}$$

$$y''' = 3 \cdot 2c_3 + 4 \cdot 3 \cdot 2c_4 z + 5 \cdot 4 \cdot 3c_5 z^2 + \ldots = \sum_{k=3}^{\infty} k(k-1)(k-2) c_k z^{k-3}. \tag{B7c}$$

After substituting the derivatives into differential equation (B5), the equation

$$3 \cdot 2c_3 \quad + 4 \cdot 3 \cdot 2c_4 z \quad + 5 \cdot 4 \cdot 3c_5 z^2 \quad + 6 \cdot 5 \cdot 4c_6 z^3 \quad + \ldots$$

$$+ \frac{7.837}{H^3} \alpha c_1 z \quad + \frac{7.837}{H^3} \alpha 2 c_2 z^2 \quad + \frac{7.837}{H^3} \alpha 3 c_3 z^3 \quad + \ldots \tag{B8}$$

$$- \frac{7.837}{H^2} \beta c_1 \quad - \frac{7.837}{H^2} \beta 2 c_2 z \quad - \frac{7.837}{H^2} \beta 3 c_3 z^2 \quad - \frac{7.837}{H^2} \beta 4 c_4 z^3 \quad - \ldots = 0$$

is obtained. This equation can only hold if the sum of the coefficients of each power of z equals zero. This condition, for z^2 for example, leads to

$$5 \cdot 4 \cdot 3 c_5 + \frac{7.837}{H^3} \alpha 2 c_2 - \frac{7.837}{H^2} \beta 3 c_3 = 0 \tag{B9}$$

or, in general form, to

$$k(k-1)(k-2) c_k + \frac{7.837}{H^3} \alpha (k-3) c_{k-3} - \frac{7.837}{H^2} \beta (k-2) c_{k-2} = 0. \tag{B10}$$

Rearranging equation (B10) results in a recursion formula for the unknown coefficients for $k \geq 3$:

$$c_k = \frac{7.837\beta}{k(k-1)H^2}c_{k-2} - \frac{7.837\alpha(k-3)}{k(k-1)(k-2)H^3}c_{k-3}. \quad (B11)$$

The initial values needed for the calculation are obtained from boundary conditions (B2a) and (B2c), after making use of the power series and its derivatives, as

$$c_0 = 0, \quad (B12a)$$

$$c_2 = 0. \quad (B12b)$$

Boundary condition (B2b), yet unused, leads to the equation

$$f(\beta,\alpha) = c_1 + \sum_{k=3}^{\infty} kc_k H^{k-1} = 0. \quad (B13)$$

Polynomial $f(\beta,\alpha)$ only vanishes when the eigenvalues of the problem $(\alpha_1, \alpha_2, \ldots \alpha_n)$, i.e. the critical loads, are substituted for α (Fig. B1). It follows that the smallest root of polynomial $f(\beta,\alpha)$ yields the first eigenvalue of the problem, which is of practical importance, being the smallest critical load of the structure.

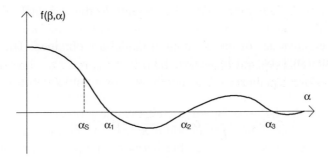

Fig. B1 Polynomial with eigenvalues $\alpha_S, \alpha_1, \alpha_2, \alpha_3$.

The original eigenvalue problem (B1) has been transformed into a simple problem of finding the smallest root of a polynomial; this is a much simpler task than the original problem, as finding the smallest root of a polynomial is a routine exercise. The procedure can be easily monitored

282 Appendix B

and when problems related to slow convergence due to ill-conditioned stiffness parameters emerge, the parameters governing the solution can be adjusted to achieve reliable results. Once a specific stiffness region is covered, the eigenvalues can be tabulated for future use.

It should be noted here that coefficient c_1 appears in each term of c_k, and consequently in each term of equation (B13), so that its value does not affect the value of the eigenvalue. For this reason, any suitable value for c_1, for example $c_1 = 1$, can be used for the actual calculation of the eigenvalue. Similarly, powers of H appear both in the formula of c_k (in the denominator) and in the formula of $f(\beta, \alpha)$ (in the numerator) in the same manner, so that the value of H does not affect the eigenvalue either. Again, a suitable value, e.g. $H = 1$ can be used for the calculation of the eigenvalue.

The determination of the smallest root is made easier as the shape of the polynomial is known: due to the physical nature of the problem – only structures with non-negative geometrical and stiffness characteristics are dealt with – the polynomial assumes a positive value at $\alpha = 0$ (Fig. B1). It is also helpful that a lower bound is known to the smallest eigenvalue: Southwell's summation theorem results in a conservative estimate of the exact eigenvalue. The Southwell solution

$$\alpha_s = 1 + \beta \tag{B14}$$

can conveniently be used as a starting value in the iteration process for the solution of equation (B13) (Fig. B1).

After solving equation (B13) representing the original eigenvalue problem (B1) for different stiffness ratios β and making use of formula (B3), the eigenvalues of the continuum model are obtained. These values are tabulated in Table 9.1 in section 9.3.2.

The procedure outlined above can be applied to any type of eigenvalue problem – it was used to solve all of the stability problems in Chapters 3 and 9. It can also be used for solving eigenvalue problems which have the eigenvalue in the boundary conditions, e.g. as is the case with the sandwich model in section 9.3.3. Furthermore, the procedure can even be applied to 3-dimensional eigenvalue problems.

Appendix C

Mode coupling parameter κ

Table C1 Mode coupling parameter κ for $\tau_X = 0.0$

r_2	$r_1=0$	$r_1=0.2$	$r_1=0.4$	$r_1=0.6$	$r_1=0.8$	$r_1=1$	$r_1=2$	$r_1=3$	$r_1=4$	$r_1=5$	$r_1=10$	τ_Y
0, 0.2, 0.4, 0.6, 0.8, 1	1.000	1.000	1.000	1.000	1.000	1.000	0.500	0.333	0.250	0.200	0.100	0.0
	1.000	0.998	0.993	0.986	0.967	0.909	0.495	0.332	0.249	0.200	0.100	0.1
	1.000	0.990	0.975	0.950	0.905	0.833	0.482	0.327	0.247	0.198	0.100	0.2
	1.000	0.979	0.948	0.904	0.843	0.769	0.464	0.320	0.243	0.196	0.099	0.3
	1.000	0.963	0.915	0.856	0.787	0.714	0.444	0.311	0.238	0.193	0.098	0.4
	1.000	0.945	0.880	0.809	0.736	0.667	0.423	0.301	0.232	0.189	0.097	0.5
	1.000	0.925	0.850	0.766	0.692	0.625	0.402	0.291	0.226	0.185	0.096	0.6
	1.000	0.903	0.810	0.726	0.652	0.588	0.383	0.280	0.220	0.181	0.095	0.7
	1.000	0.880	0.776	0.689	0.616	0.556	0.365	0.270	0.213	0.176	0.094	0.8
	1.000	0.857	0.744	0.656	0.584	0.526	0.349	0.260	0.207	0.171	0.092	0.9
	1.000	0.833	0.714	0.625	0.556	0.500	0.333	0.250	0.200	0.167	0.091	1.0
1.5	0.667	0.667	0.667	0.667	0.667	0.667	0.500	0.333	0.250	0.200	0.100	0.0
	0.667	0.667	0.667	0.667	0.667	0.667	0.495	0.332	0.249	0.200	0.100	0.1
	0.667	0.667	0.667	0.667	0.667	0.667	0.482	0.327	0.247	0.198	0.100	0.2
	0.667	0.667	0.667	0.667	0.667	0.667	0.464	0.320	0.243	0.196	0.099	0.3
	0.667	0.667	0.667	0.667	0.667	0.667	0.444	0.311	0.238	0.193	0.098	0.4
	0.667	0.667	0.667	0.667	0.667	0.667	0.423	0.301	0.232	0.189	0.097	0.5
	0.667	0.667	0.667	0.667	0.667	0.625	0.402	0.291	0.226	0.185	0.096	0.6
	0.667	0.667	0.667	0.667	0.652	0.588	0.383	0.280	0.220	0.181	0.095	0.7
	0.667	0.667	0.667	0.667	0.616	0.556	0.365	0.270	0.213	0.176	0.094	0.8
	0.667	0.667	0.667	0.656	0.584	0.526	0.349	0.260	0.207	0.171	0.092	0.9
	0.667	0.667	0.667	0.625	0.556	0.500	0.333	0.250	0.200	0.167	0.091	1.0
2	0.500	0.500	0.500	0.500	0.500	0.500	0.500	0.333	0.250	0.200	0.100	0.0
	0.500	0.500	0.500	0.500	0.500	0.500	0.495	0.332	0.249	0.200	0.100	0.1
	0.500	0.500	0.500	0.500	0.500	0.500	0.482	0.327	0.247	0.198	0.100	0.2
	0.500	0.500	0.500	0.500	0.500	0.500	0.464	0.320	0.243	0.196	0.099	0.3
	0.500	0.500	0.500	0.500	0.500	0.500	0.444	0.311	0.238	0.193	0.098	0.4
	0.500	0.500	0.500	0.500	0.500	0.500	0.423	0.301	0.232	0.189	0.097	0.5
	0.500	0.500	0.500	0.500	0.500	0.500	0.402	0.291	0.226	0.185	0.096	0.6
	0.500	0.500	0.500	0.500	0.500	0.500	0.383	0.280	0.220	0.181	0.095	0.7
	0.500	0.500	0.500	0.500	0.500	0.500	0.365	0.270	0.213	0.176	0.094	0.8
	0.500	0.500	0.500	0.500	0.500	0.500	0.349	0.260	0.207	0.171	0.092	0.9
	0.500	0.500	0.500	0.500	0.500	0.500	0.333	0.250	0.200	0.167	0.091	1.0

284 Appendix C

Table C1 Continued. Mode coupling parameter κ for $\tau_X = 0.0$

r_2	$r_1=0$	$r_1=0.2$	$r_1=0.4$	$r_1=0.6$	$r_1=0.8$	$r_1=1$	$r_1=2$	$r_1=3$	$r_1=4$	$r_1=5$	$r_1=10$	τ_Y
3	0.333	0.333	0.333	0.333	0.333	0.333	0.333	0.333	0.250	0.200	0.100	0.0
	0.333	0.333	0.333	0.333	0.333	0.333	0.333	0.332	0.249	0.200	0.100	0.1
	0.333	0.333	0.333	0.333	0.333	0.333	0.333	0.327	0.247	0.198	0.100	0.2
	0.333	0.333	0.333	0.333	0.333	0.333	0.333	0.320	0.243	0.196	0.099	0.3
	0.333	0.333	0.333	0.333	0.333	0.333	0.333	0.311	0.238	0.193	0.098	0.4
	0.333	0.333	0.333	0.333	0.333	0.333	0.333	0.301	0.232	0.189	0.097	0.5
	0.333	0.333	0.333	0.333	0.333	0.333	0.333	0.291	0.226	0.185	0.096	0.6
	0.333	0.333	0.333	0.333	0.333	0.333	0.333	0.280	0.220	0.181	0.095	0.7
	0.333	0.333	0.333	0.333	0.333	0.333	0.333	0.270	0.213	0.176	0.094	0.8
	0.333	0.333	0.333	0.333	0.333	0.333	0.333	0.260	0.207	0.171	0.092	0.9
	0.333	0.333	0.333	0.333	0.333	0.333	0.333	0.250	0.200	0.167	0.091	1.0
4	0.250	0.250	0.250	0.250	0.250	0.250	0.250	0.250	0.250	0.200	0.100	0.0
	0.250	0.250	0.250	0.250	0.250	0.250	0.250	0.250	0.249	0.200	0.100	0.1
	0.250	0.250	0.250	0.250	0.250	0.250	0.250	0.250	0.247	0.198	0.100	0.2
	0.250	0.250	0.250	0.250	0.250	0.250	0.250	0.250	0.243	0.196	0.099	0.3
	0.250	0.250	0.250	0.250	0.250	0.250	0.250	0.250	0.238	0.193	0.098	0.4
	0.250	0.250	0.250	0.250	0.250	0.250	0.250	0.250	0.232	0.189	0.097	0.5
	0.250	0.250	0.250	0.250	0.250	0.250	0.250	0.250	0.226	0.185	0.096	0.6
	0.250	0.250	0.250	0.250	0.250	0.250	0.250	0.250	0.220	0.181	0.095	0.7
	0.250	0.250	0.250	0.250	0.250	0.250	0.250	0.250	0.213	0.176	0.094	0.8
	0.250	0.250	0.250	0.250	0.250	0.250	0.250	0.250	0.207	0.171	0.092	0.9
	0.250	0.250	0.250	0.250	0.250	0.250	0.250	0.250	0.200	0.167	0.091	1.0
5	0.200	0.200	0.200	0.200	0.200	0.200	0.200	0.200	0.200	0.200	0.100	0.0
	0.200	0.200	0.200	0.200	0.200	0.200	0.200	0.200	0.200	0.200	0.100	0.1
	0.200	0.200	0.200	0.200	0.200	0.200	0.200	0.200	0.200	0.198	0.100	0.2
	0.200	0.200	0.200	0.200	0.200	0.200	0.200	0.200	0.200	0.196	0.099	0.3
	0.200	0.200	0.200	0.200	0.200	0.200	0.200	0.200	0.200	0.193	0.098	0.4
	0.200	0.200	0.200	0.200	0.200	0.200	0.200	0.200	0.200	0.189	0.097	0.5
	0.200	0.200	0.200	0.200	0.200	0.200	0.200	0.200	0.200	0.185	0.096	0.6
	0.200	0.200	0.200	0.200	0.200	0.200	0.200	0.200	0.200	0.181	0.095	0.7
	0.200	0.200	0.200	0.200	0.200	0.200	0.200	0.200	0.200	0.176	0.094	0.8
	0.200	0.200	0.200	0.200	0.200	0.200	0.200	0.200	0.200	0.171	0.092	0.9
	0.200	0.200	0.200	0.200	0.200	0.200	0.200	0.200	0.200	0.167	0,091	1.0
10	0.100	0.100	0.100	0.100	0.100	0.100	0.100	0.100	0.100	0.100	0.100	0.0
	0.100	0.100	0.100	0.100	0.100	0.100	0.100	0.100	0.100	0.100	0.100	0.1
	0.100	0.100	0.100	0.100	0.100	0.100	0.100	0.100	0.100	0.100	0.100	0.2
	0.100	0.100	0.100	0.100	0.100	0.100	0.100	0.100	0.100	0.100	0.099	0.3
	0.100	0.100	0.100	0.100	0.100	0.100	0.100	0.100	0.100	0.100	0.098	0.4
	0.100	0.100	0.100	0.100	0.100	0.100	0.100	0.100	0.100	0.100	0.097	0.5
	0.100	0.100	0.100	0.100	0.100	0.100	0.100	0.100	0.100	0.100	0.096	0.6
	0.100	0.100	0.100	0.100	0.100	0.100	0.100	0.100	0.100	0.100	0.095	0.7
	0.100	0.100	0.100	0.100	0.100	0.100	0.100	0.100	0.100	0.100	0.094	0.8
	0.100	0.100	0.100	0.100	0.100	0.100	0.100	0.100	0.100	0.100	0.092	0.9
	0.100	0.100	0.100	0.100	0.100	0.100	0.100	0.100	0.100	0.100	0.091	1.0

Table C2 Mode coupling parameter κ for $\tau_X = 0.1$

r_2	$r_1=0$	$r_1=0.2$	$r_1=0.4$	$r_1=0.6$	$r_1=0.8$	$r_1=1$	$r_1=2$	$r_1=3$	$r_1=4$	$r_1=5$	$r_1=10$	τ_Y
	1.000	1.000	1.000	1.000	1.000	1.000	0.500	0.333	0.250	0.200	0.100	0.0
	1.000	1.000	0.993	0.986	0.967	0.909	0.495	0.332	0.249	0.200	0.100	0.1
	1.000	0.990	0.975	0.950	0.905	0.833	0.482	0.327	0.247	0.198	0.100	0.2
	1.000	0.979	0.948	0.904	0.843	0.769	0.464	0.320	0.243	0.196	0.099	0.3
	1.000	0.963	0.915	0.856	0.787	0.714	0.444	0.311	0.238	0.193	0.098	0.4
0	1.000	0.945	0.880	0.809	0.736	0.667	0.423	0.301	0.232	0.189	0.097	0.5
	1.000	0.925	0.845	0.766	0.692	0.625	0.402	0.291	0.226	0.185	0.096	0.6
	1.000	0.903	0.810	0.726	0.652	0.588	0.383	0.280	0.220	0.181	0.095	0.7
	1.000	0.880	0.776	0.689	0.616	0.556	0.365	0.270	0.213	0.176	0.094	0.8
	1.000	0.857	0.744	0.656	0.584	0.526	0.349	0.260	0.207	0.171	0.092	0.9
	1.000	0.833	0.714	0.625	0.556	0.500	0.333	0.250	0.200	0.167	0.091	1.0
	1.000	0.998	0.998	0.998	0.998	0.998	0.500	0.333	0.250	0.200	0.100	0.0
	1.000	0.995	0.991	0.983	0.965	0.908	0.495	0.332	0.249	0.200	0.100	0.1
	1.000	0.988	0.973	0.948	0.904	0.833	0.482	0.327	0.247	0.198	0.100	0.2
	1.000	0.976	0.946	0.902	0.842	0.769	0.464	0.320	0.243	0.196	0.099	0.3
	1.000	0.961	0.914	0.854	0.786	0.714	0.444	0.311	0.238	0.193	0.098	0.4
0.2	1.000	0.943	0.879	0.808	0.736	0.666	0.423	0.301	0.232	0.189	0.097	0.5
	1.000	0.923	0.844	0.765	0.691	0.625	0.402	0.291	0.226	0.185	0.096	0.6
	1.000	0.901	0.809	0.725	0.652	0.588	0.383	0.280	0.220	0.181	0.095	0.7
	1.000	0.878	0.775	0.689	0.616	0.555	0.365	0.270	0.213	0.176	0.094	0.8
	1.000	0.855	0.744	0.655	0.584	0.526	0.349	0.260	0.207	0.171	0.092	0.9
	1.000	0.832	0.714	0.625	0.555	0.500	0.333	0.250	0.200	0.167	0.091	1.0
	0.993	0.993	0.993	0.993	0.993	0.993	0.500	0.333	0.250	0.200	0.100	0.0
	0.993	0.991	0.987	0.980	0.962	0.907	0.495	0.332	0.249	0.200	0.100	0.1
	0.993	0.984	0.969	0.945	0.902	0.832	0.482	0.327	0.247	0.198	0.100	0.2
	0.993	0.973	0.943	0.900	0.841	0.768	0.464	0.320	0.243	0.196	0.099	0.3
	0.993	0.958	0.911	0.853	0.785	0.713	0.443	0.311	0.238	0.193	0.098	0.4
0.4	0.993	0.940	0.877	0.807	0.735	0.666	0.423	0.301	0.232	0.189	0.097	0.5
	0.993	0.920	0.842	0.764	0.691	0.624	0.402	0.291	0.226	0.185	0.096	0.6
	0.993	0.899	0.807	0.724	0.651	0.588	0.383	0.280	0.220	0.181	0.095	0.7
	0.993	0.876	0.774	0.688	0.616	0.555	0.365	0.270	0.213	0.176	0.094	0.8
	0.993	0.853	0.743	0.655	0.584	0.526	0.349	0.260	0.207	0.171	0.092	0.9
	0.993	0.831	0.713	0.624	0.555	0.500	0.333	0.250	0.200	0.167	0.091	1.0
	0.986	0.986	0.986	0.986	0.986	0.986	0.500	0.333	0.250	0.200	0.100	0.0
	0.986	0.983	0.980	0.973	0.956	0.904	0.495	0.332	0.249	0.200	0.100	0.1
	0.986	0.977	0.963	0.939	0.898	0.830	0.482	0.327	0.247	0.198	0.100	0.2
	0.986	0.966	0.937	0.896	0.838	0.767	0.464	0.320	0.243	0.196	0.099	0.3
	0.986	0.952	0.907	0.850	0.783	0.712	0.443	0.311	0.238	0.193	0.098	0.4
0.6	0.986	0.934	0.873	0.805	0.734	0.665	0.423	0.301	0.232	0.189	0.097	0.5
	0.986	0.915	0.839	0.762	0.690	0.624	0.402	0.291	0.226	0.185	0.096	0.6
	0.986	0.894	0.805	0.723	0.650	0.587	0.383	0.280	0.220	0.181	0.095	0.7
	0.986	0.872	0.772	0.687	0.615	0.555	0.365	0.270	0.213	0.176	0.094	0.8
	0.986	0.850	0.741	0.654	0.583	0.526	0.349	0.260	0.207	0.171	0.092	0.9
	0.986	0.828	0.712	0.624	0.555	0.500	0.333	0.250	0.200	0.167	0.091	1.0

286 Appendix C

Table C2 Continued. Mode coupling parameter κ for $\tau_X = 0.1$

r_2	$r_1=0$	$r_1=0.2$	$r_1=0.4$	$r_1=0.6$	$r_1=0.8$	$r_1=1$	$r_1=2$	$r_1=3$	$r_1=4$	$r_1=5$	$r_1=10$	τ_Y
	0.967	0.967	0.967	0.967	0.967	0.967	0.500	0.333	0.250	0.200	0.100	0.0
	0.967	0.965	0.962	0.956	0.942	0.898	0.495	0.332	0.249	0.200	0.100	0.1
	0.967	0.959	0.947	0.927	0.890	0.827	0.482	0.327	0.247	0.198	0.100	0.2
	0.967	0.950	0.925	0.887	0.833	0.765	0.464	0.320	0.243	0.196	0.099	0.3
	0.967	0.937	0.897	0.844	0.780	0.711	0.443	0.311	0.238	0.193	0.098	0.4
0.8	0.967	0.922	0.866	0.801	0.732	0.664	0.423	0.301	0.232	0.189	0.097	0.5
	0.967	0.904	0.833	0.759	0.688	0.623	0.402	0.291	0.226	0.185	0.096	0.6
	0.967	0.885	0.801	0.721	0.649	0.587	0.383	0.280	0.220	0.181	0.095	0.7
	0.967	0.865	0.769	0.685	0.614	0.554	0.365	0.270	0.213	0.176	0.094	0.8
	0.967	0.844	0.739	0.653	0.583	0.525	0.349	0.260	0.207	0.171	0.092	0.9
	0.967	0.822	0.710	0.623	0.554	0.499	0.333	0.250	0.200	0.167	0.091	1.0
	0.909	0.909	0.909	0.909	0.909	0.909	0.500	0.333	0.250	0.200	0.100	0.0
	0.909	0.908	0.907	0.904	0.898	0.876	0.495	0.332	0.249	0.200	0.100	0.1
	0.909	0.905	0.899	0.889	0.866	0.817	0.482	0.327	0.247	0.198	0.100	0.2
	0.909	0.901	0.887	0.863	0.821	0.760	0.464	0.320	0.243	0.196	0.099	0.3
	0.909	0.894	0.869	0.829	0.773	0.708	0.443	0.311	0.238	0.193	0.098	0.4
1	0.909	0.885	0.846	0.791	0.727	0.662	0.422	0.301	0.232	0.189	0.097	0.5
	0.909	0.873	0.819	0.753	0.686	0.622	0.402	0.291	0.226	0.185	0.096	0.6
	0.909	0.860	0.791	0.717	0.647	0.586	0.383	0.280	0.220	0.181	0.095	0.7
	0.909	0.844	0.762	0.683	0.613	0.554	0.365	0.270	0.213	0.176	0.094	0.8
	0.909	0.827	0.733	0.651	0.582	0.525	0.349	0.260	0.207	0.171	0.092	0.9
	0.909	0.809	0.706	0.621	0.553	0.499	0.333	0.250	0.200	0.167	0.091	1.0
	0.654	0.654	0.654	0.654	0.654	0.654	0.500	0.333	0.250	0.200	0.100	0.0
	0.654	0.654	0.654	0.654	0.654	0.654	0.495	0.332	0.249	0.200	0.100	0.1
	0.654	0.654	0.654	0.654	0.653	0.652	0.482	0.327	0.247	0.198	0.100	0.2
	0.654	0.654	0.654	0.653	0.652	0.649	0.463	0.320	0.243	0.196	0.099	0.3
	0.654	0.654	0.653	0.652	0.649	0.643	0.443	0.311	0.238	0.193	0.098	0.4
1.5	0.654	0.653	0.652	0.650	0.644	0.629	0.422	0.301	0.232	0.189	0.097	0.5
	0.654	0.653	0.651	0.646	0.635	0.606	0.402	0.290	0.226	0.185	0.096	0.6
	0.654	0.653	0.649	0.641	0.620	0.577	0.383	0.280	0.220	0.181	0.095	0.7
	0.654	0.652	0.647	0.633	0.598	0.548	0.365	0.270	0.213	0.176	0.094	0.8
	0.654	0.651	0.643	0.620	0.573	0.521	0.348	0.260	0.206	0.171	0.092	0.9
	0.654	0.650	0.638	0.602	0.548	0.496	0.333	0.250	0.200	0.167	0.091	1.0
	0.495	0.495	0.495	0.495	0.495	0.495	0.495	0.333	0.250	0.200	0.100	0.0
	0.495	0.495	0.495	0.495	0.495	0.495	0.491	0.332	0.249	0.200	0.100	0.1
	0.495	0.495	0.495	0.495	0.495	0.495	0.478	0.327	0.247	0.198	0.100	0.2
	0.495	0.495	0.495	0.495	0.495	0.495	0.461	0.320	0.243	0.196	0.099	0.3
	0.495	0.495	0.495	0.495	0.495	0.494	0.441	0.311	0.238	0.193	0.098	0.4
2	0.495	0.495	0.495	0.495	0.494	0.494	0.421	0.301	0.232	0.189	0.097	0.5
	0.495	0.495	0.495	0.494	0.494	0.493	0.401	0.290	0.226	0.185	0.096	0.6
	0.495	0.495	0.495	0.494	0.493	0.491	0.382	0.280	0.220	0.181	0.095	0.7
	0.495	0.495	0.494	0.494	0.492	0.489	0.364	0.270	0.213	0.176	0.094	0.8
	0.495	0.495	0.494	0.493	0.491	0.484	0.348	0.259	0.206	0.171	0.092	0.9
	0.495	0.495	0.493	0.492	0.488	0.476	0.333	0.250	0.200	0.167	0.091	1.0

Table C2 Continued. Mode coupling parameter κ for $\tau_X = 0.1$

r_2	$r_1=0$	$r_1=0.2$	$r_1=0.4$	$r_1=0.6$	$r_1=0.8$	$r_1=1$	$r_1=2$	$r_1=3$	$r_1=4$	$r_1=5$	$r_1=10$	τ_Y
3	0.332	0.332	0.332	0.332	0.332	0.332	0.332	0.332	0.250	0.200	0.100	0.0
	0.332	0.332	0.332	0.332	0.332	0.332	0.332	0.330	0.249	0.200	0.100	0.1
	0.332	0.332	0.332	0.332	0.332	0.332	0.332	0.326	0.247	0.198	0.100	0.2
	0.332	0.332	0.332	0.332	0.332	0.332	0.332	0.319	0.243	0.196	0.099	0.3
	0.332	0.332	0.332	0.332	0.332	0.332	0.331	0.310	0.238	0.193	0.098	0.4
	0.332	0.332	0.332	0.332	0.332	0.332	0.331	0.300	0.232	0.189	0.097	0.5
	0.332	0.332	0.332	0.332	0.332	0.332	0.331	0.290	0.226	0.185	0.096	0.6
	0.332	0.332	0.332	0.332	0.332	0.332	0.330	0.279	0.220	0.181	0.095	0.7
	0.332	0.332	0.332	0.332	0.332	0.331	0.329	0.269	0.213	0.176	0.094	0.8
	0.332	0.332	0.332	0.332	0.331	0.331	0.327	0.259	0.206	0.171	0.092	0.9
	0.332	0.332	0.332	0.332	0.331	0.331	0.322	0.250	0.200	0.167	0.091	1.0
4	0.249	0.249	0.249	0.249	0.249	0.249	0.249	0.249	0.249	0.200	0.100	0.0
	0.249	0.249	0.249	0.249	0.249	0.249	0.249	0.249	0.248	0.200	0.100	0.1
	0.249	0.249	0.249	0.249	0.249	0.249	0.249	0.249	0.246	0.198	0.100	0.2
	0.249	0.249	0.249	0.249	0.249	0.249	0.249	0.249	0.242	0.196	0.099	0.3
	0.249	0.249	0.249	0.249	0.249	0.249	0.249	0.249	0.237	0.193	0.098	0.4
	0.249	0.249	0.249	0.249	0.249	0.249	0.249	0.249	0.232	0.189	0.097	0.5
	0.249	0.249	0.249	0.249	0.249	0.249	0.249	0.249	0.226	0.185	0.096	0.6
	0.249	0.249	0.249	0.249	0.249	0.249	0.249	0.248	0.219	0.180	0.095	0.7
	0.249	0.249	0.249	0.249	0.249	0.249	0.249	0.248	0.213	0.176	0.094	0.8
	0.249	0.249	0.249	0.249	0.249	0.249	0.249	0.247	0.206	0.171	0.092	0.9
	0.249	0.249	0.249	0.249	0.249	0.249	0.249	0.244	0.200	0.167	0.091	1.0
5	0.200	0.200	0.200	0.200	0.200	0.200	0.200	0.200	0.200	0.200	0.100	0.0
	0.200	0.200	0.200	0.200	0.200	0.200	0.200	0.200	0.200	0.199	0.100	0.1
	0.200	0.200	0.200	0.200	0.200	0.200	0.200	0.200	0.200	0.198	0.100	0.2
	0.200	0.200	0.200	0.200	0.200	0.200	0.200	0.200	0.200	0.195	0.099	0.3
	0.200	0.200	0.200	0.200	0.200	0.200	0.200	0.200	0.199	0.192	0.098	0.4
	0.200	0.200	0.200	0.200	0.200	0.200	0.200	0.200	0.199	0.189	0.097	0.5
	0.200	0.200	0.200	0.200	0.200	0.200	0.200	0.199	0.199	0.185	0.096	0.6
	0.200	0.200	0.200	0.200	0.200	0.200	0.200	0.199	0.199	0.180	0.095	0.7
	0.200	0.200	0.200	0.200	0.200	0.200	0.199	0.199	0.199	0.176	0.094	0.8
	0.200	0.200	0.200	0.200	0.200	0.200	0.199	0.199	0.198	0.171	0.092	0.9
	0.200	0.200	0.200	0.200	0.200	0.200	0.199	0.199	0.196	0.166	0.091	1.0
10	0.100	0.100	0.100	0.100	0.100	0.100	0.100	0.100	0.100	0.100	0.100	0.0
	0.100	0.100	0.100	0.100	0.100	0.100	0.100	0.100	0.100	0.100	0.100	0.1
	0.100	0.100	0.100	0.100	0.100	0.100	0.100	0.100	0.100	0.100	0.100	0.2
	0.100	0.100	0.100	0.100	0.100	0.100	0.100	0.100	0.100	0.100	0.099	0.3
	0.100	0.100	0.100	0.100	0.100	0.100	0.100	0.100	0.100	0.100	0.098	0.4
	0.100	0.100	0.100	0.100	0.100	0.100	0.100	0.100	0.100	0.100	0.097	0.5
	0.100	0.100	0.100	0.100	0.100	0.100	0.100	0.100	0.100	0.100	0.096	0.6
	0.100	0.100	0.100	0.100	0.100	0.100	0.100	0.100	0.100	0.100	0.095	0.7
	0.100	0.100	0.100	0.100	0.100	0.100	0.100	0.100	0.100	0.100	0.094	0.8
	0.100	0.100	0.100	0.100	0.100	0.100	0.100	0.100	0.100	0.100	0.092	0.9
	0.100	0.100	0.100	0.100	0.100	0.100	0.100	0.100	0.100	0.100	0.091	1.0

288 Appendix C

Table C3 Mode coupling parameter κ for $\tau_X = 0.2$

r_2	$r_1=0$	$r_1=0.2$	$r_1=0.4$	$r_1=0.6$	$r_1=0.8$	$r_1=1$	$r_1=2$	$r_1=3$	$r_1=4$	$r_1=5$	$r_1=10$	τ_Y
0	1.000	1.000	1.000	1.000	1.000	1.000	0.500	0.333	0.250	0.200	0.100	0.0
	1.000	0.998	0.993	0.986	0.967	0.909	0.495	0.332	0.249	0.200	0.100	0.1
	1.000	0.990	0.975	0.950	0.905	0.833	0.482	0.327	0.247	0.198	0.100	0.2
	1.000	0.979	0.948	0.904	0.843	0.769	0.464	0.320	0.243	0.196	0.099	0.3
	1.000	0.963	0.915	0.856	0.787	0.714	0.444	0.311	0.238	0.193	0.098	0.4
	1.000	0.945	0.880	0.809	0.736	0.667	0.423	0.301	0.232	0.189	0.097	0.5
	1.000	0.925	0.845	0.766	0.692	0.625	0.402	0.291	0.226	0.185	0.096	0.6
	1.000	0.903	0.810	0.726	0.652	0.588	0.383	0.280	0.220	0.181	0.095	0.7
	1.000	0.880	0.776	0.689	0.616	0.556	0.365	0.270	0.213	0.176	0.094	0.8
	1.000	0.857	0.744	0.656	0.584	0.526	0.349	0.260	0.207	0.171	0.092	0.9
	1.000	0.833	0.714	0.625	0.556	0.500	0.333	0.250	0.200	0.167	0.091	1.0
0.2	0.990	0.990	0.990	0.990	0.990	0.990	0.500	0.333	0.250	0.200	0.100	0.0
	0.990	0.988	0.984	0.977	0.959	0.905	0.495	0.332	0.249	0.200	0.100	0.1
	0.990	0.981	0.966	0.942	0.900	0.831	0.482	0.327	0.247	0.198	0.100	0.2
	0.990	0.970	0.940	0.898	0.839	0.767	0.464	0.320	0.243	0.196	0.099	0.3
	0.990	0.955	0.909	0.851	0.784	0.713	0.444	0.311	0.238	0.193	0.098	0.4
	0.990	0.937	0.875	0.806	0.734	0.665	0.423	0.301	0.232	0.189	0.097	0.5
	0.990	0.918	0.840	0.763	0.690	0.624	0.402	0.291	0.226	0.185	0.096	0.6
	0.990	0.896	0.806	0.723	0.650	0.587	0.383	0.280	0.220	0.181	0.095	0.7
	0.990	0.874	0.773	0.687	0.615	0.555	0.365	0.270	0.213	0.176	0.094	0.8
	0.990	0.852	0.742	0.654	0.583	0.526	0.349	0.260	0.207	0.171	0.092	0.9
	0.990	0.829	0.712	0.624	0.555	0.500	0.333	0.250	0.200	0.167	0.091	1.0
0.4	0.975	0.975	0.975	0.975	0.975	0.975	0.500	0.333	0.250	0.200	0.100	0.0
	0.975	0.973	0.969	0.963	0.947	0.899	0.495	0.332	0.249	0.200	0.100	0.1
	0.975	0.966	0.953	0.931	0.892	0.826	0.482	0.327	0.247	0.198	0.100	0.2
	0.975	0.956	0.929	0.889	0.833	0.764	0.464	0.320	0.243	0.196	0.099	0.3
	0.975	0.942	0.899	0.844	0.779	0.710	0.443	0.311	0.238	0.193	0.098	0.4
	0.975	0.926	0.867	0.800	0.731	0.663	0.422	0.301	0.232	0.189	0.097	0.5
	0.975	0.907	0.833	0.759	0.687	0.622	0.402	0.290	0.226	0.185	0.096	0.6
	0.975	0.887	0.800	0.720	0.648	0.586	0.383	0.280	0.220	0.181	0.095	0.7
	0.975	0.866	0.768	0.684	0.614	0.554	0.365	0.270	0.213	0.176	0.094	0.8
	0.975	0.844	0.738	0.652	0.582	0.525	0.349	0.260	0.207	0.171	0.092	0.9
	0.975	0.822	0.709	0.622	0.554	0.499	0.333	0.250	0.200	0.167	0.091	1.0
0.6	0.950	0.950	0.950	0.950	0.950	0.950	0.500	0.333	0.250	0.200	0.100	0.0
	0.950	0.950	0.945	0.939	0.927	0.889	0.495	0.332	0.249	0.200	0.100	0.1
	0.950	0.942	0.931	0.912	0.878	0.820	0.482	0.327	0.247	0.198	0.100	0.2
	0.950	0.933	0.910	0.875	0.824	0.759	0.464	0.320	0.243	0.196	0.099	0.3
	0.950	0.921	0.883	0.833	0.773	0.707	0.443	0.311	0.238	0.193	0.098	0.4
	0.950	0.907	0.854	0.792	0.726	0.661	0.422	0.301	0.232	0.189	0.097	0.5
	0.950	0.890	0.823	0.752	0.684	0.620	0.402	0.290	0.226	0.185	0.096	0.6
	0.950	0.872	0.792	0.715	0.646	0.585	0.383	0.280	0.220	0.181	0.095	0.7
	0.950	0.852	0.761	0.681	0.611	0.553	0.365	0.270	0.213	0.176	0.094	0.8
	0.950	0.832	0.732	0.649	0.580	0.524	0.348	0.260	0.206	0.171	0.092	0.9
	0.950	0.812	0.704	0.619	0.552	0.498	0.333	0.250	0.200	0.167	0.091	1.0

Mode coupling parameter κ 289

Table C3 Continued. Mode coupling parameter κ for $\tau_X = 0.2$

r_2	$r_1=0$	$r_1=0.2$	$r_1=0.4$	$r_1=0.6$	$r_1=0.8$	$r_1=1$	$r_1=2$	$r_1=3$	$r_1=4$	$r_1=5$	$r_1=10$	τ_Y
	0.905	0.905	0.905	0.905	0.905	0.905	0.500	0.333	0.250	0.200	0.100	0.0
	0.905	0.904	0.902	0.898	0.890	0.866	0.495	0.332	0.249	0.200	0.100	0.1
	0.905	0.900	0.892	0.878	0.853	0.807	0.482	0.327	0.247	0.198	0.100	0.2
	0.905	0.893	0.876	0.849	0.808	0.751	0.463	0.320	0.243	0.196	0.099	0.3
	0.905	0.884	0.855	0.815	0.762	0.701	0.443	0.311	0.238	0.193	0.098	0.4
0.8	0.905	0.873	0.831	0.778	0.718	0.657	0.422	0.301	0.232	0.189	0.097	0.5
	0.905	0.860	0.805	0.742	0.678	0.617	0.402	0.290	0.226	0.185	0.096	0.6
	0.905	0.845	0.777	0.707	0.642	0.582	0.383	0.280	0.220	0.181	0.095	0.7
	0.905	0.829	0.750	0.675	0.608	0.551	0.365	0.270	0.213	0.176	0.094	0.8
	0.905	0.812	0.723	0.644	0.578	0.522	0.348	0.259	0.206	0.171	0.092	0.9
	0.905	0.795	0.696	0.616	0.550	0.497	0.333	0.250	0.200	0.167	0.091	1.0
	0.833	0.833	0.833	0.833	0.833	0.833	0.500	0.333	0.250	0.200	0.100	0.0
	0.833	0.833	0.832	0.830	0.827	0.817	0.495	0.332	0.249	0.200	0.100	0.1
	0.833	0.831	0.826	0.820	0.807	0.780	0.482	0.327	0.247	0.198	0.100	0.2
	0.833	0.827	0.818	0.803	0.777	0.735	0.463	0.320	0.243	0.196	0.099	0.3
	0.833	0.822	0.805	0.780	0.742	0.691	0.442	0.311	0.238	0.193	0.098	0.4
1	0.833	0.816	0.790	0.753	0.705	0.650	0.422	0.301	0.232	0.189	0.097	0.5
	0.833	0.808	0.772	0.724	0.669	0.613	0.401	0.290	0.226	0.185	0.096	0.6
	0.833	0.799	0.751	0.694	0.635	0.579	0.382	0.280	0.220	0.181	0.095	0.7
	0.833	0.788	0.729	0.665	0.603	0.548	0.365	0.269	0.213	0.176	0.094	0.8
	0.833	0.777	0.707	0.637	0.574	0.520	0.348	0.259	0.206	0.171	0.092	0.9
	0.833	0.764	0.684	0.610	0.547	0.495	0.333	0.250	0.200	0.167	0.091	1.0
	0.625	0.625	0.625	0.625	0.625	0.625	0.500	0.333	0.250	0.200	0.100	0.0
	0.625	0.625	0.625	0.625	0.624	0.624	0.495	0.332	0.249	0.200	0.100	0.1
	0.625	0.625	0.624	0.624	0.623	0.621	0.480	0.327	0.247	0.198	0.100	0.2
	0.625	0.624	0.623	0.622	0.620	0.616	0.461	0.320	0.243	0.196	0.099	0.3
	0.625	0.624	0.622	0.619	0.615	0.606	0.441	0.311	0.238	0.193	0.098	0.4
1.5	0.625	0.623	0.620	0.615	0.607	0.593	0.420	0.301	0.232	0.189	0.097	0.5
	0.625	0.622	0.617	0.610	0.597	0.575	0.400	0.290	0.226	0.185	0.096	0.6
	0.625	0.621	0.614	0.603	0.584	0.554	0.381	0.280	0.220	0.181	0.095	0.7
	0.625	0.619	0.610	0.594	0.568	0.531	0.364	0.269	0.213	0.176	0.094	0.8
	0.625	0.618	0.605	0.583	0.549	0.509	0.347	0.259	0.206	0.171	0.092	0.9
	0.625	0.616	0.599	0.570	0.529	0.487	0.332	0.250	0.200	0.167	0.091	1.0
	0.482	0.482	0.482	0.482	0.482	0.482	0.482	0.333	0.250	0.200	0.100	0.0
	0.482	0.482	0.482	0.482	0.482	0.482	0.478	0.332	0.249	0.200	0.100	0.1
	0.482	0.482	0.482	0.482	0.482	0.482	0.467	0.327	0.247	0.198	0.100	0.2
	0.482	0.482	0.482	0.482	0.481	0.481	0.452	0.319	0.243	0.196	0.099	0.3
	0.482	0.482	0.482	0.481	0.480	0.480	0.434	0.310	0.238	0.193	0.098	0.4
2	0.482	0.482	0.481	0.480	0.479	0.478	0.415	0.300	0.232	0.190	0.097	0.5
	0.482	0.482	0.481	0.480	0.478	0.476	0.396	0.290	0.226	0.185	0.096	0.6
	0.482	0.481	0.480	0.479	0.476	0.472	0.378	0.279	0.220	0.180	0.095	0.7
	0.482	0.481	0.480	0.477	0.474	0.468	0.361	0.269	0.213	0.176	0.094	0.8
	0.482	0.481	0.479	0.476	0.470	0.461	0.345	0.259	0.206	0.171	0.092	0.9
	0.482	0.480	0.478	0.474	0.466	0.453	0.330	0.249	0.200	0.167	0.091	1.0

290 Appendix C

Table C3 Continued. Mode coupling parameter κ for $\tau_X = 0.2$

r_2	$r_1=0$	$r_1=0.2$	$r_1=0.4$	$r_1=0.6$	$r_1=0.8$	$r_1=1$	$r_1=2$	$r_1=3$	$r_1=4$	$r_1=5$	$r_1=10$	τ_Y
3	0.327	0.327	0.327	0.327	0.327	0.327	0.327	0.327	0.250	0.200	0.100	0.0
	0.327	0.327	0.327	0.327	0.327	0.327	0.327	0.326	0.249	0.200	0.100	0.1
	0.327	0.327	0.327	0.327	0.327	0.327	0.327	0.321	0.247	0.198	0.100	0.2
	0.327	0.327	0.327	0.327	0.327	0.327	0.326	0.315	0.243	0.196	0.099	0.3
	0.327	0.327	0.327	0.327	0.327	0.327	0.326	0.306	0.238	0.193	0.098	0.4
	0.327	0.327	0.327	0.327	0.327	0.327	0.325	0.297	0.232	0.189	0.097	0.5
	0.327	0.327	0.327	0.327	0.327	0.326	0.324	0.287	0.226	0.185	0.096	0.6
	0.327	0.327	0.327	0.327	0.326	0.326	0.323	0.277	0.219	0.180	0.095	0.7
	0.327	0.327	0.327	0.327	0.326	0.326	0.320	0.267	0.213	0.176	0.094	0.8
	0.327	0.327	0.327	0.326	0.326	0.326	0.317	0.258	0.206	0.171	0.092	0.9
	0.327	0.327	0.327	0.326	0.326	0.325	0.311	0.248	0.200	0.167	0.091	1.0
4	0.247	0.247	0.247	0.247	0.247	0.247	0.247	0.247	0.247	0.200	0.100	0.0
	0.247	0.247	0.247	0.247	0.247	0.247	0.247	0.247	0.246	0.200	0.100	0.1
	0.247	0.247	0.247	0.247	0.247	0.247	0.247	0.247	0.244	0.198	0.100	0.2
	0.247	0.247	0.247	0.247	0.247	0.247	0.247	0.247	0.240	0.196	0.099	0.3
	0.247	0.247	0.247	0.247	0.247	0.247	0.247	0.246	0.236	0.192	0.098	0.4
	0.247	0.247	0.247	0.247	0.247	0.247	0.247	0.246	0.230	0.189	0.097	0.5
	0.247	0.247	0.247	0.247	0.247	0.247	0.246	0.245	0.224	0.185	0.096	0.6
	0.247	0.247	0.247	0.247	0.247	0.247	0.246	0.244	0.218	0.180	0.095	0.7
	0.247	0.247	0.247	0.247	0.247	0.247	0.246	0.243	0.211	0.176	0.094	0.8
	0.247	0.247	0.247	0.247	0.247	0.247	0.246	0.241	0.205	0.171	0.092	0.9
	0.247	0.247	0.247	0.247	0.247	0.246	0.245	0.237	0.199	0.166	0.091	1.0
5	0.198	0.198	0.198	0.198	0.198	0.198	0.198	0.198	0.198	0.198	0.100	0.0
	0.198	0.198	0.198	0.198	0.198	0.198	0.198	0.198	0.198	0.198	0.100	0.1
	0.198	0.198	0.198	0.198	0.198	0.198	0.198	0.198	0.198	0.196	0.100	0.2
	0.198	0.198	0.198	0.198	0.198	0.198	0.198	0.198	0.198	0.194	0.099	0.3
	0.198	0.198	0.198	0.198	0.198	0.198	0.198	0.198	0.198	0.191	0.098	0.4
	0.198	0.198	0.198	0.198	0.198	0.198	0.198	0.198	0.198	0.188	0.097	0.5
	0.198	0.198	0.198	0.198	0.198	0.198	0.198	0.198	0.197	0.184	0.096	0.6
	0.198	0.198	0.198	0.198	0.198	0.198	0.198	0.198	0.197	0.179	0.095	0.7
	0.198	0.198	0.198	0.198	0.198	0.198	0.198	0.198	0.196	0.175	0.094	0.8
	0.198	0.198	0.198	0.198	0.198	0.198	0.198	0.197	0.194	0.170	0.092	0.9
	0.198	0.198	0.198	0.198	0.198	0.198	0.198	0.197	0.192	0.166	0.091	1.0
10	0.100	0.100	0.100	0.100	0.100	0.100	0.100	0.100	0.100	0.100	0.100	0.0
	0.100	0.100	0.100	0.100	0.100	0.100	0.100	0.100	0.100	0.100	0.100	0.1
	0.100	0.100	0.100	0.100	0.100	0.100	0.100	0.100	0.100	0.100	0.099	0.2
	0.100	0.100	0.100	0.100	0.100	0.100	0.100	0.100	0.100	0.100	0.099	0.3
	0.100	0.100	0.100	0.100	0.100	0.100	0.100	0.100	0.100	0.100	0.098	0.4
	0.100	0.100	0.100	0.100	0.100	0.100	0.100	0.100	0.100	0.100	0.097	0.5
	0.100	0.100	0.100	0.100	0.100	0.100	0.100	0.100	0.100	0.100	0.096	0.6
	0.100	0.100	0.100	0.100	0.100	0.100	0.100	0.100	0.100	0.100	0.095	0.7
	0.100	0.100	0.100	0.100	0.100	0.100	0.100	0.100	0.100	0.100	0.093	0.8
	0.100	0.100	0.100	0.100	0.100	0.100	0.100	0.100	0.100	0.100	0.092	0.9
	0.100	0.100	0.100	0.100	0.100	0.100	0.100	0.100	0.100	0.100	0.091	1.0

Table C4 Mode coupling parameter κ for $\tau_X = 0.3$

r_2	$r_1=0$	$r_1=0.2$	$r_1=0.4$	$r_1=0.6$	$r_1=0.8$	$r_1=1$	$r_1=2$	$r_1=3$	$r_1=4$	$r_1=5$	$r_1=10$	τ_Y
	1.000	1.000	1.000	1.000	1.000	1.000	0.500	0.333	0.250	0.200	0.100	0.0
	1.000	1.000	0.993	0.986	0.967	0.909	0.495	0.332	0.249	0.200	0.100	0.1
	1.000	0.990	0.975	0.950	0.905	0.833	0.482	0.327	0.247	0.198	0.100	0.2
	1.000	0.979	0.948	0.904	0.843	0.769	0.464	0.320	0.243	0.196	0.099	0.3
	1.000	0.963	0.915	0.856	0.787	0.714	0.444	0.311	0.238	0.193	0.098	0.4
0	1.000	0.945	0.880	0.809	0.736	0.667	0.423	0.301	0.232	0.189	0.097	0.5
	1.000	0.925	0.845	0.766	0.692	0.625	0.402	0.291	0.226	0.185	0.096	0.6
	1.000	0.903	0.810	0.726	0.652	0.588	0.383	0.280	0.220	0.181	0.095	0.7
	1.000	0.880	0.776	0.689	0.616	0.556	0.365	0.270	0.213	0.176	0.094	0.8
	1.000	0.857	0.744	0.656	0.584	0.526	0.349	0.260	0.207	0.171	0.092	0.9
	1.000	0.833	0.714	0.625	0.556	0.500	0.333	0.250	0.200	0.167	0.091	1.0
	0.979	0.979	0.979	0.979	0.979	0.979	0.500	0.333	0.250	0.200	0.100	0.0
	0.979	0.976	0.973	0.966	0.950	0.901	0.495	0.332	0.249	0.200	0.100	0.1
	0.979	0.970	0.956	0.933	0.893	0.827	0.482	0.327	0.247	0.198	0.100	0.2
	0.979	0.959	0.931	0.891	0.834	0.764	0.464	0.320	0.243	0.196	0.099	0.3
	0.979	0.945	0.901	0.845	0.780	0.711	0.443	0.311	0.238	0.193	0.098	0.4
0.2	0.979	0.928	0.868	0.801	0.731	0.664	0.422	0.301	0.232	0.189	0.097	0.5
	0.979	0.909	0.835	0.759	0.688	0.623	0.402	0.290	0.226	0.185	0.096	0.6
	0.979	0.889	0.801	0.720	0.649	0.586	0.383	0.280	0.220	0.181	0.095	0.7
	0.979	0.867	0.769	0.685	0.614	0.554	0.365	0.270	0.213	0.176	0.094	0.8
	0.979	0.845	0.738	0.652	0.582	0.525	0.349	0.260	0.207	0.171	0.092	0.9
	0.979	0.823	0.709	0.622	0.554	0.499	0.333	0.250	0.200	0.167	0.091	1.0
	0.948	0.948	0.948	0.948	0.948	0.948	0.500	0.333	0.250	0.200	0.100	0.0
	0.948	0.946	0.943	0.937	0.925	0.887	0.495	0.332	0.249	0.200	0.100	0.1
	0.948	0.940	0.929	0.910	0.876	0.818	0.482	0.327	0.247	0.198	0.100	0.2
	0.948	0.931	0.907	0.872	0.822	0.758	0.463	0.320	0.243	0.196	0.099	0.3
	0.948	0.919	0.880	0.831	0.771	0.705	0.443	0.311	0.238	0.193	0.098	0.4
0.4	0.948	0.904	0.851	0.790	0.724	0.660	0.422	0.301	0.232	0.189	0.097	0.5
	0.948	0.887	0.820	0.750	0.682	0.619	0.402	0.290	0.226	0.185	0.096	0.6
	0.948	0.869	0.789	0.713	0.644	0.584	0.383	0.280	0.220	0.181	0.095	0.7
	0.948	0.849	0.759	0.679	0.610	0.552	0.365	0.270	0.213	0.176	0.094	0.8
	0.948	0.829	0.730	0.647	0.579	0.523	0.348	0.259	0.206	0.171	0.092	0.9
	0.948	0.809	0.702	0.618	0.551	0.497	0.333	0.250	0.200	0.167	0.091	1.0
	0.904	0.904	0.904	0.904	0.904	0.904	0.500	0.333	0.250	0.200	0.100	0.0
	0.904	0.902	0.900	0.896	0.887	0.863	0.495	0.332	0.249	0.200	0.100	0.1
	0.904	0.898	0.889	0.875	0.849	0.803	0.482	0.327	0.247	0.198	0.100	0.2
	0.904	0.891	0.872	0.844	0.803	0.747	0.463	0.320	0.243	0.196	0.099	0.3
	0.904	0.881	0.850	0.809	0.757	0.698	0.442	0.311	0.238	0.193	0.098	0.4
0.6	0.904	0.869	0.825	0.773	0.714	0.654	0.422	0.301	0.232	0.189	0.097	0.5
	0.904	0.855	0.799	0.737	0.674	0.615	0.401	0.290	0.226	0.185	0.096	0.6
	0.904	0.840	0.772	0.703	0.638	0.580	0.382	0.280	0.220	0.181	0.095	0.7
	0.904	0.824	0.744	0.671	0.605	0.549	0.364	0.269	0.213	0.176	0.094	0.8
	0.904	0.806	0.717	0.640	0.575	0.521	0.348	0.259	0.206	0.171	0.092	0.9
	0.904	0.789	0.692	0.612	0.548	0.495	0.333	0.250	0.200	0.167	0.091	1.0

292   Appendix C

Table C4 Continued. Mode coupling parameter κ for $\tau_X = 0.3$

r_2	$r_1=0$	$r_1=0.2$	$r_1=0.4$	$r_1=0.6$	$r_1=0.8$	$r_1=1$	$r_1=2$	$r_1=3$	$r_1=4$	$r_1=5$	$r_1=10$	τ_Y
0.8	0.843	0.843	0.843	0.843	0.843	0.843	0.500	0.333	0.250	0.200	0.100	0.0
	0.843	0.842	0.841	0.838	0.833	0.821	0.495	0.332	0.249	0.200	0.100	0.1
	0.843	0.839	0.833	0.824	0.808	0.777	0.481	0.327	0.247	0.198	0.100	0.2
	0.843	0.834	0.822	0.803	0.774	0.730	0.463	0.320	0.243	0.196	0.099	0.3
	0.843	0.828	0.807	0.777	0.736	0.686	0.442	0.311	0.238	0.193	0.098	0.4
	0.843	0.820	0.788	0.748	0.699	0.645	0.421	0.301	0.232	0.189	0.097	0.5
	0.843	0.810	0.768	0.718	0.663	0.608	0.401	0.290	0.226	0.185	0.096	0.6
	0.843	0.799	0.746	0.688	0.630	0.575	0.382	0.280	0.220	0.181	0.095	0.7
	0.843	0.786	0.723	0.659	0.599	0.545	0.364	0.269	0.213	0.176	0.094	0.8
	0.843	0.773	0.700	0.631	0.570	0.518	0.347	0.259	0.206	0.171	0.092	0.9
	0.843	0.759	0.677	0.605	0.544	0.493	0.332	0.250	0.200	0.167	0.091	1.0
1	0.769	0.769	0.769	0.769	0.769	0.769	0.500	0.333	0.250	0.200	0.100	0.0
	0.769	0.769	0.767	0.767	0.765	0.760	0.495	0.332	0.249	0.200	0.100	0.1
	0.769	0.767	0.764	0.759	0.751	0.735	0.481	0.327	0.247	0.198	0.100	0.2
	0.769	0.764	0.758	0.747	0.730	0.702	0.462	0.320	0.243	0.196	0.099	0.3
	0.769	0.761	0.748	0.731	0.704	0.667	0.441	0.311	0.238	0.193	0.098	0.4
	0.769	0.756	0.737	0.711	0.675	0.632	0.420	0.301	0.232	0.189	0.097	0.5
	0.769	0.750	0.724	0.689	0.646	0.599	0.400	0.290	0.226	0.185	0.096	0.6
	0.769	0.743	0.708	0.665	0.617	0.568	0.381	0.280	0.220	0.180	0.095	0.7
	0.769	0.735	0.691	0.641	0.589	0.539	0.363	0.269	0.213	0.176	0.094	0.8
	0.769	0.726	0.674	0.617	0.562	0.513	0.347	0.259	0.206	0.171	0.092	0.9
	0.769	0.717	0.655	0.594	0.538	0.489	0.332	0.250	0.200	0.167	0.091	1.0
1.5	0.590	0.590	0.590	0.590	0.590	0.590	0.500	0.333	0.250	0.200	0.100	0.0
	0.590	0.590	0.590	0.590	0.590	0.589	0.494	0.332	0.249	0.200	0.100	0.1
	0.590	0.590	0.589	0.589	0.587	0.586	0.478	0.327	0.247	0.198	0.100	0.2
	0.590	0.589	0.588	0.586	0.584	0.580	0.458	0.319	0.243	0.196	0.099	0.3
	0.590	0.589	0.586	0.583	0.579	0.571	0.437	0.310	0.238	0.193	0.098	0.4
	0.590	0.588	0.584	0.579	0.571	0.560	0.417	0.300	0.232	0.189	0.097	0.5
	0.590	0.586	0.581	0.574	0.562	0.545	0.397	0.290	0.226	0.185	0.096	0.6
	0.590	0.585	0.578	0.567	0.551	0.528	0.378	0.279	0.219	0.180	0.095	0.7
	0.590	0.583	0.573	0.559	0.538	0.510	0.361	0.269	0.213	0.176	0.094	0.8
	0.590	0.581	0.568	0.549	0.523	0.491	0.345	0.259	0.206	0.171	0.092	0.9
	0.590	0.579	0.563	0.539	0.508	0.473	0.330	0.249	0.200	0.167	0.091	1.0
2	0.464	0.464	0.464	0.464	0.464	0.464	0.464	0.333	0.250	0.200	0.100	0.0
	0.464	0.464	0.464	0.464	0.464	0.464	0.461	0.332	0.249	0.200	0.100	0.1
	0.464	0.464	0.464	0.464	0.463	0.463	0.452	0.326	0.247	0.198	0.100	0.2
	0.464	0.464	0.464	0.463	0.463	0.462	0.438	0.319	0.243	0.196	0.099	0.3
	0.464	0.464	0.463	0.463	0.462	0.461	0.460	0.423	0.310	0.238	0.193	0.098	0.4
	0.464	0.463	0.462	0.461	0.460	0.458	0.406	0.299	0.232	0.189	0.097	0.5
	0.464	0.463	0.462	0.460	0.458	0.455	0.389	0.289	0.226	0.185	0.096	0.6
	0.464	0.463	0.461	0.459	0.455	0.451	0.372	0.278	0.219	0.180	0.095	0.7
	0.464	0.462	0.460	0.457	0.453	0.445	0.356	0.268	0.213	0.176	0.094	0.8
	0.464	0.462	0.459	0.454	0.448	0.439	0.341	0.258	0.206	0.171	0.092	0.9
	0.464	0.461	0.457	0.452	0.443	0.431	0.327	0.249	0.200	0.167	0.091	1.0

Table C4 Continued. Mode coupling parameter κ for $\tau_X = 0.3$

r_2	$r_1=0$	$r_1=0.2$	$r_1=0.4$	$r_1=0.6$	$r_1=0.8$	$r_1=1$	$r_1=2$	$r_1=3$	$r_1=4$	$r_1=5$	$r_1=10$	τ_Y
3	0.320	0.320	0.320	0.320	0.320	0.320	0.320	0.320	0.250	0.200	0.100	0.0
	0.320	0.320	0.320	0.320	0.320	0.320	0.320	0.319	0.249	0.200	0.100	0.1
	0.320	0.320	0.320	0.320	0.320	0.320	0.319	0.315	0.247	0.198	0.100	0.2
	0.320	0.320	0.320	0.320	0.320	0.320	0.319	0.309	0.242	0.196	0.099	0.3
	0.320	0.320	0.320	0.320	0.320	0.319	0.318	0.301	0.237	0.192	0.098	0.4
	0.320	0.320	0.320	0.320	0.319	0.319	0.317	0.292	0.231	0.189	0.097	0.5
	0.320	0.320	0.320	0.319	0.319	0.319	0.315	0.283	0.225	0.185	0.096	0.6
	0.320	0.320	0.319	0.319	0.319	0.318	0.313	0.274	0.219	0.180	0.095	0.7
	0.320	0.320	0.319	0.319	0.318	0.318	0.310	0.264	0.212	0.176	0.094	0.8
	0.320	0.320	0.319	0.319	0.318	0.317	0.306	0.255	0.205	0.171	0.092	0.9
	0.320	0.319	0.319	0.318	0.318	0.317	0.301	0.246	0.199	0.166	0.091	1.0
4	0.243	0.243	0.243	0.243	0.243	0.243	0.243	0.243	0.243	0.200	0.100	0.0
	0.243	0.243	0.243	0.243	0.243	0.243	0.243	0.243	0.242	0.200	0.100	0.1
	0.243	0.243	0.243	0.243	0.243	0.243	0.243	0.243	0.240	0.198	0.100	0.2
	0.243	0.243	0.243	0.243	0.243	0.243	0.243	0.242	0.237	0.195	0.099	0.3
	0.243	0.243	0.243	0.243	0.243	0.243	0.243	0.242	0.232	0.192	0.098	0.4
	0.243	0.243	0.243	0.243	0.243	0.243	0.243	0.241	0.227	0.188	0.097	0.5
	0.243	0.243	0.243	0.243	0.243	0.243	0.242	0.240	0.222	0.184	0.096	0.6
	0.243	0.243	0.243	0.243	0.243	0.243	0.242	0.239	0.216	0.180	0.095	0.7
	0.243	0.243	0.243	0.243	0.243	0.243	0.242	0.237	0.210	0.175	0.094	0.8
	0.243	0.243	0.243	0.243	0.243	0.242	0.241	0.234	0.203	0.171	0.092	0.9
	0.243	0.243	0.243	0.243	0.242	0.242	0.240	0.231	0.197	0.166	0.091	1.0
5	0.196	0.196	0.196	0.196	0.196	0.196	0.196	0.196	0.196	0.196	0.100	0.0
	0.196	0.196	0.196	0.196	0.196	0.196	0.196	0.196	0.196	0.195	0.100	0.1
	0.196	0.196	0.196	0.196	0.196	0.196	0.196	0.196	0.196	0.194	0.100	0.2
	0.196	0.196	0.196	0.196	0.196	0.196	0.196	0.196	0.195	0.192	0.099	0.3
	0.196	0.196	0.196	0.196	0.196	0.196	0.196	0.196	0.195	0.189	0.098	0.4
	0.196	0.196	0.196	0.196	0.196	0.196	0.196	0.195	0.195	0.186	0.097	0.5
	0.196	0.196	0.196	0.196	0.196	0.196	0.196	0.195	0.194	0.182	0.096	0.6
	0.196	0.196	0.196	0.196	0.196	0.196	0.195	0.195	0.193	0.178	0.095	0.7
	0.196	0.196	0.196	0.196	0.196	0.196	0.195	0.195	0.192	0.173	0.094	0.8
	0.196	0.196	0.196	0.196	0.196	0.196	0.195	0.194	0.190	0.169	0.092	0.9
	0.196	0.196	0.196	0.196	0.196	0.196	0.195	0.194	0.187	0.165	0.091	1.0
10	0.099	0.099	0.099	0.099	0.099	0.099	0.099	0.099	0.099	0.099	0.099	0.0
	0.099	0.099	0.099	0.099	0.099	0.099	0.099	0.099	0.099	0.099	0.099	0.1
	0.099	0.099	0.099	0.099	0.099	0.099	0.099	0.099	0.099	0.099	0.099	0.2
	0.099	0.099	0.099	0.099	0.099	0.099	0.099	0.099	0.099	0.099	0.098	0.3
	0.099	0.099	0.099	0.099	0.099	0.099	0.099	0.099	0.099	0.099	0.097	0.4
	0.099	0.099	0.099	0.099	0.099	0.099	0.099	0.099	0.099	0.099	0.097	0.5
	0.099	0.099	0.099	0.099	0.099	0.099	0.099	0.099	0.099	0.099	0.096	0.6
	0.099	0.099	0.099	0.099	0.099	0.099	0.099	0.099	0.099	0.099	0.094	0.7
	0.099	0.099	0.099	0.099	0.099	0.099	0.099	0.099	0.099	0.099	0.093	0.8
	0.099	0.099	0.099	0.099	0.099	0.099	0.099	0.099	0.099	0.099	0.092	0.9
	0.099	0.099	0.099	0.099	0.099	0.099	0.099	0.099	0.099	0.099	0.090	1.0

294 Appendix C

Table C5 Mode coupling parameter κ for $\tau_X = 0.4$

r_2	$r_1=0$	$r_1=0.2$	$r_1=0.4$	$r_1=0.6$	$r_1=0.8$	$r_1=1$	$r_1=2$	$r_1=3$	$r_1=4$	$r_1=5$	$r_1=10$	τ_Y
0	1.000	1.000	1.000	1.000	1.000	1.000	0.500	0.333	0.250	0.200	0.100	0.0
	1.000	0.998	0.993	0.986	0.967	0.909	0.495	0.332	0.249	0.200	0.100	0.1
	1.000	0.990	0.975	0.950	0.905	0.833	0.482	0.327	0.247	0.198	0.100	0.2
	1.000	0.979	0.948	0.904	0.843	0.769	0.464	0.320	0.243	0.196	0.099	0.3
	1.000	0.963	0.915	0.856	0.787	0.714	0.444	0.311	0.238	0.193	0.098	0.4
	1.000	0.945	0.880	0.809	0.736	0.667	0.423	0.301	0.232	0.189	0.097	0.5
	1.000	0.925	0.845	0.766	0.692	0.625	0.402	0.291	0.226	0.185	0.096	0.6
	1.000	0.903	0.810	0.726	0.652	0.588	0.383	0.280	0.220	0.181	0.095	0.7
	1.000	0.880	0.776	0.689	0.616	0.556	0.365	0.270	0.213	0.176	0.094	0.8
	1.000	0.857	0.744	0.656	0.584	0.526	0.349	0.260	0.207	0.171	0.092	0.9
	1.000	0.833	0.714	0.625	0.556	0.500	0.333	0.250	0.200	0.167	0.091	1.0
0.2	0.963	0.963	0.963	0.963	0.963	0.963	0.500	0.333	0.250	0.200	0.100	0.0
	0.963	0.961	0.958	0.952	0.937	0.894	0.495	0.332	0.249	0.200	0.100	0.1
	0.963	0.955	0.942	0.921	0.884	0.822	0.482	0.327	0.247	0.198	0.100	0.2
	0.963	0.945	0.919	0.881	0.828	0.761	0.464	0.320	0.243	0.196	0.099	0.3
	0.963	0.932	0.890	0.838	0.775	0.708	0.443	0.311	0.238	0.193	0.098	0.4
	0.963	0.916	0.859	0.795	0.727	0.661	0.422	0.301	0.232	0.189	0.097	0.5
	0.963	0.898	0.827	0.754	0.684	0.621	0.402	0.290	0.226	0.185	0.096	0.6
	0.963	0.878	0.795	0.716	0.646	0.585	0.383	0.280	0.220	0.181	0.095	0.7
	0.963	0.858	0.763	0.681	0.612	0.553	0.365	0.270	0.213	0.176	0.094	0.8
	0.963	0.837	0.733	0.649	0.580	0.524	0.348	0.260	0.206	0.171	0.092	0.9
	0.963	0.816	0.705	0.620	0.552	0.498	0.333	0.250	0.200	0.167	0.091	1.0
0.4	0.915	0.915	0.915	0.915	0.915	0.915	0.500	0.333	0.250	0.200	0.100	0.0
	0.915	0.914	0.911	0.907	0.897	0.869	0.495	0.331	0.249	0.200	0.100	0.1
	0.915	0.909	0.899	0.883	0.855	0.805	0.482	0.327	0.247	0.198	0.100	0.2
	0.915	0.901	0.880	0.850	0.807	0.748	0.463	0.320	0.243	0.196	0.099	0.3
	0.915	0.890	0.857	0.813	0.759	0.698	0.442	0.311	0.238	0.193	0.098	0.4
	0.915	0.877	0.831	0.776	0.715	0.654	0.421	0.301	0.232	9.189	0.097	0.5
	0.915	0.863	0.803	0.739	0.675	0.615	0.401	0.290	0.226	9.185	0.096	0.6
	0.915	0.846	0.774	0.704	0.638	0.580	0.382	0.280	0.220	0.181	0.095	0.7
	0.915	0.829	0.746	0.671	0.605	0.549	0.364	0.269	0.213	0.176	0.094	0.8
	0.915	0.811	0.719	0.641	0.575	0.521	0.348	0.259	0.206	0.171	0.092	0.9
	0.915	0.792	0.692	0.612	0.548	0.495	0.332	0.250	0.200	0.167	0.091	1.0
0.6	0.856	0.856	0.856	0.856	0.856	0.856	0.500	0.333	0.250	0.200	0.100	0.0
	0.856	0.854	0.853	0.850	0.844	0.829	0.495	0.332	0.249	0.200	0.100	0.1
	0.856	0.851	0.844	0.833	0.815	0.780	0.481	0.327	0.248	0.198	0.100	0.2
	0.856	0.845	0.831	0.809	0.777	0.731	0.462	0.320	0.243	0.196	0.099	0.3
	0.856	0.838	0.813	0.780	0.737	0.685	0.442	0.311	0.238	0.193	0.098	0.4
	0.856	0.828	0.793	0.749	0.698	0.644	0.421	0.301	0.232	0.189	0.097	0.5
	0.856	0.817	0.771	0.718	0.662	0.607	0.400	0.290	0.226	0.185	0.096	0.6
	0.856	0.804	0.747	0.687	0.628	0.574	0.381	0.280	0.220	0.180	0.095	0.7
	0.856	0.791	0.723	0.657	0.597	0.544	0.364	0.269	0.213	0.176	0.094	0.8
	0.856	0.776	0.699	0.630	0.569	0.516	0.347	0.259	0.206	0.171	0.092	0.9
	0.856	0.761	0.676	0.603	0.542	0.492	0.332	0.250	0.200	0.167	0.091	1.0

Table C5 Continued. Mode coupling parameter κ for $\tau_X = 0.4$

r_2	r_1=0	r_1=0.2	r_1=0.4	r_1=0.6	r_1=0.8	r_1=1	r_1=2	r_1=3	r_1=4	r_1=5	r_1=10	τ_Y
	0.787	0.787	0.787	0.787	0.787	0.787	0.500	0.333	0.250	0.200	0.100	0.0
	0.787	0.786	0.785	0.783	0.780	0.773	0.495	0.332	0.249	0.200	0.100	0.1
	0.787	0.784	0.779	0.773	0.762	0.742	0.480	0.327	0.247	0.198	0.100	0.2
	0.787	0.780	0.771	0.757	0.736	0.704	0.461	0.320	0.243	0.196	0.099	0.3
	0.787	0.775	0.759	0.737	0.706	0.666	0.440	0.310	0.238	0.193	0.098	0.4
0.8	0.787	0.769	0.745	0.714	0.675	0.630	0.420	0.300	0.232	0.189	0.097	0.5
	0.787	0.761	0.729	0.689	0.644	0.596	0.399	0.290	0.226	0.185	0.096	0.6
	0.787	0.753	0.711	0.664	0.614	0.565	0.380	0.279	0.220	0.180	0.095	0.7
	0.787	0.743	0.693	0.639	0.586	0.537	0.363	0.269	0.213	0.176	0.094	0.8
	0.787	0.732	0.673	0.614	0.560	0.511	0.344	0.259	0.206	0.171	0.092	0.9
	0.787	0.721	0.654	0.591	0.535	0.487	0.331	0.249	0.200	0.167	0.091	1.0
	0.714	0.714	0.714	0.714	0.714	0.714	0.500	0.333	0.250	0.200	0.100	0.0
	0.714	0.714	0.713	0.712	0.711	0.708	0.494	0.332	0.249	0.200	0.100	0.1
	0.714	0.713	0.710	0.707	0.701	0.691	0.480	0.327	0.247	0.198	0.100	0.2
	0.714	0.711	0.705	0.698	0.686	0.667	0.460	0.319	0.243	0.196	0.099	0.3
	0.714	0.708	0.698	0.685	0.666	0.639	0.439	0.310	0.238	0.193	0.098	0.4
1	0.714	0.704	0.690	0.670	0.644	0.610	0.418	0.300	0.232	0.189	0.097	0.5
	0.714	0.699	0.679	0.653	0.620	0.581	0.398	0.290	0.226	0.185	0.096	0.6
	0.714	0.694	0.667	0.634	0.595	0.554	0.379	0.279	0.219	0.180	0.095	0.7
	0.714	0.688	0.654	0.614	0.571	0.528	0.362	0.269	0.213	0.176	0.094	0.8
	0.714	0.681	0.640	0.594	0.548	0.504	0.345	0.259	0.206	0.171	0.092	0.9
	0.714	0.673	0.625	0.574	0.526	0.482	0.330	0.249	0.200	0.167	0.091	1.0
	0.556	0.556	0.556	0.556	0.556	0.556	0.500	0.333	0.250	0.200	0.100	0.0
	0.556	0.556	0.555	0.555	0.555	0.555	0.492	0.332	0.249	0.200	0.100	0.1
	0.556	0.555	0.555	0.554	0.553	0.551	0.474	0.327	0.247	0.198	0.100	0.2
	0.556	0.555	0.553	0.552	0.549	0.546	0.453	0.319	0.243	0.196	0.099	0.3
	0.556	0.554	0.552	0.549	0.545	0.538	0.432	0.310	0.238	0.193	0.098	0.4
1.5	0.556	0.553	0.549	0.545	0.538	0.529	0.412	0.299	0.232	0.189	0.097	0.5
	0.556	0.552	0.547	0.540	0.530	0.517	0.393	0.289	0.226	0.185	0.096	0.6
	0.556	0.550	0.543	0.534	0.521	0.503	0.375	0.278	0.219	0.180	0.095	0.7
	0.556	0.549	0.539	0.527	0.510	0.488	0.358	0.268	0.213	0.176	0.094	0.8
	0.556	0.547	0.535	0.519	0.498	0.473	0.342	0.258	0.206	0.171	0.092	0.9
	0.556	0.545	0.530	0.510	0.485	0.457	0.328	0.249	0.199	0.166	0.091	1.0
	0.444	0.444	0.444	0.444	0.444	0.444	0.444	0.333	0.250	0.200	0.100	0.0
	0.444	0.443	0.443	0.443	0.443	0.443	0.441	0.331	0.249	0.200	0.100	0.1
	0.444	0.443	0.443	0.443	0.443	0.442	0.434	0.326	0.247	0.198	0.100	0.2
	0.444	0.443	0.443	0.442	0.442	0.441	0.423	0.318	0.243	0.196	0.099	0.3
	0.444	0.443	0.442	0.442	0.440	0.439	0.409	0.309	0.238	0.192	0.098	0.4
2	0.444	0.443	0.442	0.440	0.439	0.436	0.395	0.298	0.232	0.189	0.097	0.5
	0.444	0.442	0.441	0.439	0.436	0.433	0.379	0.288	0.225	0.184	0.096	0.6
	0.444	0.442	0.440	0.437	0.434	0.429	0.364	0.277	0.219	0.180	0.095	0.7
	0.444	0.441	0.439	0.435	0.430	0.424	0.350	0.267	0.212	0.175	0.094	0.8
	0.444	0.441	0.437	0.433	0.426	0.418	0.336	0.260	0.206	0.171	0.092	0.9
	0.444	0.440	0.436	0.430	0.422	0.411	0.322	0.248	0.199	0.166	0.091	1.0

296 Appendix C

Table C5 Continued. Mode coupling parameter κ for $\tau_X = 0.4$

r_2	$r_1=0$	$r_1=0.2$	$r_1=0.4$	$r_1=0.6$	$r_1=0.8$	$r_1=1$	$r_1=2$	$r_1=3$	$r_1=4$	$r_1=5$	$r_1=10$	τ_Y
3	0.311	0.311	0.311	0.311	0.311	0.311	0.311	0.311	0.250	0.200	0.100	0.0
	0.311	0.311	0.311	0.311	0.311	0.311	0.311	0.310	0.249	0.200	0.100	0.1
	0.311	0.311	0.311	0.311	0.311	0.311	0.310	0.306	0.246	0.198	0.100	0.2
	0.311	0.311	0.311	0.311	0.311	0.311	0.310	0.301	0.242	0.196	0.099	0.3
	0.311	0.311	0.311	0.311	0.310	0.310	0.309	0.294	0.237	0.192	0.098	0.4
	0.311	0.311	0.311	0.310	0.310	0.310	0.307	0.286	0.231	0.189	0.097	0.5
	0.311	0.311	0.310	0.310	0.310	0.309	0.305	0.278	0.224	0.184	0.096	0.6
	0.311	0.311	0.310	0.310	0.309	0.309	0.302	0.269	0.218	0.180	0.095	0.7
	0.311	0.311	0.310	0.310	0.309	0.308	0.299	0.260	0.211	0.175	0.094	0.8
	0.311	0.310	0.310	0.309	0.308	0.307	0.295	0.251	0.205	0.171	0.092	0.9
	0.311	0.310	0.310	0.309	0.308	0.306	0.290	0.243	0.198	0.166	0.091	1.0
4	0.238	0.238	0.238	0.238	0.238	0.238	0.238	0.238	0.238	0.200	0.100	0.0
	0.238	0.238	0.238	0.238	0.238	0.238	0.238	0.238	0.237	0.199	0.100	0.1
	0.238	0.238	0.238	0.238	0.238	0.238	0.238	0.238	0.236	0.198	0.100	0.2
	0.238	0.238	0.238	0.238	0.238	0.238	0.238	0.237	0.232	0.195	0.099	0.3
	0.238	0.238	0.238	0.238	0.238	0.238	0.238	0.237	0.228	0.192	0.098	0.4
	0.238	0.238	0.238	0.238	0.238	0.238	0.237	0.236	0.224	0.188	0.097	0.5
	0.238	0.238	0.238	0.238	0.238	0.238	0.237	0.234	0.218	0.184	0.096	0.6
	0.238	0.238	0.238	0.238	0.238	0.238	0.236	0.233	0.213	0.179	0.095	0.7
	0.238	0.238	0.238	0.238	0.238	0.237	0.236	0.231	0.207	0.174	0.094	0.8
	0.238	0.238	0.238	0.238	0.237	0.237	0.235	0.228	0.201	0.170	0.092	0.9
	0.238	0.238	0.238	0.238	0.237	0.237	0.234	0.224	0.195	0.165	0.091	1.0
5	0.193	0.193	0.193	0.193	0.193	0.193	0.193	0.193	0.193	0.193	0.100	0.0
	0.193	0.193	0.193	0.193	0.193	0.193	0.193	0.193	0.193	0.192	0.100	0.1
	0.193	0.193	0.193	0.193	0.193	0.193	0.193	0.193	0.192	0.191	0.100	0.2
	0.193	0.193	0.193	0.193	0.193	0.193	0.193	0.192	0.192	0.189	0.099	0.3
	0.193	0.193	0.193	0.193	0.193	0.193	0.193	0.192	0.192	0.186	0.098	0.4
	0.193	0.193	0.193	0.193	0.193	0.193	0.192	0.192	0.191	0.183	0.097	0.5
	0.193	0.193	0.193	0.193	0.193	0.193	0.192	0.192	0.190	0.180	0.096	0.6
	0.193	0.193	0.193	0.193	0.193	0.193	0.192	0.191	0.189	0.176	0.095	0.7
	0.193	0.193	0.193	0.193	0.193	0.192	0.192	0.191	0.188	0.172	0.094	0.8
	0.193	0.193	0.193	0.193	0.192	0.192	0.192	0.190	0.186	0.167	0.092	0.9
	0.193	0.193	0.193	0.192	0.192	0.192	0.192	0.190	0.183	0.163	0.091	1.0
10	0.098	0.098	0.098	0.098	0.098	0.098	0.098	0.098	0.098	0.098	0.098	0.0
	0.098	0.098	0.098	0.098	0.098	0.098	0.098	0.098	0.098	0.098	0.098	0.1
	0.098	0.098	0.098	0.098	0.098	0.098	0.098	0.098	0.098	0.098	0.098	0.2
	0.098	0.098	0.098	0.098	0.098	0.098	0.098	0.098	0.098	0.098	0.097	0.3
	0.098	0.098	0.098	0.098	0.098	0.098	0.098	0.098	0.098	0.098	0.097	0.4
	0.098	0.098	0.098	0.098	0.098	0.098	0.098	0.098	0.098	0.098	0.096	0.5
	0.098	0.098	0.098	0.098	0.098	0.098	0.098	0.098	0.098	0.098	0.095	0.6
	0.098	0.098	0.098	0.098	0.098	0.098	0.098	0.098	0.098	0.098	0.094	0.7
	0.098	0.098	0.098	0.098	0.098	0.098	0.098	0.098	0.098	0.098	0.093	0.8
	0.098	0.098	0.098	0.098	0.098	0.098	0.098	0.098	0.098	0.098	0.091	0.9
	0.098	0.098	0.098	0.098	0.098	0.098	0.098	0.098	0.098	0.098	0.090	1.0

Table C6 Mode coupling parameter κ for $\tau_X = 0.5$

r_2	$r_1=0$	$r_1=0.2$	$r_1=0.4$	$r_1=0.6$	$r_1=0.8$	$r_1=1$	$r_1=2$	$r_1=3$	$r_1=4$	$r_1=5$	$r_1=10$	τ_Y
	1.000	1.000	1.000	1.000	1.000	1.000	0.500	0.333	0.250	0.200	0.100	0.0
	1.000	0.998	0.993	0.986	0.967	0.909	0.495	0.332	0.249	0.200	0.100	0.1
	1.000	0.990	0.975	0.950	0.905	0.833	0.482	0.327	0.247	0.198	0.100	0.2
	1.000	0.979	0.948	0.904	0.843	0.769	0.464	0.320	0.243	0.196	0.099	0.3
	1.000	0.963	0.915	0.856	0.787	0.714	0.444	0.311	0.238	0.193	0.098	0.4
0	1.000	0.945	0.880	0.809	0.736	0.667	0.423	0.301	0.232	0.189	0.097	0.5
	1.000	0.925	0.845	0.766	0.692	0.625	0.402	0.291	0.226	0.185	0.096	0.6
	1.000	0.903	0.810	0.726	0.652	0.588	0.383	0.280	0.220	0.181	0.095	0.7
	1.000	0.880	0.776	0.689	0.616	0.556	0.365	0.270	0.213	0.176	0.094	0.8
	1.000	0.857	0.744	0.656	0.584	0.526	0.349	0.260	0.207	0.171	0.092	0.9
	1.000	0.833	0.714	0.625	0.556	0.500	0.333	0.250	0.200	0.167	0.091	1.0
	0.945	0.945	0.945	0.945	0.945	0.945	0.500	0.333	0.250	0.200	0.100	0.0
	0.945	0.943	0.940	0.934	0.932	0.885	0.495	0.332	0.249	0.200	0.100	0.1
	0.945	0.937	0.926	0.907	0.873	0.816	0.482	0.327	0.247	0.198	0.100	0.2
	0.945	0.928	0.904	0.869	0.820	0.756	0.463	0.320	0.243	0.196	0.099	0.3
	0.945	0.916	0.877	0.828	0.769	0.704	0.443	0.311	0.238	0.193	0.098	0.4
0.2	0.945	0.901	0.848	0.787	0.722	0.658	0.422	0.301	0.232	0.189	0.097	0.5
	0.945	0.884	0.817	0.748	0.681	0.618	0.402	0.290	0.226	0.185	0.096	0.6
	0.945	0.866	0.786	0.711	0.643	0.583	0.382	0.280	0.220	0.181	0.095	0.7
	0.945	0.847	0.756	0.677	0.609	0.551	0.365	0.269	0.213	0.176	0.094	0.8
	0.945	0.827	0.727	0.646	0.578	0.522	0.348	0.259	0.206	0.171	0.092	0.9
	0.945	0.806	0.700	0.617	0.550	0.497	0.333	0.250	0.200	0.167	0.091	1.0
	0.880	0.880	0.880	0.880	0.880	0.880	0.500	0.333	0.250	0.200	0.100	0.0
	0.880	0.879	0.877	0.873	0.866	0.846	0.495	0.332	0.249	0.200	0.100	0.1
	0.880	0.875	0.867	0.854	0.831	0.790	0.481	0.327	0.247	0.198	0.100	0.2
	0.880	0.868	0.851	0.825	0.788	0.737	0.462	0.320	0.243	0.196	0.099	0.3
	0.880	0.859	0.831	0.793	0.745	0.690	0.442	0.311	0.238	0.193	0.098	0.4
0.4	0.880	0.848	0.807	0.759	0.704	0.647	0.421	0.301	0.232	0.189	0.097	0.5
	0.880	0.835	0.783	0.725	0.666	0.609	0.401	0.290	0.226	0.185	0.096	0.6
	0.880	0.821	0.757	0.693	0.631	0.575	0.381	0.280	0.220	0.180	0.095	0.7
	0.880	0.805	0.731	0.662	0.599	0.545	0.364	0.269	0.213	0.176	0.094	0.8
	0.880	0.789	0.706	0.633	0.570	0.517	0.347	0.259	0.206	0.171	0.092	0.9
	0.880	0.773	0.681	0.606	0.544	0.492	0.332	0.250	0.200	0.167	0.091	1.0
	0.809	0.809	0.809	0.809	0.809	0.809	0.500	0.333	0.250	0.200	0.100	0.0
	0.809	0.808	0.807	0.805	0.801	0.791	0.495	0.332	0.249	0.200	0.100	0.1
	0.809	0.806	0.800	0.792	0.778	0.753	0.480	0.327	0.247	0.198	0.100	0.2
	0.809	0.801	0.790	0.773	0.748	0.711	0.461	0.320	0.243	0.196	0.099	0.3
	0.809	0.795	0.776	0.749	0.714	0.670	0.440	0.310	0.238	0.193	0.098	0.4
0.6	0.809	0.787	0.759	0.723	0.680	0.632	0.419	0.300	0.232	0.189	0.097	0.5
	0.809	0.778	0.740	0.696	0.647	0.597	0.399	0.290	0.226	0.185	0.096	0.6
	0.809	0.768	0.720	0.687	0.616	0.566	0.380	0.279	0.219	0.180	0.095	0.7
	0.809	0.757	0.700	0.642	0.587	0.537	0.363	0.269	0.213	0.176	0.094	0.8
	0.809	0.744	0.679	0.616	0.560	0.511	0.346	0.259	0.206	0.171	0.092	0.9
	0.809	0.732	0.658	0.592	0.535	0.487	0.331	0.249	0.200	0.167	0.091	1.0

298 Appendix C

Table C6 Continued. Mode coupling parameter κ for $\tau_X = 0.5$

r_2	$r_1=0$	$r_1=0.2$	$r_1=0.4$	$r_1=0.6$	$r_1=0.8$	$r_1=1$	$r_1=2$	$r_1=3$	$r_1=4$	$r_1=5$	$r_1=10$	τ_Y
	0.736	0.736	0.736	0.736	0.736	0.736	0.500	0.333	0.250	0.200	0.100	0.0
	0.736	0.736	0.735	0.734	0.732	0.727	0.494	0.332	0.249	0.200	0.100	0.1
	0.736	0.734	0.731	0.726	0.718	0.705	0.479	0.327	0.247	0.198	0.100	0.2
	0.736	0.731	0.724	0.714	0.699	0.675	0.460	0.319	0.243	0.196	0.099	0.3
	0.736	0.727	0.715	0.698	0.675	0.644	0.439	0.310	0.238	0.193	0.098	0.4
0.8	0.736	0.722	0.704	0.680	0.649	0.612	0.418	0.300	0.232	0.189	0.097	0.5
	0.736	0.716	0.691	0.660	0.623	0.582	0.398	0.290	0.226	0.185	0.096	0.6
	0.736	0.710	0.677	0.639	0.597	0.554	0.379	0.279	0.219	0.180	0.095	0.7
	0.736	0.702	0.662	0.617	0.572	0.527	0.361	0.269	0.213	0.176	0.094	0.8
	0.736	0.694	0.646	0.596	0.548	0.503	0.345	0.259	0.206	0.171	0.092	0.9
	0.736	0.684	0.629	0.575	0.525	0.481	0.330	0.249	0.200	0.167	0.091	1.0
	0.667	0.667	0.667	0.667	0.667	0.667	0.500	0.333	0.250	0.200	0.100	0.0
	0.667	0.666	0.666	0.665	0.664	0.662	0.494	0.332	0.249	0.200	0.100	0.1
	0.667	0.665	0.663	0.661	0.657	0.650	0.478	0.327	0.247	0.198	0.100	0.2
	0.667	0.664	0.660	0.654	0.645	0.632	0.458	0.319	0.243	0.196	0.099	0.3
	0.667	0.661	0.654	0.644	0.630	0.610	0.436	0.310	0.238	0.193	0.098	0.4
1	0.667	0.658	0.647	0.632	0.612	0.586	0.416	0.300	0.232	0.189	0.097	0.5
	0.667	0.655	0.639	0.618	0.593	0.562	0.396	0.289	0.226	0.185	0.096	0.6
	0.667	0.650	0.629	0.603	0.572	0.538	0.377	0.279	0.219	0.180	0.095	0.7
	0.667	0.645	0.619	0.587	0.552	0.515	0.360	0.268	0.213	0.176	0.094	0.8
	0.667	0.640	0.607	0.570	0.531	0.493	0.344	0.258	0.206	0.171	0.092	0.9
	0.667	0.634	0.595	0.553	0.512	0.472	0.329	0.249	0.200	0.166	0.091	1.0
	0.523	0.523	0.523	0.523	0.523	0.523	0.500	0.333	0.250	0.200	0.100	0.0
	0.523	0.523	0.523	0.523	0.523	0.522	0.487	0.332	0.249	0.200	0.100	0.1
	0.523	0.523	0.522	0.522	0.521	0.519	0.467	0.326	0.247	0.198	0.100	0.2
	0.523	0.522	0.521	0.520	0.518	0.515	0.446	0.318	0.243	0.196	0.099	0.3
	0.523	0.522	0.520	0.517	0.513	0.508	0.425	0.309	0.238	0.193	0.098	0.4
1.5	0.523	0.521	0.518	0.513	0.508	0.500	0.406	0.299	0.232	0.189	0.097	0.5
	0.523	0.520	0.515	0.509	0.501	0.490	0.387	0.288	0.226	0.185	0.096	0.6
	0.523	0.518	0.512	0.504	0.493	0.479	0.370	0.277	0.219	0.180	0.095	0.7
	0.523	0.517	0.508	0.498	0.484	0.467	0.354	0.267	0.212	0.176	0.094	0.8
	0.523	0.515	0.505	0.491	0.474	0.454	0.338	0.257	0.206	0.171	0.092	0.9
	0.523	0.513	0.500	0.484	0.463	0.440	0.324	0.248	0.199	0.166	0.091	1.0
	0.423	0.423	0.423	0.423	0.423	0.423	0.423	0.333	0.250	0.200	0.100	0.0
	0.423	0.423	0.423	0.423	0.423	0.422	0.421	0.331	0.249	0.200	0.100	0.1
	0.423	0.423	0.422	0.422	0.422	0.422	0.415	0.325	0.247	0.198	0.100	0.2
	0.423	0.422	0.422	0.422	0.421	0.420	0.406	0.317	0.242	0.196	0.099	0.3
	0.423	0.422	0.421	0.421	0.420	0.418	0.395	0.307	0.237	0.192	0.098	0.4
2	0.423	0.422	0.421	0.419	0.418	0.416	0.382	0.296	0.231	0.189	0.097	0.5
	0.423	0.421	0.420	0.418	0.415	0.412	0.369	0.286	0.225	0.185	0.096	0.6
	0.423	0.421	0.419	0.416	0.413	0.408	0.355	0.275	0.218	0.180	0.095	0.7
	0.423	0.420	0.418	0.414	0.410	0.404	0.342	0.265	0.212	0.175	0.094	0.8
	0.423	0.420	0.416	0.412	0.406	0.398	0.329	0.255	0.205	0.171	0.092	0.9
	0.423	0.419	0.415	0.409	0.401	0.392	0.317	0.246	0199	0.166	0.091	1.0

Mode coupling parameter κ

Table C6 Continued. Mode coupling parameter κ for $\tau_X = 0.5$

r_2	r_1=0	r_1=0.2	r_1=0.4	r_1=0.6	r_1=0.8	r_1=1	r_1=2	r_1=3	r_1=4	r_1=5	r_1=10	τ_Y
3	0.301	0.301	0.301	0.301	0.301	0.301	0.301	0.301	0.250	0.200	0.100	0.0
	0.301	0.301	0.301	0.301	0.301	0.301	0.301	0.300	0.250	0.200	0.100	0.1
	0.301	0.301	0.301	0.301	0.301	0.301	0.300	0.297	0.246	0.198	0.100	0.2
	0.301	0.301	0.301	0.301	0.301	0.301	0.299	0.292	0.241	0.195	0.099	0.3
	0.301	0.301	0.301	0.301	0.300	0.300	0.298	0.286	0.236	0.192	0.098	0.4
	0.301	0.301	0.301	0.300	0.300	0.300	0.296	0.279	0.230	0.188	0.097	0.5
	0.301	0.301	0.300	0.300	0.300	0.299	0.294	0.272	0.223	0.184	0.096	0.6
	0.301	0.301	0.300	0.300	0.299	0.299	0.292	0.264	0.216	0.180	0.095	0.7
	0.301	0.301	0.300	0.299	0.299	0.298	0.288	0.255	0.210	0.175	0.094	0.8
	0.301	0.300	0.300	0.299	0.298	0.297	0.284	0.247	0.203	0.170	0.092	0.9
	0.301	0.300	0.299	0.298	0.297	0.296	0.280	0.239	0.197	0.166	0.091	1.0
4	0.232	0.232	0.232	0.232	0.232	0.232	0.232	0.232	0.232	0.200	0.100	0.0
	0.232	0.232	0.232	0.232	0.232	0.232	0.232	0.232	0.232	0.199	0.100	0.1
	0.232	0.232	0.232	0.232	0.232	0.232	0.232	0.232	0.230	0.198	0.100	0.2
	0.232	0.232	0.232	0.232	0.232	0.232	0.232	0.231	0.227	0.195	0.099	0.3
	0.232	0.232	0.232	0.232	0.232	0.232	0.232	0.231	0.224	0.191	0.098	0.4
	0.232	0.232	0.232	0.232	0.232	0.232	0.232	0.230	0.219	0.187	0.097	0.5
	0.232	0.232	0.232	0.232	0.232	0.232	0.231	0.228	0.214	0.183	0.096	0.6
	0.232	0.232	0.232	0.232	0.232	0.232	0.230	0.226	0.209	0.178	0.095	0.7
	0.232	0.232	0.232	0.232	0.232	0.232	0.230	0.224	0.204	0.174	0.094	0.8
	0.232	0.232	0.232	0.232	0.232	0.231	0.229	0.221	0.198	0.169	0.092	0.9
	0.232	0.232	0.232	0.232	0.231	0.231	0.228	0.218	0.193	0.164	0.091	1.0
5	0.189	0.189	0.189	0.189	0.189	0.189	0.189	0.189	0.189	0.189	0.100	0.0
	0.189	0.189	0.189	0.189	0.189	0.189	0.189	0.189	0.189	0.189	0.100	0.1
	0.189	0.189	0.189	0.189	0.189	0.189	0.189	0.189	0.189	0.188	0.100	0.2
	0.189	0.189	0.189	0.189	0.189	0.189	0.189	0.189	0.188	0.186	0.099	0.3
	0.189	0.189	0.189	0.189	0.189	0.189	0.189	0.189	0.188	0.183	0.098	0.4
	0.189	0.189	0.189	0.189	0.189	0.189	0.189	0.188	0.187	0.180	0.097	0.5
	0.189	0.189	0.189	0.189	0.189	0.189	0.189	0.188	0.186	0.177	0.096	0.6
	0.189	0.189	0.189	0.189	0.189	0.189	0.188	0.188	0.185	0.173	0.095	0.7
	0.189	0.189	0.189	0.189	0.189	0.189	0.188	0.187	0.183	0.169	0.094	0.8
	0.189	0.189	0.189	0.189	0.189	0.189	0.188	0.186	0.181	0.165	0.092	0.9
	0.189	0.189	0.189	0.189	0.189	0.188	0.188	0.185	0.179	0.161	0.091	1.0
10	0.097	0.097	0.097	0.097	0.097	0.097	0.097	0.097	0.097	0.097	0.097	0.0
	0.097	0.097	0.097	0.097	0.097	0.097	0.097	0.097	0.097	0.097	0.097	0.1
	0.097	0.097	0.097	0.097	0.097	0.097	0.097	0.097	0.097	0.097	0.097	0.2
	0.097	0.097	0.097	0.097	0.097	0.097	0.097	0.097	0.097	0.097	0.097	0.3
	0.097	0.097	0.097	0.097	0.097	0.097	0.097	0.097	0.097	0.097	0.096	0.4
	0.097	0.097	0.097	0.097	0.097	0.097	0.097	0.097	0.097	0.097	0.095	0.5
	0.097	0.097	0.097	0.097	0.097	0.097	0.097	0.097	0.097	0.097	0.094	0.6
	0.097	0.097	0.097	0.097	0.097	0.097	0.097	0.097	0.097	0.097	0.093	0.7
	0.097	0.097	0.097	0.097	0.097	0.097	0.097	0.097	0.097	0.097	0.092	0.8
	0.097	0.097	0.097	0.097	0.097	0.097	0.097	0.097	0.097	0.097	0.091	0.9
	0.097	0.097	0.097	0.097	0.097	0.097	0.097	0.097	0.097	0.097	0.089	1.0

300 Appendix C

Table C7 Mode coupling parameter κ for $\tau_X = 0.6$

r_2	$r_1=0$	$r_1=0.2$	$r_1=0.4$	$r_1=0.6$	$r_1=0.8$	$r_1=1$	$r_1=2$	$r_1=3$	$r_1=4$	$r_1=5$	$r_1=10$	τ_Y
0	1.000	1.000	1.000	1.000	1.000	1.000	0.500	0.333	0.250	0.200	0.100	0.0
	1.000	0.998	0.993	0.986	0.967	0.909	0.495	0.332	0.249	0.200	0.100	0.1
	1.000	0.990	0.975	0.950	0.905	0.833	0.482	0.327	0.247	0.198	0.100	0.2
	1.000	0.979	0.948	0.904	0.843	0.769	0.464	0.320	0.243	0.196	0.099	0.3
	1.000	0.963	0.915	0.856	0.787	0.714	0.444	0.311	0.238	0.193	0.098	0.4
	1.000	0.945	0.880	0.809	0.736	0.667	0.423	0.301	0.232	0.189	0.097	0.5
	1.000	0.925	0.845	0.766	0.692	0.625	0.402	0.291	0.226	0.185	0.096	0.6
	1.000	0.903	0.810	0.726	0.652	0.588	0.383	0.280	0.220	0.181	0.095	0.7
	1.000	0.880	0.776	0.689	0.616	0.556	0.365	0.270	0.213	0.176	0.094	0.8
	1.000	0.857	0.744	0.656	0.584	0.526	0.349	0.260	0.207	0.171	0.092	0.9
	1.000	0.833	0.714	0.625	0.556	0.500	0.333	0.250	0.200	0.167	0.091	1.0
0.2	0.925	0.925	0.925	0.925	0.925	0.925	0.500	0.333	0.250	0.200	0.100	0.0
	0.925	0.923	0.920	0.915	0.904	0.873	0.495	0.332	0.249	0.200	0.100	0.1
	0.925	0.918	0.907	0.890	0.860	0.808	0.482	0.327	0.247	0.198	0.100	0.2
	0.925	0.909	0.887	0.855	0.810	0.750	0.463	0.320	0.243	0.196	0.099	0.3
	0.925	0.898	0.863	0.817	0.761	0.699	0.442	0.311	0.238	0.193	0.098	0.4
	0.925	0.884	0.835	0.778	0.716	0.655	0.421	0.301	0.232	0.189	0.097	0.5
	0.925	0.869	0.806	0.741	0.676	0.615	0.401	0.290	0.226	0.185	0.096	0.6
	0.925	0.852	0.777	0.705	0.639	0.580	0.382	0.280	0.220	0.181	0.095	0.7
	0.925	0.833	0.748	0.672	0.606	0.549	0.364	0.269	0.213	0.176	0.094	0.8
	0.925	0.815	0.720	0.641	0.575	0.521	0.348	0.259	0.206	0.171	0.092	0.9
	0.925	0.795	0.694	0.613	0.548	0.495	0.332	0.250	0.200	0.167	0.091	1.0
0.4	0.845	0.845	0.845	0.845	0.845	0.845	0.500	0.333	0.250	0.200	0.100	0.0
	0.845	0.844	0.842	0.839	0.833	0.819	0.495	0.332	0.249	0.200	0.100	0.1
	0.845	0.840	0.833	0.823	0.805	0.772	0.481	0.327	0.247	0.198	0.100	0.2
	0.845	0.835	0.820	0.799	0.768	0.724	0.462	0.320	0.243	0.196	0.099	0.3
	0.845	0.827	0.803	0.771	0.729	0.679	0.441	0.310	0.238	0.193	0.098	0.4
	0.845	0.817	0.783	0.740	0.691	0.639	0.420	0.300	0.232	0.189	0.097	0.5
	0.845	0.806	0.761	0.710	0.656	0.603	0.400	0.290	0.226	0.185	0.096	0.6
	0.845	0.794	0.738	0.680	0.623	0.570	0.381	0.279	0.220	0.180	0.095	0.7
	0.845	0.780	0.714	0.651	0.592	0.540	0.363	0.269	0.213	0.176	0.094	0.8
	0.845	0.766	0.691	0.624	0.564	0.513	0.347	0.259	0.206	0.171	0.092	0.9
	0.845	0.751	0.668	0.598	0.539	0.489	0.331	0.249	0.200	0.167	0.091	1.0
0.6	0.766	0.766	0.766	0.766	0.766	0.766	0.500	0.333	0.250	0.200	0.100	0.0
	0.766	0.765	0.764	0.762	0.759	0.753	0.494	0.332	0.249	0.200	0.100	0.1
	0.766	0.763	0.759	0.752	0.742	0.724	0.480	0.327	0.247	0.198	0.100	0.2
	0.766	0.759	0.750	0.737	0.718	0.689	0.460	0.319	0.243	0.196	0.099	0.3
	0.766	0.754	0.739	0.718	0.689	0.653	0.439	0.310	0.238	0.193	0.098	0.4
	0.766	0.748	0.725	0.696	0.660	0.618	0.418	0.300	0.232	0.189	0.097	0.5
	0.766	0.741	0.710	0.673	0.631	0.586	0.398	0.290	0.226	0.185	0.096	0.6
	0.766	0.732	0.693	0.649	0.603	0.557	0.379	0.279	0.219	0.180	0.095	0.7
	0.766	0.723	0.675	0.625	0.576	0.530	0.361	0.269	0.213	0.176	0.094	0.8
	0.766	0.713	0.657	0.602	0.551	0.505	0.345	0.259	0.206	0.171	0.092	0.9
	0.766	0.702	0.638	0.580	0.527	0.482	0.330	0.249	0.200	0.167	0.091	1.0

Table C7 Continued. Mode coupling parameter κ for $\tau_X = 0.6$

r_2	$r_1=0$	$r_1=0.2$	$r_1=0.4$	$r_1=0.6$	$r_1=0.8$	$r_1=1$	$r_1=2$	$r_1=3$	$r_1=4$	$r_1=5$	$r_1=10$	τ_Y
0.8	0.692	0.692	0.692	0.692	0.692	0.692	0.500	0.333	0.250	0.200	0.100	0.0
	0.692	0.691	0.691	0.690	0.688	0.686	0.494	0.332	0.249	0.200	0.100	0.1
	0.692	0.690	0.687	0.684	0.678	0.669	0.478	0.327	0.247	0.198	0.100	0.2
	0.692	0.688	0.682	0.674	0.663	0.646	0.458	0.319	0.243	0.196	0.099	0.3
	0.692	0.684	0.675	0.662	0.644	0.620	0.436	0.310	0.238	0.193	0.098	0.4
	0.692	0.681	0.666	0.647	0.623	0.593	0.415	0.300	0.232	0.189	0.097	0.5
	0.692	0.676	0.656	0.631	0.600	0.566	0.396	0.289	0.226	0.185	0.096	0.6
	0.692	0.670	0.644	0.613	0.578	0.540	0.377	0.279	0.219	0.180	0.095	0.7
	0.692	0.664	0.631	0.595	0.556	0.516	0.359	0.268	0.213	0.176	0.094	0.8
	0.692	0.657	0.618	0.576	0.534	0.494	0.343	0.258	0.206	0.171	0.092	0.9
	0.692	0.650	0.604	0.558	0.513	0.473	0.329	0.249	0.200	0.166	0.091	1.0
1	0.625	0.625	0.625	0.625	0.625	0.625	0.500	0.333	0.250	0.200	0.100	0.0
	0.625	0.625	0.624	0.624	0.623	0.622	0.493	0.332	0.249	0.200	0.100	0.1
	0.625	0.624	0.622	0.620	0.617	0.613	0.476	0.326	0.247	0.198	0.100	0.2
	0.625	0.623	0.619	0.615	0.608	0.599	0.455	0.319	0.243	0.196	0.099	0.3
	0.625	0.621	0.615	0.607	0.596	0.581	0.433	0.309	0.238	0.193	0.098	0.4
	0.625	0.618	0.609	0.597	0.582	0.562	0.412	0.299	0.232	0.189	0.097	0.5
	0.625	0.615	0.603	0.586	0.566	0.541	0.393	0.289	0.226	0.185	0.096	0.6
	0.625	0.612	0.595	0.574	0.549	0.520	0.374	0.278	0.219	0.180	0.095	0.7
	0.625	0.608	0.586	0.561	0.531	0.500	0.357	0.268	0.212	0.176	0.094	0.8
	0.625	0.603	0.577	0.547	0.514	0.480	0.341	0.258	0.206	0.171	0.092	0.9
	0.625	0.598	0.567	0.532	0.497	0.462	0.327	0.248	0.200	0.166	0.091	1.0
1.5	0.494	0.494	0.494	0.494	0.494	0.494	0.494	0.333	0.250	0.200	0.100	0.0
	0.494	0.493	0.493	0.493	0.493	0.493	0.476	0.331	0.249	0.200	0.100	0.1
	0.494	0.493	0.493	0.493	0.491	0.490	0.456	0.326	0.247	0.198	0.100	0.2
	0.494	0.493	0.492	0.490	0.489	0.486	0.436	0.318	0.243	0.196	0.099	0.3
	0.494	0.492	0.490	0.488	0.485	0.481	0.417	0.308	0.237	0.192	0.098	0.4
	0.494	0.491	0.488	0.485	0.480	0.474	0.398	0.297	0.232	0.189	0.097	0.5
	0.494	0.490	0.486	0.481	0.474	0.466	0.381	0.287	0.225	0.185	0.096	0.6
	0.494	0.489	0.484	0.477	0.468	0.456	0.364	0.276	0.219	0.180	0.095	0.7
	0.494	0.488	0.481	0.471	0.460	0.446	0.349	0.266	0.212	0.175	0.094	0.8
	0.494	0.486	0.477	0.466	0.452	0.435	0.334	0.256	0.205	0.171	0.092	0.9
	0.494	0.484	0.473	0.459	0.443	0.424	0.321	0.247	0.199	0.166	0.091	1.0
2	0.402	0.402	0.402	0.402	0.402	0.402	0.402	0.333	0.250	0.200	0.100	0.0
	0.402	0.402	0.402	0.402	0.402	0.402	0.401	0.331	0.249	0.200	0.100	0.1
	0.402	0.402	0.402	0.402	0.402	0.401	0.396	0.324	0.246	0.198	0.100	0.2
	0.402	0.402	0.402	0.401	0.401	0.400	0.389	0.315	0.242	0.196	0.099	0.3
	0.402	0.402	0.401	0.400	0.399	0.398	0.379	0.305	0.237	0.192	0.098	0.4
	0.402	0.402	0.401	0.399	0.398	0.396	0.369	0.294	0.231	0.189	0.097	0.5
	0.402	0.401	0.400	0.398	0.396	0.393	0.357	0.284	0.224	0.184	0.096	0.6
	0.402	0.401	0.399	0.396	0.393	0.389	0.345	0.273	0.218	0.180	0.095	0.7
	0.402	0.400	0.397	0.394	0.390	0.385	0.333	0.263	0.211	0.175	0.094	0.8
	0.402	0.400	0.396	0.392	0.386	0.380	0.322	0.254	0.205	0.171	0.092	0.9
	0.402	0.399	0.395	0.389	0.383	0.374	0.310	0.244	0.198	0.166	0.091	1.0

Table C7 Continued. Mode coupling parameter κ for $\tau_X = 0.6$

r_2	$r_1=0$	$r_1=0.2$	$r_1=0.4$	$r_1=0.6$	$r_1=0.8$	$r_1=1$	$r_1=2$	$r_1=3$	$r_1=4$	$r_1=5$	$r_1=10$	τ_Y
3	0.291	0.291	0.291	0.291	0.291	0.291	0.291	0.291	0.250	0.200	0.100	0.0
	0.291	0.291	0.291	0.291	0.291	0.291	0.290	0.290	0.249	0.199	0.100	0.1
	0.291	0.291	0.290	0.290	0.290	0.290	0.290	0.287	0.245	0.198	0.100	0.2
	0.291	0.290	0.290	0.290	0.290	0.290	0.289	0.283	0.240	0.195	0.099	0.3
	0.291	0.290	0.290	0.290	0.290	0.290	0.288	0.278	0.234	0.192	0.098	0.4
	0.291	0.290	0.290	0.290	0.290	0.289	0.286	0.272	0.228	0.188	0.097	0.5
	0.291	0.290	0.290	0.290	0.290	0.289	0.284	0.265	0.222	0.184	0.096	0.6
	0.291	0.290	0.290	0.289	0.289	0.288	0.281	0.258	0.215	0.179	0.095	0.7
	0.291	0.290	0.289	0.289	0.288	0.287	0.278	0.250	0.208	0.174	0.094	0.8
	0.291	0.290	0.289	0.288	0.287	0.286	0.274	0.243	0.202	0.170	0.092	0.9
	0.291	0.290	0.289	0.288	0.286	0.285	0.270	0.235	0.196	0.165	0.091	1.0
4	0.226	0.226	0.226	0.226	0.226	0.226	0.226	0.226	0.226	0.200	0.100	0.0
	0.226	0.226	0.226	0.226	0.226	0.226	0.226	0.226	0.226	0.199	0.100	0.1
	0.226	0.226	0.226	0.226	0.226	0.226	0.226	0.226	0.224	0.197	0.100	0.2
	0.226	0.226	0.226	0.226	0.226	0.226	0.226	0.225	0.222	0.194	0.099	0.3
	0.226	0.226	0.226	0.226	0.226	0.226	0.225	0.224	0.218	0.190	0.098	0.4
	0.226	0.226	0.226	0.226	0.226	0.226	0.225	0.223	0.214	0.186	0.097	0.5
	0.226	0.226	0.226	0.226	0.226	0.226	0.224	0.222	0.210	0.182	0.096	0.6
	0.226	0.226	0.226	0.226	0.226	0.225	0.224	0.220	0.205	0.177	0.095	0.7
	0.226	0.226	0.226	0.226	0.225	0.225	0.223	0.217	0.200	0.172	0.094	0.8
	0.226	0.226	0.226	0.226	0.225	0.225	0.222	0.215	0.195	0.168	0.092	0.9
	0.226	0.226	0.226	0.225	0.225	0.225	0.221	0.212	0.190	0.163	0.091	1.0
5	0.185	0.185	0.185	0.185	0.185	0.185	0.185	0.185	0.185	0.185	0.100	0.0
	0.185	0.185	0.185	0.185	0.185	0.185	0.185	0.185	0.185	0.185	0.100	0.1
	0.185	0.185	0.185	0.185	0.185	0.185	0.185	0.185	0.185	0.184	0.100	0.2
	0.185	0.185	0.185	0.185	0.185	0.185	0.185	0.185	0.184	0.182	0.099	0.3
	0.185	0.185	0.185	0.185	0.185	0.185	0.185	0.184	0.184	0.180	0.098	0.4
	0.185	0.185	0.185	0.185	0.185	0.185	0.185	0.184	0.183	0.177	0.097	0.5
	0.185	0.185	0.185	0.185	0.185	0.185	0.184	0.184	0.182	0.174	0.096	0.6
	0.185	0.185	0.185	0.185	0.185	0.185	0.184	0.183	0.180	0.170	0.095	0.7
	0.185	0.185	0.185	0.185	0.185	0.185	0.184	0.182	0.179	0.167	0.094	0.8
	0.185	0.185	0.185	0.185	0.185	0.184	0.183	0.182	0.177	0.163	0.092	0.9
	0.185	0.185	0.185	0.185	0.184	0.184	0.183	0.181	0.174	0.159	0.091	1.0
10	0.096	0.096	0.096	0.096	0.096	0.096	0.096	0.096	0.096	0.096	0.096	0.0
	0.096	0.096	0.096	0.096	0.096	0.096	0.096	0.096	0.096	0.096	0.096	0.1
	0.096	0.096	0.096	0.096	0.096	0.096	0.096	0.096	0.096	0.096	0.096	0.2
	0.096	0.096	0.096	0.096	0.096	0.096	0.096	0.096	0.096	0.096	0.096	0.3
	0.096	0.096	0.096	0.096	0.096	0.096	0.096	0.096	0.096	0.096	0.095	0.4
	0.096	0.096	0.096	0.096	0.096	0.096	0.096	0.096	0.096	0.096	0.094	0.5
	0.096	0.096	0.096	0.096	0.096	0.096	0.096	0.096	0.096	0.096	0.093	0.6
	0.096	0.096	0.096	0.096	0.096	0.096	0.096	0.096	0.096	0.096	0.092	0.7
	0.096	0.096	0.096	0.096	0.096	0.096	0.096	0.096	0.096	0.096	0.091	0.8
	0.096	0.096	0.096	0.096	0.096	0.096	0.096	0.096	0.096	0.096	0.090	0.9
	0.096	0.096	0.096	0.096	0.096	0.096	0.096	0.096	0.096	0.096	0.088	1.0

Mode coupling parameter κ 303

Table C8 Mode coupling parameter κ for $\tau_X = 0.7$

r_2	$r_1=0$	$r_1=0.2$	$r_1=0.4$	$r_1=0.6$	$r_1=0.8$	$r_1=1$	$r_1=2$	$r_1=3$	$r_1=4$	$r_1=5$	$r_1=10$	τ_Y
0	1.000	1.000	1.000	1.000	1.000	1.000	0.500	0.333	0.250	0.200	0.100	0.0
	1.000	0.998	0.993	0.986	0.967	0.909	0.495	0.332	0.249	0.200	0.100	0.1
	1.000	0.990	0.975	0.950	0.905	0.833	0.482	0.327	0.247	0.198	0.100	0.2
	1.000	0.979	0.948	0.904	0.843	0.769	0.464	0.320	0.243	0.196	0.099	0.3
	1.000	0.963	0.915	0.856	0.787	0.714	0.444	0.311	0.238	0.193	0.098	0.4
	1.000	0.945	0.880	0.809	0.736	0.667	0.423	0.301	0.232	0.189	0.097	0.5
	1.000	0.925	0.845	0.766	0.692	0.625	0.402	0.291	0.226	0.185	0.096	0.6
	1.000	0.903	0.810	0.726	0.652	0.588	0.383	0.280	0.220	0.181	0.095	0.7
	1.000	0.880	0.776	0.689	0.616	0.556	0.365	0.270	0.213	0.176	0.094	0.8
	1.000	0.857	0.744	0.656	0.584	0.526	0.349	0.260	0.207	0.171	0.092	0.9
	1.000	0.833	0.714	0.625	0.556	0.500	0.333	0.250	0.200	0.167	0.091	1.0
0.2	0.903	0.903	0.903	0.903	0.903	0.903	0.500	0.333	0.250	0.200	0.100	0.0
	0.903	0.901	0.899	0.894	0.885	0.860	0.495	0.332	0.249	0.200	0.100	0.1
	0.903	0.896	0.887	0.872	0.845	0.799	0.481	0.327	0.247	0.198	0.100	0.2
	0.903	0.889	0.869	0.840	0.799	0.743	0.463	0.320	0.243	0.196	0.099	0.3
	0.903	0.878	0.846	0.804	0.753	0.694	0.442	0.311	0.238	0.193	0.098	0.4
	0.903	0.866	0.821	0.768	0.710	0.650	0.421	0.301	0.232	0.189	0.097	0.5
	0.903	0.852	0.794	0.732	0.670	0.612	0.401	0.290	0.226	0.185	0.096	0.6
	0.903	0.836	0.766	0.698	0.634	0.577	0.382	0.280	0.220	0.181	0.095	0.7
	0.903	0.819	0.739	0.666	0.602	0.546	0.364	0.269	0.213	0.176	0.094	0.8
	0.903	0.801	0.712	0.636	0.572	0.519	0.347	0.259	0.206	0.171	0.092	0.9
	0.903	0.783	0.687	0.609	0.545	0.493	0.332	0.250	0.200	0.167	0.091	1.0
0.4	0.810	0.810	0.810	0.810	0.810	0.810	0.500	0.333	0.250	0.200	0.100	0.0
	0.810	0.809	0.807	0.805	0.801	0.791	0.495	0.332	0.249	0.200	0.100	0.1
	0.810	0.806	0.800	0.792	0.777	0.751	0.480	0.327	0.247	0.198	0.100	0.2
	0.810	0.801	0.789	0.772	0.746	0.708	0.461	0.319	0.243	0.196	0.099	0.3
	0.810	0.795	0.774	0.747	0.711	0.667	0.440	0.310	0.238	0.193	0.098	0.4
	0.810	0.786	0.757	0.720	0.677	0.629	0.419	0.300	0.232	0.189	0.097	0.5
	0.810	0.777	0.738	0.693	0.644	0.595	0.399	0.290	0.226	0.185	0.096	0.6
	0.810	0.766	0.717	0.665	0.613	0.563	0.380	0.279	0.219	0.180	0.095	0.7
	0.810	0.754	0.696	0.639	0.584	0.535	0.362	0.269	0.213	0.176	0.094	0.8
	0.810	0.742	0.675	0.613	0.558	0.509	0.346	0.259	0.206	0.171	0.092	0.9
	0.810	0.729	0.654	0.589	0.533	0.485	0.331	0.249	0.200	0.167	0.091	1.0
0.6	0.726	0.726	0.726	0.726	0.726	0.726	0.500	0.333	0.250	0.200	0.100	0.0
	0.726	0.725	0.724	0.723	0.791	0.717	0.494	0.332	0.249	0.200	0.100	0.1
	0.726	0.723	0.720	0.715	0.707	0.694	0.479	0.327	0.247	0.198	0.100	0.2
	0.726	0.720	0.713	0.703	0.688	0.665	0.459	0.319	0.243	0.196	0.099	0.3
	0.726	0.716	0.704	0.687	0.664	0.634	0.437	0.310	0.238	0.193	0.098	0.4
	0.726	0.711	0.693	0.669	0.639	0.603	0.416	0.300	0.232	0.189	0.097	0.5
	0.726	0.705	0.680	0.649	0.613	0.574	0.396	0.289	0.226	0.185	0.096	0.6
	0.726	0.698	0.665	0.628	0.588	0.547	0.377	0.279	0.219	0.180	0.095	0.7
	0.726	0.690	0.650	0.607	0.563	0.521	0.360	0.268	0.213	0.176	0.094	0.8
	0.726	0.682	0.634	0.586	0.540	0.497	0.344	0.258	0.206	0.171	0.092	0.9
	0.726	0.673	0.618	0.566	0.518	0.476	0.329	0.249	0.200	0.166	0.091	1.0

304 Appendix C

Table C8 Continued. Mode coupling parameter κ for $\tau_X = 0.7$

r_2	$r_1=0$	$r_1=0.2$	$r_1=0.4$	$r_1=0.6$	$r_1=0.8$	$r_1=1$	$r_1=2$	$r_1=3$	$r_1=4$	$r_1=5$	$r_1=10$	τ_Y
	0.652	0.652	0.652	0.652	0.652	0.652	0.500	0.333	0.250	0.200	0.100	0.0
	0.652	0.652	0.651	0.650	0.649	0.647	0.493	0.332	0.249	0.200	0.100	0.1
	0.652	0.650	0.648	0.646	0.642	0.635	0.476	0.326	0.247	0.198	0.100	0.2
	0.652	0.649	0.644	0.638	0.630	0.617	0.455	0.319	0.243	0.196	0.099	0.3
	0.652	0.646	0.638	0.628	0.614	0.595	0.434	0.309	0.238	0.193	0.098	0.4
0.8	0.652	0.643	0.631	0.616	0.597	0.572	0.413	0.299	0.232	0.189	0.097	0.5
	0.652	0.639	0.623	0.603	0.578	0.549	0.393	0.289	0.226	0.185	0.096	0.6
	0.652	0.634	0.613	0.588	0.558	0.526	0.374	0.278	0.219	0.180	0.095	0.7
	0.652	0.629	0.603	0.572	0.539	0.504	0.357	0.268	0.212	0.176	0.094	0.8
	0.652	0.624	0.591	0.556	0.520	0.484	0.342	0.258	0.206	0.171	0.092	0.9
	0.652	0.618	0.580	0.540	0.501	0.464	0.327	0.248	0.199	0.166	0.091	1.0
	0.588	0.588	0.588	0.588	0.588	0.588	0.500	0.333	0.250	0.200	0.100	0.0
	0.588	0.588	0.588	0.587	0.587	0.586	0.491	0.332	0.249	0.200	0.100	0.1
	0.588	0.587	0.586	0.585	0.582	0.579	0.472	0.326	0.247	0.198	0.100	0.2
	0.588	0.586	0.584	0.580	0.575	0.568	0.451	0.318	0.243	0.196	0.099	0.3
	0.588	0.585	0.580	0.574	0.565	0.554	0.429	0.309	0.238	0.193	0.098	0.4
1	0.588	0.583	0.575	0.566	0.554	0.538	0.408	0.299	0.232	0.189	0.097	0.5
	0.588	0.580	0.570	0.557	0.540	0.520	0.389	0.288	0.225	0.185	0.096	0.6
	0.588	0.577	0.563	0.547	0.526	0.503	0.371	0.277	0.219	0.180	0.095	0.7
	0.588	0.574	0.556	0.535	0.511	0.485	0.354	0.267	0.212	0.176	0.094	0.8
	0.588	0.570	0.549	0.524	0.496	0.467	0.339	0.257	0.206	0.171	0.092	0.9
	0.588	0.566	0.540	0.511	0.481	0.450	0.324	0.247	0.199	0.166	0.091	1.0
	0.467	0.467	0.467	0.467	0.467	0.467	0.467	0.333	0.250	0.200	0.100	0.0
	0.467	0.467	0.466	0.466	0.466	0.466	0.458	0.331	0.249	0.200	0.100	0.1
	0.467	0.466	0.466	0.465	0.465	0.464	0.442	0.325	0.246	0.198	0.100	0.2
	0.467	0.466	0.465	0.464	0.462	0.461	0.424	0.317	0.242	0.196	0.099	0.3
	0.467	0.465	0.464	0.462	0.459	0.456	0.406	0.307	0.237	0.192	0.098	0.4
1.5	0.467	0.465	0.462	0.459	0.455	0.450	0.389	0.296	0.231	0.189	0.097	0.5
	0.467	0.464	0.460	0.456	0.450	0.443	0.373	0.285	0.225	0.184	0.096	0.6
	0.467	0.463	0.458	0.452	0.444	0.435	0.358	0.275	0.218	0.180	0.095	0.7
	0.467	0.462	0.455	0.447	0.438	0.426	0.343	0.265	0.211	0.175	0.094	0.8
	0.467	0.460	0.452	0.443	0.431	0.417	0.329	0.255	0.205	0.171	0.092	0.9
	0.467	0.459	0.449	0.437	0.423	0.407	0.316	0.245	0.198	0.166	0.091	1.0
	0.383	0.383	0.383	0.383	0.383	0.383	0.383	0.333	0.250	0.200	0.100	0.0
	0.383	0.383	0.383	0.383	0.383	0.383	0.382	0.330	0.249	0.200	0.100	0.1
	0.383	0.383	0.383	0.383	0.383	0.382	0.378	0.323	0.246	0.198	0.100	0.2
	0.383	0.383	0.383	0.383	0.382	0.382	0.381	0.372	0.313	0.242	0.195	0.099	0.3
	0.383	0.383	0.382	0.381	0.380	0.379	0.364	0.302	0.236	0.192	0.098	0.4
2	0.383	0.382	0.381	0.380	0.379	0.377	0.355	0.292	0.230	0.188	0.097	0.5
	0.383	0.382	0.380	0.379	0.377	0.374	0.345	0.281	0.224	0.184	0.096	0.6
	0.383	0.382	0.380	0.377	0.374	0.371	0.335	0.271	0.217	0.180	0.095	0.7
	0.383	0.381	0.379	0.376	0.372	0.367	0.324	0.261	0.210	0.175	0.094	0.8
	0.383	0.381	0.377	0.373	0.369	0.363	0.314	0.251	0.204	0.170	0.092	0.9
	0.383	0.380	0.376	0.371	0.365	0.358	0.303	0.242	0.197	0.166	0.091	1.0

Mode coupling parameter κ 305

Table C8 Continued. Mode coupling parameter κ for $\tau_X = 0.7$

r_2	$r_1=0$	$r_1=0.2$	$r_1=0.4$	$r_1=0.6$	$r_1=0.8$	$r_1=1$	$r_1=2$	$r_1=3$	$r_1=4$	$r_1=5$	$r_1=10$	τ_Y
3	0.280	0.280	0.280	0.280	0.280	0.280	0.280	0.280	0.250	0.200	0.100	0.0
	0.280	0.280	0.280	0.280	0.280	0.280	0.280	0.279	0.248	0.199	0.100	0.1
	0.280	0.280	0.280	0.280	0.280	0.280	0.279	0.277	0.244	0.198	0.100	0.2
	0.280	0.280	0.280	0.280	0.280	0.280	0.278	0.274	0.239	0.195	0.099	0.3
	0.280	0.280	0.280	0.280	0.279	0.279	0.277	0.269	0.233	0.191	0.098	0.4
	0.280	0.280	0.280	0.279	0.279	0.279	0.275	0.264	0.226	0.187	0.097	0.5
	0.280	0.280	0.279	0.279	0.279	0.278	0.273	0.258	0.220	0.183	0.096	0.6
	0.280	0.280	0.279	0.279	0.278	0.277	0.271	0.251	0.213	0.178	0.095	0.7
	0.280	0.279	0.279	0.278	0.277	0.276	0.268	0.244	0.207	0.174	0.094	0.8
	0.280	0.279	0.279	0.278	0.277	0.275	0.264	0.237	0.200	0.169	0.092	0.9
	0.280	0.279	0.278	0.277	0.276	0.274	0.261	0.231	0.194	0.164	0.091	1.0
4	0.220	0.220	0.220	0.220	0.220	0.220	0.220	0.220	0.220	0.200	0.100	0.0
	0.220	0.220	0.220	0.220	0.220	0.220	0.220	0.220	0.219	0.199	0.100	0.1
	0.220	0.220	0.220	0.220	0.220	0.220	0.220	0.219	0.218	0.197	0.100	0.2
	0.220	0.220	0.220	0.220	0.220	0.220	0.219	0.219	0.216	0.193	0.099	0.3
	0.220	0.220	0.220	0.220	0.220	0.219	0.219	0.218	0.213	0.189	0.098	0.4
	0.220	0.220	0.220	0.219	0.219	0.219	0.218	0.216	0.209	0.185	0.097	0.5
	0.220	0.220	0.220	0.219	0.219	0.219	0.218	0.215	0.205	0.180	0.096	0.6
	0.220	0.220	0.219	0.219	0.219	0.219	0.217	0.213	0.201	0.176	0.095	0.7
	0.220	0.220	0.219	0.219	0.219	0.219	0.216	0.211	0.196	0.171	0.094	0.8
	0.220	0.219	0.219	0.219	0.219	0.218	0.215	0.208	0.191	0.167	0.092	0.9
	0.220	0.219	0.219	0.219	0.218	0.218	0.214	0.206	0.186	0.162	0.091	1.0
5	0.181	0.181	0.181	0.181	0.181	0.181	0.181	0.181	0.181	0.181	0.100	0.0
	0.181	0.181	0.181	0.181	0.181	0.181	0.181	0.181	0.180	0.180	0.100	0.1
	0.181	0.181	0.181	0.181	0.181	0.181	0.180	0.180	0.180	0.179	0.100	0.2
	0.181	0.181	0.181	0.181	0.181	0.180	0.180	0.180	0.180	0.178	0.099	0.3
	0.181	0.181	0.181	0.180	0.180	0.180	0.180	0.180	0.179	0.176	0.098	0.4
	0.181	0.181	0.180	0.180	0.180	0.180	0.180	0.180	0.178	0.173	0.097	0.5
	0.181	0.181	0.180	0.180	0.180	0.180	0.180	0.179	0.177	0.170	0.096	0.6
	0.181	0.181	0.180	0.180	0.180	0.180	0.180	0.178	0.176	0.167	0.095	0.7
	0.181	0.180	0.180	0.180	0.180	0.180	0.179	0.178	0.174	0.164	0.094	0.8
	0.181	0.180	0.180	0.180	0.180	0.180	0.179	0.177	0.172	0.160	0.092	0.9
	0.181	0.180	0.180	0.180	0.180	0.180	0.178	0.176	0.170	0.157	0.091	1.0
10	0.095	0.095	0.095	0.095	0.095	0.095	0.095	0.095	0.095	0.095	0.095	0.0
	0.095	0.095	0.095	0.095	0.095	0.095	0.095	0.095	0.095	0.095	0.095	0.1
	0.095	0.095	0.095	0.095	0.095	0.095	0.095	0.095	0.095	0.095	0.095	0.2
	0.095	0.095	0.095	0.095	0.095	0.095	0.095	0.095	0.095	0.095	0.094	0.3
	0.095	0.095	0.095	0.095	0.095	0.095	0.095	0.095	0.095	0.095	0.094	0.4
	0.095	0.095	0.095	0.095	0.095	0.095	0.095	0.095	0.095	0.095	0.093	0.5
	0.095	0.095	0.095	0.095	0.095	0.095	0.095	0.095	0.095	0.095	0.092	0.6
	0.095	0.095	0.095	0.095	0.095	0.095	0.095	0.095	0.095	0.095	0.091	0.7
	0.095	0.095	0.095	0.095	0.095	0.095	0.095	0.095	0.095	0.095	0.090	0.8
	0.095	0.095	0.095	0.095	0.095	0.095	0.095	0.095	0.095	0.095	0.089	0.9
	0.095	0.095	0.095	0.095	0.095	0.095	0.095	0.095	0.095	0.095	0.088	1.0

306 Appendix C

Table C9 Mode coupling parameter κ for $\tau_X = 0.8$

r_2	$r_1=0$	$r_1=0.2$	$r_1=0.4$	$r_1=0.6$	$r_1=0.8$	$r_1=1$	$r_1=2$	$r_1=3$	$r_1=4$	$r_1=5$	$r_1=10$	τ_Y
0	1.000	1.000	1.000	1.000	1.000	1.000	0.500	0.333	0.250	0.200	0.100	0.0
	1.000	0.998	0.993	0.986	0.967	0.909	0.495	0.332	0.249	0.200	0.100	0.1
	1.000	0.990	0.975	0.950	0.905	0.833	0.482	0.327	0.247	0.198	0.100	0.2
	1.000	0.979	0.948	0.904	0.843	0.769	0.464	0.320	0.243	0.196	0.099	0.3
	1.000	0.963	0.915	0.856	0.787	0.714	0.444	0.311	0.238	0.193	0.098	0.4
	1.000	0.945	0.880	0.809	0.736	0.667	0.423	0.301	0.232	0.189	0.097	0.5
	1.000	0.925	0.845	0.766	0.692	0.625	0.402	0.291	0.226	0.185	0.096	0.6
	1.000	0.903	0.810	0.726	0.652	0.588	0.383	0.280	0.220	0.181	0.095	0.7
	1.000	0.880	0.776	0.689	0.616	0.556	0.365	0.270	0.213	0.176	0.094	0.8
	1.000	0.857	0.744	0.656	0.584	0.526	0.349	0.260	0.207	0.171	0.092	0.9
	1.000	0.833	0.714	0.625	0.556	0.500	0.333	0.250	0.200	0.167	0.091	1.0
0.2	0.880	0.880	0.880	0.880	0.880	0.880	0.500	0.333	0.250	0.200	0.100	0.0
	0.880	0.878	0.876	0.872	0.865	0.844	0.495	0.332	0.249	0.200	0.100	0.1
	0.880	0.874	0.866	0.852	0.829	0.788	0.481	0.327	0.247	0.198	0.100	0.2
	0.880	0.867	0.849	0.824	0.786	0.735	0.462	0.320	0.243	0.196	0.099	0.3
	0.880	0.858	0.830	0.791	0.743	0.688	0.441	0.311	0.238	0.193	0.098	0.4
	0.880	0.847	0.805	0.757	0.702	0.645	0.420	0.301	0.232	0.189	0.097	0.5
	0.880	0.833	0.780	0.723	0.664	0.608	0.400	0.290	0.226	0.185	0.096	0.6
	0.880	0.819	0.754	0.690	0.629	0.574	0.381	0.279	0.220	0.180	0.095	0.7
	0.880	0.803	0.729	0.660	0.598	0.544	0.363	0.269	0.213	0.176	0.094	0.8
	0.880	0.787	0.703	0.631	0.569	0.516	0.347	0.259	0.206	0.171	0.092	0.9
	0.880	0.770	0.679	0.604	0.542	0.491	0.332	0.250	0.200	0.167	0.091	1.0
0.4	0.776	0.776	0.776	0.776	0.776	0.776	0.500	0.333	0.250	0.200	0.100	0.0
	0.776	0.775	0.774	0.772	0.769	0.762	0.494	0.332	0.249	0.200	0.100	0.1
	0.776	0.0773	0.768	0.761	0.750	0.729	0.480	0.327	0.247	0.198	0.100	0.2
	0.776	0.769	0.759	0.744	0.723	0.691	0.460	0.319	0.243	0.196	0.099	0.3
	0.776	0.763	0.746	0.723	0.693	0.654	0.439	0.310	0.238	0.193	0.098	0.4
	0.776	0.756	0.731	0.700	0.662	0.619	0.418	0.300	0.232	0.189	0.097	0.5
	0.776	0.748	0.714	0.675	0.631	0.586	0.397	0.289	0.226	0.185	0.096	0.6
	0.776	0.739	0.696	0.650	0.603	0.556	0.379	0.279	0.219	0.180	0.095	0.7
	0.776	0.729	0.678	0.626	0.576	0.529	0.361	0.269	0.213	0.176	0.094	0.8
	0.776	0.718	0.659	0.602	0.550	0.504	0.345	0.258	0.206	0.171	0.092	0.9
	0.776	0.706	0.640	0.579	0.527	0.481	0.330	0.249	0.200	0.167	0.091	1.0
0.6	0.689	0.689	0.689	0.689	0.689	0.689	0.500	0.333	0.250	0.200	0.100	0.0
	0.689	0.689	0.688	0.687	0.685	0.683	0.494	0.332	0.249	0.200	0.100	0.1
	0.689	0.687	0.684	0.681	0.675	0.665	0.477	0.327	0.247	0.198	0.100	0.2
	0.689	0.685	0.679	0.671	0.659	0.641	0.457	0.319	0.243	0.196	0.099	0.3
	0.689	0.681	0.671	0.657	0.639	0.614	0.435	0.310	0.238	0.193	0.098	0.4
	0.689	0.677	0.662	0.642	0.617	0.587	0.414	0.299	0.232	0.189	0.097	0.5
	0.689	0.672	0.651	0.625	0.595	0.561	0.394	0.289	0.226	0.185	0.096	0.6
	0.689	0.666	0.639	0.607	0.572	0.535	0.376	0.278	0.219	0.180	0.095	0.7
	0.689	0.660	0.626	0.589	0.550	0.512	0.358	0.268	0.212	0.176	0.094	0.8
	0.689	0.652	0.612	0.570	0.529	0.490	0.342	0.258	0.206	0.171	0.092	0.9
	0.689	0.645	0.598	0.552	0.509	0.469	0.327	0.248	0.199	0.166	0.091	1.0

Table C9 Continued. Mode coupling parameter κ for $\tau_X = 0.8$

r_2	$r_1=0$	$r_1=0.2$	$r_1=0.4$	$r_1=0.6$	$r_1=0.8$	$r_1=1$	$r_1=2$	$r_1=3$	$r_1=4$	$r_1=5$	$r_1=10$	τ_Y
	0.616	0.616	0.616	0.616	0.616	0.616	0.500	0.333	0.250	0.200	0.100	0.0
	0.616	0.616	0.616	0.615	0.614	0.613	0.492	0.332	0.249	0.200	0.100	0.1
	0.616	0.615	0.614	0.611	0.608	0.603	0.474	0.326	0.247	0.198	0.100	0.2
	0.616	0.614	0.610	0.605	0.599	0.589	0.452	0.315	0.243	0.196	0.099	0.3
	0.616	0.612	0.605	0.597	0.586	0.571	0.430	0.309	0.238	0.193	0.098	0.4
0.8	0.616	0.609	0.599	0.587	0.572	0.552	0.410	0.299	0.232	0.189	0.097	0.5
	0.616	0.606	0.592	0.576	0.556	0.531	0.390	0.288	0.225	0.185	0.096	0.6
	0.616	0.602	0.584	0.563	0.539	0.511	0.372	0.277	0.219	0.180	0.095	0.7
	0.616	0.598	0.576	0.550	0.522	0.492	0.355	0.267	0.212	0.176	0.094	0.8
	0.616	0.593	0.566	0.536	0.505	0.473	0.339	0.257	0.206	0.171	0.092	0.9
	0.616	0.588	0.556	0.522	0.488	0.455	0.325	0.248	0.199	0.166	0.091	1.0
	0.556	0.556	0.556	0.556	0.556	0.556	0.500	0.333	0.250	0.200	0.100	0.0
	0.556	0.555	0.555	0.555	0.554	0.554	0.489	0.331	0.249	0.200	0.100	0.1
	0.556	0.555	0.554	0.553	0.551	0.548	0.468	0.326	0.247	0.198	0.100	0.2
	0.556	0.554	0.552	0.549	0.545	0.539	0.445	0.318	0.243	0.196	0.099	0.3
	0.556	0.553	0.549	0.544	0.537	0.528	0.424	0.308	0.237	0.192	0.098	0.4
1	0.556	0.551	0.545	0.537	0.527	0.515	0.404	0.298	0.232	0.189	0.097	0.5
	0.556	0.549	0.540	0.530	0.516	0.500	0.385	0.287	0.225	0.185	0.096	0.6
	0.556	0.546	0.535	0.521	0.504	0.485	0.367	0.276	0.219	0.180	0.095	0.7
	0.556	0.544	0.529	0.512	0.492	0.469	0.351	0.266	0.212	0.175	0.094	0.8
	0.556	0.540	0.522	0.502	0.479	0.454	0.336	0.256	0.205	0.171	0.092	0.9
	0.556	0.537	0.515	0.491	0.465	0.439	0.322	0.247	0.199	0.166	0.091	1.0
	0.442	0.442	0.442	0.442	0.442	0.442	0.500	0.333	0.250	0.200	0.100	0.0
	0.442	0.442	0.442	0.442	0.442	0.442	0.437	0.331	0.249	0.200	0.100	0.1
	0.442	0.442	0.442	0.441	0.441	0.440	0.425	0.325	0.246	0.198	0.100	0.2
	0.442	0.442	0.441	0.440	0.439	0.437	0.410	0.316	0.242	0.195	0.099	0.3
	0.442	0.441	0.440	0.438	0.436	0.433	0.395	0.305	0.237	0.192	0.098	0.4
1.5	0.442	0.441	0.438	0.436	0.432	0.428	0.380	0.294	0.231	0.188	0.097	0.5
	0.442	0.440	0.437	0.433	0.428	0.422	0.365	0.284	0.224	0.184	0.096	0.6
	0.442	0.439	0.435	0.429	0.423	0.416	0.350	0.273	0.218	0.180	0.095	0.7
	0.442	0.438	0.432	0.426	0.418	0.408	0.337	0.263	0.211	0.175	0.094	0.8
	0.442	0.437	0.430	0.421	0.412	0.400	0.324	0.253	0.204	0.170	0.092	0.9
	0.442	0.435	0.427	0.417	0.405	0.392	0.311	0.244	0.198	0.166	0.091	1.0
	0.365	0.365	0.365	0.365	0.365	0.365	0.365	0.333	0.250	0.200	0.100	0.0
	0.365	0.365	0.365	0.365	0.365	0.365	0.364	0.329	0.249	0.199	0.100	0.1
	0.365	0.365	0.365	0.365	0.365	0.365	0.361	0.320	0.246	0.198	0.100	0.2
	0.365	0.365	0.365	0.364	0.364	0.363	0.356	0.310	0.242	0.195	0.099	0.3
	0.365	0.365	0.364	0.364	0.363	0.362	0.350	0.299	0.236	0.192	0.098	0.4
2	0.365	0.365	0.364	0.363	0.361	0.360	0.342	0.288	0.230	0.188	0.097	0.5
	0.365	0.364	0.363	0.361	0.359	0.357	0.333	0.278	0.223	0.184	0.096	0.6
	0.365	0.364	0.362	0.360	0.357	0.354	0.324	0.268	0.216	0.179	0.095	0.7
	0.365	0.363	0.361	0.358	0.355	0.351	0.315	0.258	0.210	0.175	0.094	0.8
	0.365	0.363	0.360	0.356	0.352	0.347	0.305	0.249	0.203	0.170	0.092	0.9
	0.365	0.362	0.359	0.354	0.349	0.343	0.296	0.240	0.197	0.165	0.091	1.0

308 Appendix C

Table C9 Continued. Mode coupling parameter κ for $\tau_X = 0.8$

r_2	r_1=0	r_1=0.2	r_1=0.4	r_1=0.6	r_1=0.8	r_1=1	r_1=2	r_1=3	r_1=4	r_1=5	r_1=10	τ_Y
3	0.270	0.270	0.270	0.270	0.270	0.270	0.270	0.270	0.250	0.200	0.100	0.0
	0.270	0.270	0.270	0.270	0.270	0.270	0.270	0.269	0.248	0.199	0.100	0.1
	0.270	0.270	0.270	0.270	0.270	0.269	0.269	0.267	0.243	0.198	0.100	0.2
	0.270	0.270	0.270	0.269	0.269	0.269	0.268	0.264	0.237	0.195	0.099	0.3
	0.270	0.270	0.270	0.269	0.269	0.269	0.267	0.260	0.231	0.191	0.098	0.4
	0.270	0.269	0.269	0.269	0.269	0.268	0.265	0.255	0.224	0.187	0.097	0.5
	0.270	0.269	0.269	0.269	0.268	0.268	0.263	0.250	0.217	0.182	0.096	0.6
	0.270	0.269	0.269	0.268	0.268	0.267	0.261	0.244	0.211	0.178	0.095	0.7
	0.270	0.269	0.269	0.268	0.267	0.266	0.258	0.238	0.205	0.173	0.094	0.8
	0.270	0.269	0.269	0.267	0.266	0.265	0.255	0.232	0.198	0.168	0.092	0.9
	0.270	0.269	0.269	0.267	0.265	0.264	0.251	0.226	0.192	0.164	0.091	1.0
4	0.213	0.213	0.213	0.213	0.213	0.213	0.213	0.213	0.213	0.200	0.100	0.0
	0.213	0.213	0.213	0.213	0.213	0.213	0.213	0.213	0.213	0.199	0.100	0.1
	0.213	0.213	0.213	0.213	0.213	0.213	0.213	0.213	0.211	0.196	0.100	0.2
	0.213	0.213	0.213	0.213	0.213	0.213	0.213	0.212	0.210	0.192	0.099	0.3
	0.213	0.213	0.213	0.213	0.213	0.213	0.212	0.211	0.207	0.188	0.098	0.4
	0.213	0.213	0.213	0.213	0.213	0.213	0.212	0.210	0.204	0.183	0.097	0.5
	0.213	0.213	0.213	0.213	0.213	0.212	0.211	0.208	0.200	0.179	0.096	0.6
	0.213	0.213	0.213	0.213	0.212	0.212	0.210	0.207	0.196	0.174	0.095	0.7
	0.213	0.213	0.213	0.212	0.212	0.212	0.210	0.205	0.192	0.170	0.094	0.8
	0.213	0.213	0.213	0.212	0.212	0.212	0.209	0.202	0.187	0.165	0.092	0.9
	0.213	0.213	0.212	0.212	0.212	0.211	0.207	0.199	0.183	0.161	0.091	1.0
5	0.176	0.176	0.176	0.176	0.176	0.176	0.176	0.176	0.176	0.176	0.100	0.0
	0.176	0.176	0.176	0.176	0.176	0.176	0.176	0.176	0.176	0.176	0.100	0.1
	0.176	0.176	0.176	0.176	0.176	0.176	0.176	0.176	0.176	0.175	0.100	0.2
	0.176	0.176	0.176	0.176	0.176	0.176	0.176	0.176	0.175	0.173	0.099	0.3
	0.176	0.176	0.176	0.176	0.176	0.176	0.176	0.175	0.174	0.172	0.098	0.4
	0.176	0.176	0.176	0.176	0.176	0.176	0.175	0.175	0.174	0.169	0.097	0.5
	0.176	0.176	0.176	0.176	0.176	0.176	0.175	0.174	0.172	0.167	0.096	0.6
	0.176	0.176	0.176	0.176	0.176	0.176	0.175	0.174	0.171	0.164	0.095	0.7
	0.176	0.176	0.176	0.176	0.176	0.175	0.175	0.173	0.170	0.161	0.094	0.8
	0.176	0.176	0.176	0.176	0.175	0.175	0.174	0.172	0.168	0.157	0.092	0.9
	0.176	0.176	0.176	0.176	0.175	0.175	0.174	0.171	0.166	0.154	0.091	1.0
10	0.094	0.094	0.094	0.094	0.094	0.094	0.094	0.094	0.094	0.094	0.094	0.0
	0.094	0.094	0.094	0.094	0.094	0.094	0.094	0.094	0.094	0.094	0.094	0.1
	0.094	0.094	0.094	0.094	0.094	0.094	0.094	0.094	0.094	0.094	0.093	0.2
	0.094	0.094	0.094	0.094	0.094	0.094	0.094	0.094	0.094	0.094	0.093	0.3
	0.094	0.094	0.094	0.094	0.094	0.094	0.094	0.094	0.094	0.094	0.093	0.4
	0.094	0.094	0.094	0.094	0.094	0.094	0.094	0.094	0.094	0.094	0.092	0.5
	0.094	0.094	0.094	0.094	0.094	0.094	0.094	0.094	0.094	0.094	0.091	0.6
	0.094	0.094	0.094	0.094	0.094	0.094	0.094	0.094	0.094	0.094	0.090	0.7
	0.094	0.094	0.094	0.094	0.094	0.094	0.094	0.094	0.094	0.094	0.089	0.8
	0.094	0.094	0.094	0.094	0.094	0.094	0.094	0.094	0.094	0.093	0.088	0.9
	0.094	0.094	0.094	0.094	0.094	0.094	0.094	0.094	0.093	0.093	0.087	1.0

Table C10 Mode coupling parameter κ for $\tau_X = 0.9$

r_2	$r_1=0$	$r_1=0.2$	$r_1=0.4$	$r_1=0.6$	$r_1=0.8$	$r_1=1$	$r_1=2$	$r_1=3$	$r_1=4$	$r_1=5$	$r_1=10$	τ_Y
0	1.000	1.000	1.000	1.000	1.000	1.000	0.500	0.333	0.250	0.200	0.100	0.0
	1.000	0.998	0.993	0.986	0.967	0.909	0.495	0.332	0.249	0.200	0.100	0.1
	1.000	0.990	0.975	0.950	0.905	0.833	0.482	0.327	0.247	0.198	0.100	0.2
	1.000	0.979	0.948	0.904	0.843	0.769	0.464	0.320	0.243	0.196	0.099	0.3
	1.000	0.963	0.915	0.856	0.787	0.714	0.444	0.311	0.238	0.193	0.098	0.4
	1.000	0.945	0.880	0.809	0.736	0.667	0.423	0.301	0.232	0.189	0.097	0.5
	1.000	0.925	0.845	0.766	0.692	0.625	0.402	0.291	0.226	0.185	0.096	0.6
	1.000	0.903	0.810	0.726	0.652	0.588	0.383	0.280	0.220	0.181	0.095	0.7
	1.000	0.880	0.776	0.689	0.616	0.556	0.365	0.270	0.213	0.176	0.094	0.8
	1.000	0.857	0.744	0.656	0.584	0.526	0.349	0.260	0.207	0.171	0.092	0.9
	1.000	0.833	0.714	0.625	0.556	0.500	0.333	0.250	0.200	0.167	0.091	1.0
0.2	0.857	0.857	0.857	0.857	0.857	0.857	0.500	0.333	0.250	0.200	0.100	0.0
	0.857	0.855	0.853	0.850	0.844	0.827	0.495	0.332	0.249	0.200	0.100	0.1
	0.857	0.852	0.844	0.832	0.812	0.777	0.481	0.327	0.247	0.198	0.100	0.2
	0.857	0.845	0.829	0.806	0.773	0.726	0.462	0.320	0.243	0.196	0.099	0.3
	0.857	0.837	0.811	0.776	0.732	0.681	0.441	0.310	0.238	0.193	0.098	0.4
	0.857	0.827	0.789	0.744	0.694	0.640	0.420	0.300	0.232	0.189	0.097	0.5
	0.857	0.815	0.766	0.713	0.657	0.603	0.400	0.290	0.226	0.185	0.096	0.6
	0.857	0.801	0.742	0.682	0.624	0.570	0.381	0.279	0.219	0.180	0.095	0.7
	0.857	0.787	0.718	0.652	0.593	0.540	0.363	0.269	0.213	0.176	0.094	0.8
	0.857	0.772	0.694	0.625	0.565	0.514	0.346	0.259	0.206	0.171	0.092	0.9
	0.857	0.756	0.670	0.599	0.539	0.489	0.331	0.249	0.200	0.167	0.091	1.0
0.4	0.744	0.744	0.744	0.744	0.744	0.744	0.500	0.333	0.250	0.200	0.100	0.0
	0.744	0.744	0.743	0.741	0.739	0.733	0.494	0.332	0.249	0.200	0.100	0.1
	0.744	0.742	0.738	0.732	0.723	0.707	0.479	0.327	0.247	0.198	0.100	0.2
	0.744	0.738	0.730	0.717	0.700	0.674	0.459	0.319	0.243	0.196	0.099	0.3
	0.744	0.733	0.719	0.699	0.673	0.640	0.437	0.310	0.238	0.193	0.098	0.4
	0.744	0.727	0.706	0.679	0.646	0.607	0.416	0.300	0.232	0.189	0.097	0.5
	0.744	0.720	0.691	0.657	0.618	0.577	0.396	0.289	0.226	0.185	0.096	0.6
	0.744	0.712	0.675	0.634	0.591	0.549	0.377	0.279	0.219	0.180	0.095	0.7
	0.744	0.703	0.659	0.612	0.566	0.522	0.360	0.268	0.213	0.176	0.094	0.8
	0.744	0.694	0.641	0.590	0.542	0.498	0.344	0.258	0.206	0.171	0.092	0.9
	0.744	0.684	0.624	0.569	0.520	0.476	0.329	0.249	0.199	0.166	0.091	1.0
0.6	0.656	0.656	0.656	0.656	0.656	0.656	0.500	0.333	0.250	0.200	0.100	0.0
	0.656	0.655	0.655	0.654	0.653	0.651	0.493	0.332	0.249	0.200	0.100	0.1
	0.656	0.654	0.652	0.649	0.644	0.637	0.476	0.326	0.247	0.198	0.100	0.2
	0.656	0.652	0.647	0.640	0.631	0.617	0.454	0.319	0.243	0.196	0.099	0.3
	0.656	0.649	0.641	0.630	0.614	0.594	0.433	0.309	0.238	0.193	0.098	0.4
	0.656	0.646	0.633	0.616	0.596	0.570	0.412	0.299	0.232	0.189	0.097	0.5
	0.656	0.641	0.624	0.602	0.576	0.547	0.392	0.288	0.226	0.185	0.096	0.6
	0.656	0.636	0.613	0.586	0.556	0.524	0.373	0.278	0.219	0.180	0.095	0.7
	0.656	0.631	0.602	0.570	0.536	0.502	0.356	0.267	0.212	0.176	0.094	0.8
	0.656	0.625	0.590	0.554	0.517	0.481	0.341	0.257	0.206	0.171	0.092	0.9
	0.656	0.618	0.578	0.537	0.498	0.462	0.326	0.248	0.199	0.166	0.091	1.0

310 Appendix C

Table C10 Continued. Mode coupling parameter κ for $\tau_X = 0.9$

r_2	r_1=0	r_1=0.2	r_1=0.4	r_1=0.6	r_1=0.8	r_1=1	r_1=2	r_1=3	r_1=4	r_1=5	r_1=10	τ_Y
0.8	0.584	0.584	0.584	0.584	0.584	0.584	0.500	0.333	0.250	0.200	0.100	0.0
	0.584	0.584	0.584	0.583	0.583	0.582	0.491	0.331	0.249	0.200	0.100	0.1
	0.584	0.583	0.582	0.580	0.578	0.574	0.470	0.326	0.247	0.198	0.100	0.2
	0.584	0.582	0.579	0.575	0.570	0.562	0.448	0.318	0.243	0.196	0.099	0.3
	0.584	0.580	0.575	0.569	0.560	0.548	0.426	0.308	0.237	0.192	0.098	0.4
	0.584	0.578	0.570	0.560	0.548	0.531	0.406	0.298	0.232	0.189	0.097	0.5
	0.584	0.575	0.564	0.551	0.534	0.514	0.386	0.287	0.225	0.185	0.096	0.6
	0.584	0.572	0.558	0.540	0.520	0.496	0.369	0.277	0.219	0.180	0.095	0.7
	0.584	0.569	0.550	0.529	0.505	0.479	0.352	0.266	0.212	0.175	0.094	0.8
	0.584	0.565	0.542	0.517	0.490	0.461	0.337	0.256	0.205	0.171	0.092	0.9
	0.584	0.560	0.534	0.505	0.475	0.445	0.323	0.247	0.199	0.166	0.091	1.0
1	0.526	0.526	0.526	0.526	0.526	0.526	0.500	0.333	0.250	0.200	0.100	0.0
	0.526	0.526	0.526	0.526	0.525	0.525	0.484	0.331	0.249	0.200	0.100	0.1
	0.526	0.526	0.525	0.524	0.522	0.520	0.461	0.326	0.247	0.198	0.100	0.2
	0.526	0.525	0.523	0.521	0.518	0.513	0.439	0.317	0.242	0.196	0.099	0.3
	0.526	0.524	0.521	0.516	0.511	0.504	0.418	0.307	0.237	0.192	0.098	0.4
	0.526	0.522	0.517	0.511	0.503	0.493	0.398	0.297	0.231	0.189	0.097	0.5
	0.526	0.521	0.513	0.505	0.494	0.480	0.380	0.286	0.225	0.184	0.096	0.6
	0.526	0.519	0.509	0.497	0.484	0.467	0.363	0.275	0.218	0.180	0.095	0.7
	0.526	0.516	0.504	0.490	0.473	0.454	0.347	0.265	0.212	0.175	0.094	0.8
	0.526	0.614	0.498	0.481	0.461	0.440	0.332	0.255	0.205	0.171	0.092	0.9
	0.526	0.511	0.492	0.472	0.450	0.426	0.319	0.246	0.198	0.166	0.091	1.0
1.5	0.420	0.420	0.420	0.420	0.420	0.420	0.420	0.333	0.250	0.200	0.100	0.0
	0.420	0.420	0.420	0.420	0.420	0.420	0.417	0.331	0.249	0.200	0.100	0.1
	0.420	0.420	0.420	0.419	0.419	0.418	0.408	0.324	0.246	0.198	0.100	0.2
	0.420	0.420	0.419	0.418	0.417	0.416	0.396	0.314	0.242	0.195	0.099	0.3
	0.420	0.419	0.418	0.416	0.415	0.412	0.383	0.303	0.236	0.192	0.098	0.4
	0.420	0.419	0.417	0.414	0.412	0.408	0.369	0.292	0.230	0.188	0.097	0.5
	0.420	0.418	0.415	0.412	0.408	0.403	0.356	0.282	0.224	0.184	0.096	0.6
	0.420	0.417	0.413	0.409	0.404	0.397	0.342	0.271	0.217	0.179	0.095	0.7
	0.420	0.416	0.411	0.406	0.399	0.391	0.330	0.261	0.210	0.175	0.094	0.8
	0.420	0.415	0.409	0.402	0.394	0.384	0.318	0.251	0.204	0.170	0.092	0.9
	0.420	0.414	0.407	0.398	0.388	0.377	0.306	0.242	0.197	0.165	0.091	1.0
2	0.349	0.349	0.349	0.349	0.349	0.349	0.349	0.333	0.250	0.200	0.100	0.0
	0.349	0.349	0.349	0.349	0.349	0.349	0.348	0.327	0.249	0.199	0.100	0.1
	0.349	0.349	0.349	0.348	0.348	0.348	0.345	0.317	0.246	0.198	0.100	0.2
	0.349	0.349	0.348	0.348	0.347	0.347	0.341	0.306	0.241	0.195	0.099	0.3
	0.349	0.348	0.348	0.347	0.346	0.345	0.336	0.295	0.235	0.192	0.098	0.4
	0.349	0.348	0.347	0.346	0.345	0.344	0.329	0.284	0.229	0.188	0.097	0.5
	0.349	0.348	0.347	0.345	0.343	0.341	0.322	0.274	0.222	0.183	0.096	0.6
	0.349	0.347	0.346	0.344	0.342	0.339	0.314	0.264	0.215	0.179	0.095	0.7
	0.349	0.347	0.345	0.342	0.339	0.336	0.305	0.255	0.209	0.174	0.094	0.8
	0.349	0.346	0.344	0.341	0.337	0.332	0.297	0.246	0.202	0.169	0.092	0.9
	0.349	0.346	0.343	0.339	0.334	0.329	0.288	0.238	0.196	0.165	0.091	1.0

Table C10 Continued. Mode coupling parameter κ for $\tau_X = 0.9$

r_2	$r_1=0$	$r_1=0.2$	$r_1=0.4$	$r_1=0.6$	$r_1=0.8$	$r_1=1$	$r_1=2$	$r_1=3$	$r_1=4$	$r_1=5$	$r_1=10$	τ_Y
3	0.260	0.260	0.260	0.260	0.260	0.260	0.260	0.260	0.250	0.200	0.100	0.0
	0.260	0.260	0.260	0.260	0.260	0.260	0.259	0.259	0.247	0.199	0.100	0.1
	0.260	0.260	0.260	0.260	0.259	0.259	0.259	0.258	0.241	0.197	0.100	0.2
	0.260	0.260	0.259	0.259	0.259	0.259	0.258	0.255	0.234	0.194	0.099	0.3
	0.260	0.260	0.259	0.259	0.259	0.259	0.257	0.251	0.228	0.190	0.098	0.4
	0.260	0.259	0.259	0.259	0.259	0.258	0.255	0.247	0.221	0.186	0.097	0.5
	0.260	0.259	0.259	0.259	0.258	0.258	0.254	0.243	0.215	0.182	0.096	0.6
	0.260	0.259	0.259	0.258	0.258	0.257	0.251	0.237	0.208	0.177	0.095	0.7
	0.260	0.259	0.258	0.258	0.257	0.256	0.249	0.232	0.202	0.172	0.094	0.8
	0.260	0.259	0.258	0.257	0.256	0.255	0.246	0.226	0.196	0.167	0.092	0.9
	0.260	0.259	0.258	0.257	0.256	0.254	0.243	0.221	0.190	0.163	0.091	1.0
4	0.207	0.207	0.207	0.207	0.207	0.207	0.207	0.207	0.207	0.200	0.100	0.0
	0.207	0.207	0.207	0.207	0.207	0.207	0.206	0.206	0.206	0.198	0.100	0.1
	0.207	0.207	0.207	0.206	0.206	0.206	0.206	0.206	0.205	0.194	0.100	0.2
	0.207	0.207	0.206	0.206	0.206	0.206	0.206	0.205	0.203	0.190	0.099	0.3
	0.207	0.206	0.206	0.206	0.206	0.206	0.206	0.205	0.201	0.186	0.098	0.4
	0.207	0.206	0.206	0.206	0.206	0.206	0.205	0.203	0.198	0.181	0.097	0.5
	0.207	0.206	0.206	0.206	0.206	0.206	0.205	0.202	0.195	0.177	0.096	0.6
	0.207	0.206	0.206	0.206	0.206	0.206	0.204	0.200	0.191	0.172	0.095	0.7
	0.207	0.206	0.206	0.206	0.206	0.205	0.203	0.198	0.187	0.168	0.094	0.8
	0.207	0.206	0.206	0.206	0.205	0.205	0.202	0.196	0.183	0.163	0.092	0.9
	0.207	0.206	0.206	0.205	0.205	0.204	0.201	0.194	0.179	0.159	0.091	1.0
5	0.171	0.171	0.171	0.171	0.171	0.171	0.171	0.171	0.171	0.171	0.100	0.0
	0.171	0.171	0.171	0.171	0.171	0.171	0.171	0.171	0.171	0.171	0.100	0.1
	0.171	0.171	0.171	0.171	0.171	0.171	0.171	0.171	0.171	0.170	0.100	0.2
	0.171	0.171	0.171	0.171	0.171	0.171	0.171	0.171	0.171	0.169	0.099	0.3
	0.171	0.171	0.171	0.171	0.171	0.171	0.171	0.171	0.170	0.167	0.098	0.4
	0.171	0.171	0.171	0.171	0.171	0.171	0.171	0.170	0.169	0.165	0.097	0.5
	0.171	0.171	0.171	0.171	0.171	0.171	0.171	0.170	0.168	0.163	0.096	0.6
	0.171	0.171	0.171	0.171	0.171	0.171	0.170	0.169	0.167	0.160	0.095	0.7
	0.171	0.171	0.171	0.171	0.171	0.171	0.170	0.168	0.165	0.157	0.093	0.8
	0.171	0.171	0.171	0.171	0.171	0.171	0.169	0.167	0.163	0.154	0.092	0.9
	0.171	0.171	0.171	0.171	0.171	0.170	0.169	0.166	0.161	0.151	0.090	1.0
10	0.092	0.092	0.092	0.092	0.092	0.092	0.092	0.092	0.092	0.092	0.092	0.0
	0.092	0.092	0.092	0.092	0.092	0.092	0.092	0.092	0.092	0.092	0.092	0.1
	0.092	0.092	0.092	0.092	0.092	0.092	0.092	0.092	0.092	0.092	0.092	0.2
	0.092	0.092	0.092	0.092	0.092	0.092	0.092	0.092	0.092	0.092	0.092	0.3
	0.092	0.092	0.092	0.092	0.092	0.092	0.092	0.092	0.092	0.092	0.091	0.4
	0.092	0.092	0.092	0.092	0.092	0.092	0.092	0.092	0.092	0.092	0.091	0.5
	0.092	0.092	0.092	0.092	0.092	0.092	0.092	0.092	0.092	0.092	0.090	0.6
	0.092	0.092	0.092	0.092	0.092	0.092	0.092	0.092	0.092	0.092	0.089	0.7
	0.092	0.092	0.092	0.092	0.092	0.092	0.092	0.092	0.092	0.092	0.088	0.8
	0.092	0.092	0.092	0.092	0.092	0.092	0.092	0.092	0.092	0.092	0.087	0.9
	0.092	0.092	0.092	0.092	0.092	0.092	0.092	0.092	0.092	0.092	0.086	1.0

312 Appendix C

Table C11 Mode coupling parameter κ for $\tau_X = 1.0$

r_2	$r_1=0$	$r_1=0.2$	$r_1=0.4$	$r_1=0.6$	$r_1=0.8$	$r_1=1$	$r_1=2$	$r_1=3$	$r_1=4$	$r_1=5$	$r_1=10$	τ_Y
0	1.000	1.000	1.000	1.000	1.000	1.000	0.500	0.333	0.250	0.200	0.100	0.0
	1.000	0.998	0.993	0.986	0.967	0.909	0.495	0.332	0.249	0.200	0.100	0.1
	1.000	0.990	0..975	0.950	0.905	0.833	0.482	0.327	0.247	0.198	0.100	0.2
	1.000	0.979	0.948	0.904	0.843	0.769	0.464	0.320	0.243	0.196	0.099	0.3
	1.000	0.963	0.915	0.856	0.787	0.714	0.444	0.311	0.238	0.193	0.098	0.4
	1.000	0.945	0.880	0.809	0.736	0.667	0.423	0.301	0.232	0.189	0.097	0.5
	1.000	0.925	0.845	0.766	0.692	0.625	0.402	0.291	0.226	0.185	0.096	0.6
	1.000	0.903	0.810	0.726	0.652	0.588	0.383	0.280	0.220	0.181	0.095	0.7
	1.000	0.880	0.776	0.689	0.616	0.556	0.365	0.270	0.213	0.176	0.094	0.8
	1.000	0.857	0.744	0.656	0.584	0.526	0.349	0.260	0.207	0.171	0.092	0.9
	1.000	0.833	0.714	0.625	0.556	0.500	0.333	0.250	0.200	0.167	0.091	1.0
0.2	0.833	0.833	0.833	0.833	0.833	0.833	0.500	0.333	0.250	0.200	0.100	0.0
	0.833	0.832	0.831	0.828	0.822	0.809	0.495	0.332	0.249	0.200	0.100	0.1
	0.833	0.829	0.822	0.812	0.795	0.764	0.480	0.327	0.247	0.198	0.100	0.2
	0.833	0.823	0.809	0.789	0.759	0.717	0.461	0.319	0.243	0.196	0.099	0.3
	0.833	0.816	0.792	0.761	0.721	0.673	0.440	0.310	0.238	0.193	0.098	0.4
	0.833	0.806	0.773	0.732	0.684	0.634	0.419	0.300	0.232	0.189	0.097	0.5
	0.833	0.795	0.751	0.702	0.650	0.598	0.399	0.290	0.226	0.185	0.096	0.6
	0.833	0.783	0.729	0.673	0.618	0.566	0.380	0.279	0.219	0.180	0.095	0.7
	0.833	0.770	0.706	0.645	0.588	0.537	0.362	0.269	0.213	0.176	0.094	0.8
	0.833	0.756	0.684	0.618	0.560	0.511	0.346	0.259	0.206	0.171	0.092	0.9
	0.833	0.742	0.661	0.593	0.535	0.487	0.331	0.249	0.200	0.167	0.091	1.0
0.4	0.714	0.714	0.714	0.714	0.714	0.714	0.500	0.333	0.250	0.200	0.100	0.0
	0.714	0.714	0.713	0.712	0.710	0.706	0.494	0.332	0.249	0.200	0.100	0.1
	0.714	0.712	0.709	0.704	0.696	0.684	0.478	0.327	0.247	0.198	0.100	0.2
	0.714	0.709	0.702	0.692	0.677	0.655	0.457	0.319	0.243	0.196	0.099	0.3
	0.714	0.705	0.692	0.676	0.654	0.625	0.436	0.310	0.238	0.193	0.098	0.4
	0.714	0.700	0.681	0.658	0.629	0.595	0.415	0.299	0.232	0.189	0.097	0.5
	0.714	0.694	0.668	0.638	0.604	0.567	0.395	0.289	0.226	0.185	0.096	0.6
	0.714	0.687	0.654	0.618	0.580	0.540	0.376	0.278	0.219	0.180	0.095	0.7
	0.714	0.679	0.640	0.598	0.556	0.515	0.359	0.268	0.212	0.176	0.094	0.8
	0.714	0.670	0.624	0.578	0.534	0.492	0.343	0.258	0.206	0.171	0.092	0.9
	0.714	0.661	0.609	0.558	0.512	0.471	0.328	0.248	0.199	0.166	0.091	1.0
0.6	0.625	0.625	0.625	0.625	0.625	0.625	0.500	0.333	0.250	0.200	0.100	0.0
	0.625	0.625	0.624	0.624	0.623	0.621	0.492	0.332	0.249	0.200	0.100	0.1
	0.625	0.624	0.622	0.619	0.616	0.610	0.474	0.326	0.247	0.198	0.100	0.2
	0.625	0.622	0.618	0.612	0.605	0.594	0.452	0.318	0.243	0.196	0.099	0.3
	0.625	0.620	0.612	0.603	0.591	0.574	0.430	0.309	0.238	0.192	0.098	0.4
	0.625	0.617	0.606	0.592	0.575	0.553	0.409	0.298	0.232	0.189	0.097	0.5
	0.625	0.613	0.598	0.580	0.558	0.532	0.389	0.288	0.225	0.185	0.096	0.6
	0.625	0.609	0.589	0.566	0.540	0.511	0.371	0.277	0.219	0.180	0.095	0.7
	0.625	0.604	0.579	0.552	0.522	0.491	0.354	0.267	0.212	0.176	0.094	0.8
	0.625	0.599	0.569	0.537	0.505	0.472	0.339	0.257	0.205	0.171	0.092	0.9
	0.625	0.593	0.558	0.523	0.487	0.454	0.324	0.247	0.199	0.166	0.091	1.0

Mode coupling parameter κ 313

Table C11 Continued. Mode coupling parameter κ for $\tau_X = 1.0$

r_2	$r_1=0$	$r_1=0.2$	$r_1=0.4$	$r_1=0.6$	$r_1=0.8$	$r_1=1$	$r_1=2$	$r_1=3$	$r_1=4$	$r_1=5$	$r_1=10$	τ_Y
	0.556	0.556	0.556	0.556	0.556	0.556	0.500	0.333	0.250	0.200	0.100	0.0
	0.556	0.555	0.555	0.555	0.554	0.553	0.488	0.331	0.249	0.200	0.100	0.1
	0.556	0.555	0.554	0.552	0.550	0.547	0.466	0.326	0.247	0.198	0.100	0.2
	0.556	0.554	0.551	0.548	0.544	0.538	0.443	0.318	0.242	0.196	0.099	0.3
	0.556	0.552	0.548	0.542	0.535	0.526	0.422	0.308	0.237	0.192	0.098	0.4
0.8	0.556	0.550	0.544	0.535	0.525	0.512	0.401	0.297	0.231	0.189	0.097	0.5
	0.556	0.548	0.539	0.527	0.513	0.497	0.383	0.286	0.225	0.184	0.096	0.6
	0.556	0.545	0.533	0.518	0.501	0.481	0.365	0.276	0.218	0.180	0.095	0.7
	0.556	0.542	0.527	0.509	0.488	0.465	0.349	0.265	0.212	0.175	0.094	0.8
	0.556	0.539	0.520	0.498	0.475	0.450	0.334	0.256	0.205	0.171	0.092	0.9
	0.556	0.535	0.512	0.487	0.461	0.435	0.320	0.246	0.199	0.166	0.091	1.0
	0.500	0.500	0.500	0.500	0.500	0.500	0.500	0.333	0.250	0.200	0.100	0.0
	0.500	0.500	0.500	0.500	0.499	0.499	0.476	0.331	0.249	0.200	0.100	0.1
	0.500	0.499	0.499	0.498	0.497	0.495	0.453	0.325	0.246	0.198	0.100	0.2
	0.500	0.499	0.497	0.495	0.493	0.489	0.431	0.317	0.242	0.196	0.099	0.3
	0.500	0.498	0.495	0.492	0.487	0.482	0.411	0.306	0.237	0.192	0.098	0.4
1	0.500	0.497	0.492	0.487	0.481	0.472	0.392	0.296	0.231	0.188	0.097	0.5
	0.500	0.495	0.489	0.482	0.473	0.462	0.374	0.285	0.225	0.184	0.096	0.6
	0.500	0.493	0.485	0.476	0.464	0.450	0.358	0.274	0.218	0.180	0.095	0.7
	0.500	0.491	0.481	0.469	0.455	0.439	0.343	0.264	0.211	0.175	0.094	0.8
	0.500	0.489	0.476	0.462	0.445	0.426	0.329	0.254	0.204	0.170	0.092	0.9
	0.500	0.487	0.471	0.454	0.435	0.414	0.316	0.245	0.198	0.166	0.091	1.0
	0.400	0.400	0.400	0.400	0.400	0.400	0.400	0.333	0.250	0.200	0.100	0.0
	0.400	0.400	0.400	0.400	0.400	0.400	0.398	0.330	0.249	0.199	0.100	0.1
	0.400	0.400	0.400	0.399	0.399	0.398	0.391	0.322	0.246	0.198	0.100	0.2
	0.400	0.400	0.399	0.398	0.397	0.396	0.381	0.312	0.242	0.195	0.099	0.3
	0.400	0.399	0.398	0.397	0.395	0.393	0.370	0.301	0.236	0.192	0.098	0.4
1.5	0.400	0.399	0.397	0.395	0.393	0.390	0.358	0.290	0.230	0.188	0.097	0.5
	0.400	0.398	0.396	0.393	0.389	0.385	0.346	0.279	0.223	0.184	0.096	0.6
	0.400	0.397	0.394	0.390	0.386	0.380	0.334	0.269	0.216	0.179	0.095	0.7
	0.400	0.397	0.392	0.387	0.382	0.375	0.323	0.259	0.210	0.175	0.094	0.8
	0.400	0.396	0.390	0.384	0.377	0.369	0.311	0.250	0.203	0.170	0.092	0.9
	0.400	0.395	0.388	0.381	0.372	0.362	0.301	0.241	0.197	0.165	0.091	1.0
	0.333	0.333	0.333	0.333	0.333	0.333	0.333	0.333	0.250	0.200	0.100	0.0
	0.333	0.333	0.333	0.333	0.333	0.333	0.333	0.322	0.249	0.199	0.100	0.1
	0.333	0.333	0.333	0.333	0.333	0.333	0.330	0.311	0.245	0.198	0.100	0.2
	0.333	0.333	0.333	0.333	0.332	0.332	0.327	0.301	0.240	0.195	0.099	0.3
	0.333	0.333	0.332	0.332	0.331	0.330	0.322	0.290	0.234	0.192	0.098	0.4
2	0.333	0.333	0.332	0.331	0.330	0.329	0.317	0.280	0.228	0.188	0.097	0.5
	0.333	0.332	0.331	0.330	0.329	0.327	0.310	0.270	0.221	0.183	0.096	0.6
	0.333	0.332	0.331	0.329	0.327	0.324	0.303	0.261	0.214	0.178	0.095	0.7
	0.333	0.332	0.330	0.327	0.325	0.322	0.296	0.251	0.207	0.174	0.094	0.8
	0.333	0.331	0.329	0.326	0.323	0.319	0.288	0.243	0.201	0.169	0.092	0.9
	0.333	0.331	0.328	0.324	0.320	0.315	0.281	0.235	0.195	0.164	0.091	1.0

314 Appendix C

Table C11 Continued. Mode coupling parameter κ for $\tau_X = 1.0$

r_2	$r_1=0$	$r_1=0.2$	$r_1=0.4$	$r_1=0.6$	$r_1=0.8$	$r_1=1$	$r_1=2$	$r_1=3$	$r_1=4$	$r_1=5$	$r_1=10$	τ_Y
3	0.250	0.250	0.250	0.250	0.250	0.250	0.250	0.250	0.250	0.200	0.100	0.0
	0.250	0.250	0.250	0.250	0.250	0.250	0.250	0.250	0.244	0.199	0.100	0.1
	0.250	0.250	0.250	0.250	0.250	0.250	0.249	0.248	0.237	0.197	0.100	0.2
	0.250	0.250	0.250	0.250	0.250	0.250	0.249	0.246	0.231	0.194	0.099	0.3
	0.250	0.250	0.250	0.250	0.249	0.249	0.248	0.243	0.224	0.190	0.098	0.4
	0.250	0.250	0.250	0.249	0.249	0.249	0.246	0.239	0.218	0.185	0.097	0.5
	0.250	0.250	0.249	0.249	0.249	0.248	0.244	0.235	0.212	0.181	0.096	0.6
	0.250	0.250	0.249	0.249	0.248	0.247	0.242	0.231	0.206	0.176	0.095	0.7
	0.250	0.250	0.249	0.248	0.248	0.247	0.240	0.226	0.199	0.171	0.094	0.8
	0.250	0.249	0.249	0.248	0.247	0.246	0.238	0.221	0.194	0.166	0.092	0.9
	0.250	0.249	0.248	0.247	0.246	0.245	0.235	0.215	0.188	0.162	0.091	1.0
4	0.200	0.200	0.200	0.200	0.200	0.200	0.200	0.200	0.200	0.200	0.100	0.0
	0.200	0.200	0.200	0.200	0.200	0.200	0.200	0.200	0.200	0.196	0.100	0.1
	0.200	0.200	0.200	0.200	0.200	0.200	0.200	0.200	0.199	0.192	0.100	0.2
	0.200	0.200	0.200	0.200	0.200	0.200	0.200	0.199	0.197	0.187	0.099	0.3
	0.200	0.200	0.200	0.200	0.200	0.200	0.199	0.198	0.195	0.183	0.098	0.4
	0.200	0.200	0.200	0.200	0.200	0.200	0.199	0.197	0.193	0.179	0.097	0.5
	0.200	0.200	0.200	0.200	0.199	0.199	0.198	0.196	0.190	0.174	0.096	0.6
	0.200	0.200	0.200	0.200	0.199	0.199	0.197	0.194	0.186	0.170	0.095	0.7
	0.200	0.200	0.200	0.199	0.199	0.199	0.197	0.192	0.183	0.166	0.093	0.8
	0.200	0.200	0.199	0.199	0.199	0.198	0.196	0.190	0.179	0.161	0.092	0.9
	0.200	0.200	0.199	0.199	0.199	0.198	0.195	0.188	0.175	0.157	0.091	1.0
5	0.167	0.167	0.167	0.167	0.167	0.167	0.167	0.167	0.167	0.167	0.100	0.0
	0.167	0.167	0.167	0.167	0.167	0.167	0.167	0.167	0.167	0.166	0.100	0.1
	0.167	0.167	0.167	0.167	0.167	0.167	0.167	0.167	0.166	0.166	0.100	0.2
	0.167	0.167	0.167	0.167	0.167	0.167	0.167	0.166	0.166	0.165	0.099	0.3
	0.167	0.167	0.167	0.167	0.167	0.167	0.166	0.166	0.165	0.163	0.098	0.4
	0.167	0.167	0.167	0.167	0.167	0.166	0.166	0.166	0.164	0.161	0.097	0.5
	0.167	0.167	0.167	0.167	0.167	0.166	0.166	0.166	0.163	0.159	0.096	0.6
	0.167	0.167	0.167	0.167	0.166	0.166	0.166	0.164	0.162	0.157	0.095	0.7
	0.167	0.167	0.167	0.166	0.166	0.166	0.165	0.164	0.161	0.154	0.093	0.8
	0.167	0.167	0.166	0.166	0.166	0.166	0.165	0.163	0.159	0.151	0.092	0.9
	0.167	0.167	0.166	0.166	0.166	0.166	0.164	0.162	0.157	0.148	0.090	1.0
10	0.091	0.091	0.091	0.091	0.091	0.091	0.091	0.091	0.091	0.091	0.091	0.0
	0.091	0.091	0.091	0.091	0.091	0.091	0.091	0.091	0.091	0.091	0.091	0.1
	0.091	0.091	0.091	0.091	0.091	0.091	0.091	0.091	0.091	0.091	0.091	0.2
	0.091	0.091	0.091	0.091	0.091	0.091	0.091	0.091	0.091	0.091	0.090	0.3
	0.091	0.091	0.091	0.091	0.091	0.091	0.091	0.091	0.091	0.091	0.090	0.4
	0.091	0.091	0.091	0.091	0.091	0.091	0.091	0.091	0.091	0.091	0.089	0.5
	0.091	0.091	0.091	0.091	0.091	0.091	0.091	0.091	0.091	0.091	0.088	0.6
	0.091	0.091	0.091	0.091	0.091	0.091	0.091	0.091	0.091	0.091	0.088	0.7
	0.091	0.091	0.091	0.091	0.091	0.091	0.091	0.091	0.091	0.091	0.087	0.8
	0.091	0.091	0.091	0.091	0.091	0.091	0.091	0.091	0.091	0.090	0.086	0.9
	0.091	0.091	0.091	0.091	0.091	0.091	0.091	0.091	0.091	0.090	0.084	1.0

References

Allen, H.G. (1969) *Analysis and design of structural sandwich panels.* Oxford

Appeltauer, J. and Kollár, L. (1999) Buckling of frames. In *Structural stability in engineering practice* (ed. L. Kollár), 129–186. E & FN Spon

Asztalos, Z. (1972) Buckling analysis of multistorey, one-bay frameworks using the continuum model, taking the axial deformations of the columns into account (in Hungarian). *Magyar Építőipar*, 471–474

Barbero, E. and Tomblin, J. (1993) Euler buckling of thin-walled composite columns. *Thin-walled Structures*, **17**, 237–258

Barkan, D.D. (1962) *Dynamics of bases and foundations.* McGraw-Hill, London

Beck, H. (1956) Ein neues Berechnungsverfahren für gegliederte Scheiben, dargestellt am Beispiel des Vierendeel-Trägers. *Der Bauingenieur*, **31**, 436–443

Beck, H., König, G. and Reeh, H. (1968) Kenngrößen zur Beurteilung der Torsionssteifigkeit von Hochhäusern. *Beton und Stahlbetonbau*, **63**, 268–277

Beck, H. and Schäfer, H.G. (1969) Die Berechnung von Hochhäusern durch Zusammenfassung aller aussteifenden Bauteile zu einen Balken. *Der Bauingenieur*, **44**, 80–87

Bishop, R.E.D. and Price, W.G. (1977) Coupled bending and twisting of a Timoshenko beam. *Journal of Sound and Vibration*, **50**, (4), 469–477

Boughton, J. (1994) Dynamic tests on a small-scale building model. Manuscript. *Building Research Establishment*, Watford

Brohn, D.M. and Cowan, J. (1977) Teaching towards an improved understanding of structural behaviour. *The Structural Engineer*, **55**, (1), 9–17

Brohn, D.M. (1992) A new paradigm for structural engineering. *The Structural Engineer*, **70**, (13), 239–242

Brohn, D.M. (1996) Avoiding CAD: The Computer Aided Disaster. Symposium: Safer computing. *The Institution of Structural Engineers*, 30 January 1996

BS5950 (1990) Structural use of steelwork in buildings. *British Standards Institution*, London

Capuani, D., Savoiva, M. and Laudiero, F. (1992) A generalization of the Timoshenko beam model for coupled vibration analysis of thin-walled beams. *Earthquake Engineering and Structural Dynamics*, **21**, 859–879

CEB (1983) *Seismic design of concrete structures*, Vol. 7, Code Manual. Vienna

Chwalla, E. (1959) Die neuen Hilfstafeln zur Berechnung von Spannungs-problemen der Theorie zweiter Ordnung und von Knickproblemen. *Bauingenieur*, **34**, (4, 6 and 8), p128, p240 and p299

Cook, N.J. (1985) *Designer's guide to wind loading of building structures. Part 1: Background, damage survey, wind data and structural classification.* Butterworth, London

Cook, N.J. (1990) *Designer's guide to wind loading of building structures. Part 2: Static structures.* Butterworth, London

Coull, A. and Mukherjee, P.R. (1978) Natural vibrations of shear wall buildings on flexible supports. *Earthquake Engineering and Structural Dynamics*, **6**, 295–315

Council on Tall Buildings and Urban Habitat (1978a) *Monograph on planning and design of tall buildings. Vol. CB: Structural design of tall concrete and masonry buildings.* American Society of Civil Engineers

Council on Tall Buildings and Urban Habitat (1978b) *Monograph on planning and design of tall buildings. Vol. SB: Structural design of tall steel buildings.* American Society of Civil Engineers

Council on Tall Buildings and Urban Habitat (1978c) *Monograph on planning and design of tall buildings. Vol. SC: Tall buildings systems and concepts.* American Society of Civil Engineers, 323–340

Council on Tall Buildings and Urban Habitat (1978d) *Monograph on planning and design of tall buildings. Vol. CB: Structural design of tall concrete and masonry buildings.* American Society of Civil Engineers, 396–398

Croll, J.G.A. and Walker, A.C. (1972) *Elements of structural stability.* Macmillan, London

Csonka, P. (1956) Über proportinierte Rahmen. *Die Bautechnik*, **33**, 19–20

Csonka, P. (1965a) Simple procedure for multistory frameworks under wind load (in Hungarian). *Az MTA VI. Oszt. Közl.*, **35**, Budapest, 209–219

Csonka, P. (1965b) Maximum bending moments in multistorey frameworks of rectangular network under wind load (in Hungarian). *Az MTA VI. Oszt. Közl.*, **35**, Budapest, 271–275

Daniels, B.J. (1994) Serviceability testing at Cardington. *Proceedings of the First Cardington Conference*. 16–17 November 1994, Cardington, England

DeWolf, J.T. and Pelliccione, J.F. (1979) Cross-bracing design. *Journal of Structural Division, ASCE,* **105**, (ST7), 1379–1391

DIN 18800, Teil 2 (1990) *Stahlbauten. Stabilitätsfälle*. Knicken von Stäben und Stabwerken

Dowrick, D.J. (1976) Overall stability of structures. *The Structural Engineer*, **54**, 399–409

Dulácska, E. and Kollár, L. (1960) Angenäherte Berechnung des Momentenzuwachses und der Stabilität von gedrückten Rahmenstielen. *Die Bautechnik*, **37**, (3), 98

Dunkerley, S. (1894) On the whirling and vibration of shafts. *Philosophical Transactions of the Royal Society of London*, Ser. A, **185**, 279–360

Ellis, R.B. (1980) An assessment of the accuracy of predicting the fundamental natural frequencies of buildings. *Proceedings of ICE*, **69**, Part 2, September, 763–776

Ellis, R.B. (1984) *Dynamic soil-structure interaction in buildings*. PhD thesis. University College London

Ellis, R.B. (1986) The significance of dynamic soil-structure interaction in tall buildings. *Proceedings of ICE*, Part 2, **81**, 221–242

Ellis, R.B. and Ji, T. (1996) Dynamic testing and numerical modelling of the Cardington Steel Framed Building from construction to completion. *The Structural Engineer*, **74**, (11), 186–192

Eurocode 1 (1995) *Actions on structures*. Part 2.4: Wind actions. EN 1991. CEN/TC250/SC1 Technical Secretariat, March 1995

Eurocode 2 (1992) *Design of concrete structures*. EN 1992

Eurocode 3 (1992) *Design of steel structures*. EN 1993

Eurocode 8 (1996) *Design of structures for earthquake resistance*. EN 1998

Fintel, M. Ed. (1974) *Handbook of concrete engineering*. Van Nostrand Reinhold

Gardner, P. (1999) Computers aided engineering – the implications for education and training. *The Structural Engineer*, **77**, (17), 18–19

Garland, C.F. (1940) The normal modes of vibrations of beams having noncollinear elastic and mass axes. *Journal of Applied Mechanics*, September 1940, 97–105

Gere, J.M. (1954) Torsional vibrations of beams of thin-walled open section. *Journal of Applied Mechanics*, December 1954, 381–387

Gere, J.M. and Lin, J.K. (1958) Coupled vibrations of thin-walled beams of open cross section. *Journal of Applied Mechanics*, **25**, 373–378

Gergely, P. (1975) Design and construction of earthquake resistant buildings. Lecture at the Society of Civil Engineers, Budapest

Gjelsvik, A. (1981) *The theory of thin walled bars*. John Wiley, New York

Gjelsvik, A. (1990) Buckling of built-up columns with or without stay plates. *Journal of Engineering Mechanics, ASCE*, **116**, (5), 1142–1159

Gjelsvik, A. (1991) Stability of built-up columns. *Journal of Engineering Mechanics, ASCE*, **117**, (6), 1331–1345

Goldberg, J.E. (1973) Approximate methods for stability and frequency analysis of tall buildings. *Regional Conference on 'Planning and Design of Tall Buildings'*, Madrid, 123–146

Goschy, B. (1981) *Statics and dynamics of flat-slab buildings* (in Hungarian). Műszaki Könyvkiadó, Budapest

Goschy, B. (1990) *Design of buildings to withstand abnormal loading*. Butterworth, London

Griffel, W. (1966) *Handbook of formulas for stress and strain*. Frederik Ungar Publishing Co., New York

Halldorsson, O.P. and Wang, C.K. (1968) Stability analysis of frameworks by matrix methods. *Journal of the Structural Division, ASCE,* **94**, ST7, 1745

Hart, G.C. and Vasudevan, R. (1975) Earthquake design of buildings: damping. *Journal of Structural Division, ASCE,* **101**, (ST1), 11–30

Hegedűs, I. and Kollár, L.P. (1984) Buckling of sandwich columns with thick faces subjected to axial loads of arbitrary contribution. *Acta Technica Hung.*, **97**, 123–132

Hegedűs, I. and Kollár, L.P. (1987) Stabilitätsuntersuchung von Rahmen und Wandscheiben mit der Sandwichtheorie. *Die Bautechnik*, **64**, 420–425

Hegedűs, I. and Kollár, L.P. (1999) Application of the sandwich theory in the stability analysis of structures. In *Structural stability in engineering practice* (ed. L. Kollár), E & FN Spon, 187–241

Hegedűs, I. (1987) Comment on 'Stability of large frameworks on hinged supports', a study by K.A. Zalka. *Acta Technica Hung.*, **100**, (3–4), 385–386

Horne, M.R. (1975) An approximate method for calculating the elastic critical loads of multistorey plane frames. *The Structural Engineer*, **53**, 242–248

Hrobst, E.H. and Comrie, J. (eds) (1951) *Civil engineering reference book*. Butterworth, London

Huang, T.C. (1969) The effect of rotatory inertia and of shear deformation of the frequency and normal mode equations of uniform beams with simple end conditions. *Journal of Applied Mechanics, ASME*, **28**, 579–584

Irwin, A.W. (1984) *Design of shear wall buildings*. Construction Industry Research and Information Association, Report 102, London

ISE (1988) *Stability of buildings*. The Institution of Structural Engineers, London

Jeary, A.P. (1981) *The dynamic behaviour of tall buildings*. PhD Thesis. University College London

Jeary, A.P. and Ellis, B.R. (1981) *The accuracy of mathematikal models of structural dynamics*. Building Research Establishment. PD112/81

Kaliszky, S. (1978) Simple discrete models on elastic subgrade. *Acta Technica Hung.*, **86**, 301–316

Key, D.E. (1988) *Earthquake design practice for buildings*. Thomas Telford, London

Knowles, N.C. (1996) Some lessons from recent case studies. Symposium: Safer computing. *The Institution of Structural Engineers*, 30 January 1996

Kollár, L. (1972) *Stability of frameworks and column systems* (in Hungarian). Budapesti Városfejlesztési Tervező Vállalat, Budapest

Kollár, L. (1977) Bracing of buildings against torsional buckling (in Hungarian). *Magyar Építőipar*, Budapest, 150–154

Kollár, L. (1979) *The dynamic effects of the wind on tall structures* (in Hungarian). Műszaki Könyvkiadó, Budapest

Kollár, L. and Zalka, K.A. (1999) Bracing of buildings against torsional buckling. In *Structural stability in engineering practice* (ed. L. Kollár), E & FN Spon, 242–275

Kollár, L. and Póth, L. (1994) Stresses in shear-wall buildings (in Hungarian). *Építés- Építészettudomány*, Budapest, 3–36

Kollár, L.P. (1986) *Stability analysis of frameworks and coupled shear walls by the method of final differences and the continuum method* (in Hungarian). PhD thesis, Budapest

Kollbrunner, C.F. and Basler, K. (1969) *Torsion in structures*. Springer-Verlag, Berlin, New York

König, G. and Liphardt, S. (1990) *Hochhäuser aus Stahlbeton, Beton-Kalender*, Teil II, 474–539

Lagomarsino, S. (1993) Forecast models for damping and vibration periods of buildings. *Journal of Wind Engineering and Industrial Aerodynamics*, **48**, 221–239

Lightfoot, E., McPharlin, R.D. and Le Messurier, A.P. (1979) Framework instability analysis using Blaszkowiak's stiffness functions. *Int. J. Mech. Sci.*, **21**, (9), 547–555

Littler, J.D. (1993) An assessment of some of the different methods for estimating damping from full-scale testing. *1st European and African Regional Conference of the International Association of Wind Engineering*. Guernsey, 20–24

LUSAS Finite Element System V11 (1995) *User Manual*, FEA Ltd, Forge House, Kingston Upon Thames, Surrey, KT1 1HN, UK

Maciag, E. and Kuzniar, K. (1993) The influence of ground flexibility on the fundamental frequencies of natural vibrations of medium-height buildings with load bearing concrete walls. *Archives of Civil Engineering*, **39**, (2), 139–151

MacLeod, I.A. and Marshall, J. (1983) Elastic stability of building structures. *Proceedings of 'The Michael R. Horne Conference: Instability and plastic collapse of steel structures'* (ed. L.J. Morris), Granada, London, 75–85

MacLeod, I.A. (1990) *Analytical modelling of structural systems*. Ellis Horwood

MacLeod, I.A. (1995) A strategy for the use of computers in structural engineering. *The Structural Engineer*, **73**, (21), 366–370

MacLeod, I.A. and Zalka, K.A. (1996) The global critical load ratio approach to stability of building structures. *The Structural Engineer*, **74**, (15), 249–254

Madan, A., Reinhorn, A.M., Mander, J.B. and Valles, R.E. (1997) Modeling of masonry infill panels for structural analysis. *Journal of Structural Engineering, ASCE*, **123**, (10), 1295–1302

Mainstone, R.J. and Weeks, G.A. (1972) *The influence of a bounding frame on the racking stiffness and strength of brick walls*. Building Research Station, Current Paper 3/72

Manninger, M. and Kollár, L. (1998) The shear centre curve of bracing systems containing parallel plane members subjected to uniformly distributed lateral load. *2nd International Symposium in Civil Engineering*, Technical University of Budapest, 298–304

MSZ (1986) *Structural design of the load bearing system of buildings. Loads and actions* (in Hungarian). MSZ 15021–86, Hungarian Standards Institution, Budapest

Murray, N.W. (1984) *Introduction to the theory of thin-walled structures*. Oxford University Press
Nadjai, A. and Johnson, D. (1996) Elastic analysis of spatial shear wall systems with flexible bases. *The Structural Design of Tall Buildings*, **5**, 55–72
Newmark, N.M. and Rosenblueth, E. (1971) *Fundamentals of earthquake engineering*. Prentice Hall, Englewood Cliffs
Parmelee, R.R., Perelman, D.S. and Lee, S.L. (1969) Seismic response of multiple-story structures on flexible foundations. *Bulletin of Seismological Society of America*, **59**, (3), June
Pearce, D.J. and Matthews, D.D. (1971) *Shear walls. An appraisal of their design in box-frame structures*. Property Services Agency, Department of the Environment, London
Plantema, F.J. (1961) *Sandwich construction. The bending and buckling of sandwich beams, plates and shells*. McGraw-Hill, London
Polyakov, S.V. (1956) *Masonry in framed buildings (An investigation into the strength and stiffness of masonry infilling)*. Gosudarstvennoe izdatel'stvo literatury po stroitel'stvu i arkhitekture, Moscow. (English translation by G.L. Cairns, National Lending Library for Science and Technology, Boston, Yorkshire, England, 1963)
Prosec (1994) *Prosec: Section properties*. Version 4.05. PROKON Software Consultants, 75 Lower Richmond Road, Putney, London, SW15 1ET
Richart, F.E., Woods, R.D. and Hall, J.R. (1970) *Vibrations of soils and foundations*. Prentice-Hall, Englewood Cliffs
Roark, R.J. and Young, W.C. (1975) *Formulas for stress and strain*. 5th edn. McGraw-Hill, London
Roberts, T.M. (1985) Section properties of thin walled bars of open crosssection. *The Structural Engineer*, **63B**, (3), 63–67
Rosman, R. (1967) Faltwerke als aussteifende Systeme bei Hochbauten. *Bauingenieur*, **42**, p55
Rosman, R. (1980) Stabilität im Grundriβ unsymmetrischer Stützen- und Wandscheibensysteme. *Die Bautechnik*, **57**, (1), 21–32
Rosman, R. (1981) Buckling and vibrations of spatial building structures. *Engineering Structures*, **3**, (4), 194–202
Scarlat, A.S. (1996) *Approximate methods in structural seismic design*. E & FN Spon, London
Schueller, W. (1977) *High-rise building structures*. John Wiley and Sons
Smart, R.A. (1997) Computers in the design office: boon or bane? *The Structural Engineer*, **75**, (3), 52

Southwell, R.V. (1922) On the free transverse vibration of a uniform circular disc clamped at its centre, and on the effects of rotation. *Proceedings of the Royal Society of London*, Ser. A., **101**, 133–153

Spyrakos, C.C. and Beskos, D.E. (1986) Dynamic response of flexible foundations by boundary and finite elements. *Soil dynam. Earthquake Engng*, **5**, 84–96

Stafford Smith, B. and Coull, A. (1991) *Tall building structures. Analysis and design*. John Wiley & Sons, New York

Stafford Smith, B. and Vézina, S. (1985) Evaluation of centres of resistance in multistorey building structures. *Proceedings of ICE*, **79**, Part 2, 623–635

Steinle, A. and Hahn, V. (1988) *Bauen mit Betonfertigteilen im Hochbau*. Beton-Kalender, Teil II. Ernst & Sohn, Berlin

Stevens, L.K. (1967) Elastic stability of practical multi-storey frames. *Proceedings of ICE*, **36**, 99–117

Stevens, L.K. (1983) The practical significance of the elastic critical load in the design of frames. *Proceedings of 'The Michael R. Horne Conference: Instability and plastic collapse of steel structures'* (ed. L.J. Morris), Granada, London, 36–46

Stiller, M. (1965) Verteilung der Horizontalkräfte auf die aussteifenden Scheibensysteme von Hochhäusern. *Beton- und Stahlbetonbau*, **60**, 42–45

Stoman, S.H. (1988) Stability criteria for X-bracing systems. *Journal of Engineering Mechanics, ASCE*, **114**, (8), 1426–1434

Stoman, S.H. (1989) Effective length spectra for cross bracings. *Journal of Structural Engineering, ASCE*, **115**, (12), 3112–3122

Stüssi, F. (1965) Die Grenzlagen der Schubmittelpunktes bei Kastenträgern. *IVBH Publications*, **25**, 279–315

Szittner, A. (1979) *Small-scale tests of a framework* (in Hungarian). Technical University of Budapest. Department of Steel Structures, No. 202002/9

Szmodits, K. (1975) *Guidelines for the structural design of large panel system building* (in Hungarian). Építéstudományi Intézet, Budapest

Tarnai, T. (1996) Unsymmetrical bending of beams: a matrix formulation. *International Journal of Mechanical Engineering Education*, **24**, 144–149

Tarnai, T. (1999) Summation theorems concerning critical loads of bifurcation. In *Structural stability in engineering practice* (ed. L. Kollár), E & FN Spon, 23–58

Thevendran, V. and Wang, C.M. (1993) Stability of nonsymmetric crossbracing systems. *Journal of Structural Engineering, ASCE.* **119**, (1), 169–180

Timoshenko, S. (1928) *Vibration problems in engineering.* D. Van Nostrand Company, London

Timoshenko, S.P. (1955) *Strength of materials. Part I: Elementary theory and problems.* D. Van Nostrand Company, Princeton, New York

Timoshenko, S.P. and Gere, J. (1961) *Theory of elastic stability.* 2nd edn, New York, McGraw-Hill

Timoshenko, S.P. and Young, D.H. (1955) *Vibration problems in engineering.* 3rd edn. Van Nostrand Company, Princeton, New York

Timoshenko, S.P., Young, D.H. and Weaver, W. (1974) *Vibration problems in engineering.* 4th edn. John Wiley

Vafai, A., Estekanchi, H. and Mofid, M. (1995) The use of single diagonal bracing in improvement of the lateral response of tall buildings. *The Structural Design of Tall Buildings*, **4**, 115–126

Vértes, Gy. (1985) *Structural dynamics.* Elsevier, New York

Vlasov, V.Z. (1940) *Tonkostennye uprugie sterzhni.* Moscow. 2nd edn: *Thin-walled elastic beams.* Israeli Program for Scientific Translations, Jerusalem, 1961

Waldron, P. (1986) Sectorial properties of straight thin-walled beams. *Computers and Structures*, **24**, (1), 147–156

Wang, D.Q. and Boresi, A.P. (1992) Theoretical study of stability criteria for X-bracing systems. *Journal of Engineering Mechanics, ASCE*, **118**, (7), 1357–1364

Wolf, J.P. (1985) *Dynamic soil-structure interaction.* Englewood Cliffs: Prentice Hall

Wood, R.H. (1974) Effective lengths of columns in multi-storey buildings. *The Structural Engineer*, **52**, 235–244, 295–302 and 341–346

Wood, R.H. (1975) Effective lengths of columns in multi-storey buildings (discussion). *The Structural Engineer*, **53**, 235–241

Wood, R.H. and Marshall, W.J. (1983) The search for a unified column design. *Proceedings of 'The Michael R. Horne Conference: Instability and plastic collapse of steel structures'* (ed. L.J. Morris). Granada, London, 216–231

Zalka, K.A. (1988) LPS Building: The Hungarian experience. *Building Research and Practice. The Journal of CIB*, **16**, (2), 79–86

Zalka, K.A. and White, D.S. (1992) *Deformations and the stability of small-scale buildings models.* Building Research Establishment, Watford, N231/92

Zalka, K.A. and Armer, G.S.T. (1992) *Stability of large structures.* Butterworth-Heinemann, Oxford

Zalka, K.A. (1992) *Characteristic deformations and the stability of large frameworks.* Building Research Establishment, Watford, N214/92

Zalka, K.A. (1993) *An analytical procedure for 3-dimensional eigenvalue problems.* Building Research Establishment, Watford, N32/93

Zalka, K.A. and White, D.S. (1993) Model analysis of regular multistorey buildings. *International Conference on Tall Buildings*, Rio de Janeiro, Brazil. Proceedings: Lehigh University, Bethlehem, Pennsylvania, USA, 337–352

Zalka, K.A. (1994a) *Section properties of thin-walled cross-sections.* Building Research Establishment, Watford, N53/94

Zalka, K.A. (1994b) *Dynamic analysis of core supported buildings.* Building Research Establishment. Watford, N127/94

Zalka, K.A. (1994c) Mode coupling in the torsional-flexural buckling of regular multistorey buildings. *The International Journal of the Structural Design of Tall Buildings*, **3**, 227–245

Zalka, K.A. (1997) *SpaStab: Spatial stability, stress and frequency analyses of multistorey building structures.* Vol. **1** and Vol. **2**. Manual for the computer procedure SpaStab, Version 4.16. Manuscript

Zalka, K.A. (1998a) *Global stability of cross-bracing systems.* Building Research Establishment, Research Report, Watford

Zalka, K.A. (1998b) *Equivalent wall for frameworks for the global stability analysis.* Building Research Establishment, N33/98, Watford

Zalka, K.A. (1999) Full-height buckling of frameworks with cross-bracing. *Strucures & Buildings, Proceedings of ICE,* **134**, May, 181–191

Zbirohowski-Koscia, K. (1967) *Thin walled beams. From theory to practice.* Crosby Lockwood and Son, London

Zhang, W.J., Xu, Y.L. and Kwok, K.C.S. (1993) Torsional vibration and stability of wind-excited tall buildings with eccentricity. *Journal of Wind Engineering and Industrial Aerodynamics*, **50**, Part 2, 299–308

Further reading

Barta, T.A. (1967) On the torsional-flexural buckling of thin-walled elastic bars with monosymmetric open cross section. In *Thin-walled structures* (ed. A.H. Chilver), Chatto and Windus, London, 60–86

Beck, H. (1962) Contribution to the analysis of coupled shear walls. *Journal of the American Concrete Institution*, **59**, 1055–1069

Beck, H. and König, G. (1967) Restraining forces in the analysis of tall buildings. *Proceedings of a Symposium on Tall Buildings*, Pergamon Press, Oxford, 513–536

Beck, H. and Zilch, K. (1972) Stability (in the design of tall concrete buildings). Theme report. *International Conference on Planning and Design of Tall Buildings*, Vol. III, Lehigh University, Pennsylvania, 499–516

Biswas, J.K. and Tso, W.K. (1974) Three dimensional analysis of shear wall buildings to lateral load. *Journal of the Structural Division, ASCE*, **100**, (ST5), 1019–1036

Bleich, F. (1952) *Buckling strength of metal structures.* McGraw-Hill, New York

Blevins, R.D. (1979) *Formulas for natural frequency and mode shape.* Van Nostrand Reinhold Company, New York

Bolton, A. (1976) A simple understanding of elastic critical loads. *The Structural Engineer*, **54**, (6), 213–218

Bornscheuer, F.W. (1952) Systematische Darstellung des Biege- und Verdrehvorganges unter besonderer Berücksichtigung der Wölbkrafttorsion. *Der Stahlbau*, **21**, Heft 1, 1–9

Bowles, J.E. (1979) *Physical and geotechnical properties of soils.* McGraw-Hill, London

Bowles, J.E. (1982) *Foundation analysis and design.* 3rd edn. McGraw-Hill, London

Chen, W.F. and Atsuta, T. (1977) *Theory of beam-columns.* McGraw-Hill, London

Coull, A. (1975) Free vibrations of regular symmetrical shear wall buildings. *Building Science*, **10**, 127–133

Coull, A. and Irwin, A.W. (1971) Torsional analysis of multistorey shear wall structures. *American Concrete Institute*, SP 35–6, Detroit, Michigan. 211–238

Coull, A. and Stafford Smith, B. (Eds) (1967) *Tall buildings*. Proceedings of a Symposium on Tall Buildings with particular reference to shear wall structures. University of Southampton, Department of Civil Engineering. Pergamon Press, Oxford

Coull, A. and Wahab, A.F.A. (1993) Lateral load distribution in asymmetrical tall building structures. *Journal of Structural Engineering, ASCE*, **119**, 1032–1047

Coull, A. and Wong, Y.C. (1986) Stiffening of structural cores by floor slabs. *Journal of Structural Engineering, ASCE*, **112**, 977–994

Council on Tall Buildings (1972) *Proceedings of the International Conference on Planning and Design of Tall Buildings*. ASCE-IABSE, Lehigh University, Bethlehem, Pennsylvania

Council on Tall Buildings (1978) *Planning and Design of Tall Buildings*, a Monograph in 5 volumes. ASCE, New York

Council on Tall Buildings and Urban Habitat (1983) *Development in Tall Buildings*. Van Nostrand Reinhold, New York

Council on Tall Buildings and Urban Habitat (1986) *Advances in Tall Buildings*. Van Nostrand Reinhold, New York

Council on Tall Buildings and Urban Habitat (1986) *High-rise buildings: Recent progress*. Lehigh University, Bethlehem

Council on Tall Buildings and Urban Habitat (1988) *Second century of the skyscraper*. Van Nostrand Reinhold, New York

Council on Tall Buildings and Urban Habitat (1990) *Tall Buildings: 2000 and beyond* (eds L.S. Beedle and D.B. Rice), The 4th World Congress. November 5–9, Hong Kong

Danay, A., Gellert, M. and Glück, J. (1975) Continuum method for overall stability of tall asymmetric buildings. *Journal of the Structural Division, ASCE*, **101**, ST12, 2505–2521

Danay, A., Glück, J. and Gellert, M. (1975) A generalized continuum method for dynamic analysis of asymmetric tall buildings. *Earthquake Engineering and Structural Dynamics*, **4**, 179–203

Den Hartog, J.P. (1956) *Mechanical vibrations*. McGraw-Hill, London

Despeyroux, J. (1972) *Analyse statique et dynamique des contraventments par consoles*. Annales de l'Institut Technique du Bâtiment et des Travaux Publics. No. 290

Dicke, D. (1972) Effect of fundation rotation. *International Conference on Planning and Design of Tall Buildings*. Lehigh University, Pennsylvania, Volume III, 621–622

Eurocode (1999) *Basis of design*, EN 1990, May 1999, Draft

Fertis, D.G. (1973) *Dynamics and vibration of structures*. John Wiley & Sons, New York

Föppl, L. (1933) Über das Ausknicken von Gittermasten, insbesondere von hohen Funktürmen. *ZAMM*, **13**, 1–10

Glück, J. (1973) The vibration frequencies of an elastically supported cantilever column with variable cross-section and distributed axial load. *Earthquake Engineering and Structural Dynamics*, **1**, 371–376

Glück, J. and Gellert, M. (1971) On the stability of elastically supported cantilever with continuous lateral restraint. *International Journal of Mechanical Sciences*, **13**, 887–891

Glück, J. and Gellert, M. (1972) Buckling of lateral restrained thin-walled cantilevers of open cross-section. *Journal of the Structural Division, ASCE*, **98**, 2031–2042

Goschy, B. (1971) Spatial stability of system-buildings. *Acta Technica Hung.*, **70**, (3–4), 459–470

Goschy, B. (1978) Stability of core-supported structures. *Acta Technica Hung.*, **87**, (1–2), 59–68

Harrison, T. (1971) *The elastic behaviour of structures composed of interconnected thin-walled members*. PhD Thesis. University of Bradford

Hegedűs, I. and Kollár, L.P. (1988) Generalized bar models and their physical interpretation. *Acta Technica Hung., Civil Engineering*, **101**, 67–93

Horne, M.R. and Merchant, W. (1965) *The stability of frames*. Pergamon, Oxford

Iffland, J.S.B. (Ed.) (1979) Stability. In *Planning and design of tall buildings*. SB-4, ASCE, New York, 238–342

Jubb, J.E.M., Phillips, I.G. and Becker, H. (1975) Interrelation of structural stability, stiffness, residual stress and natural frequency. *Journal of Sound and Vibration*, **39**, (1), 121–134

Kármán, T. and Biot, M.A. (1939) *Mathematical methods in engineering*. McGraw-Hill, London

Kollár, L.P. (1991) Calculation of plane frames braced by shear walls for seismic load. *Acta Technica Hung.*, **104**, (1–3), 187–209

Ku, A.B. (1974) On the structural stability of a tall building. *Proceedings of the Regional Conference on Tall Buildings*. Bangkok, Thailand, 139–149

Lamb, H. and Southwell, R.V. (1921) The vibrations of a spinning disk. *Proceedings of the Royal Society of London*, Ser A99, 272–280

Littler, J.D. (1991) *The response of a tall building to wind loading.* PhD thesis, University College London

Nair, R.S. (1975) Overall elastic stability of multistorey buildings. *Journal of the Structural Division, ASCE*, **101**, (ST12), 2487–2503

Papkovich, P.F. (1963) *On structural mechanics in shipbuilding. 4, Stability of beams, trusses and plates* (in Russian). Sudpromgiz, Leningrad

Pearson, C.E. (1956) Remarks on the centre of shear. *Z. angew. Math. Mech.*, **36**, 94–96

Petersson, H. (1973) *Analysis of building structures.* Chalmers University of Technology, Gothenburg

Petersson, H. (1974) *Analysis of loadbearing walls in multistorey buildings.* Chalmers University of Technology, Gothenburg

Pflüger, A. (1950) *Stabilitätsprobleme der Elastostatik.* Springer, Berlin

Plaut, R.H. and Virgin, L.N. (1990) Use of frequency data to predict buckling. *Journal of Engineering Mechanics, ASCE,* **116**, (10), 2330–2335

Podolsky, D.M. (1971) Method of spatial calculation of multistorey buildings for strength, stability and vibrations. *CIB Symposium on Tall Buildings*. Moscow

Puri, R.D. (1967) *A study of the continuous connection method of analysis of coupled shear walls.* PhD thesis. University of Southampton

Qiusheng, L., Hong, C. and Guiqing, L. (1994) Analysis of free vibrations of tall buildings. *Journal of Engineering Mechanics, ASCE*, **120**, (9), 1861–1876

Rosman, R. (1972) The continuum analysis of shear-wall structures. *Proceedings of the International Conference on the Planning and Design of Tall Buildings*, 271–275. Bethlehem, Pennsylvania

Rosman, R. (1973) Dynamics and stability of shear wall building structures. *Proceedings of ICE*, Part 2, **55**, 411–423

Rosman, R. (1974) Knicklasten von Hochbauten und Schnittkräfte der Theorie II. Ordnung unter Berücksichtigung der Verformung der Grundkörperunterlage. *Beton- und Stahlbetonbau,* H. **5**

Rosman, R. (1974) Stability and dynamics of shear-wall frame structures. *Building Science*, **9**, 55–63

Rutenberg, A. (1975) Approximate natural frequencies of coupled shear walls. *Earthquake Engineering and Structural Dynamics*, **4**, 95–100

Rutenberg, A. (1979) An accurate approximate formula for the natural frequencies of sandwich beams. *Computers and Structures*, **10**, 875–878

Rutenberg, A., Glück, J. and Reinhorn, D.A. (1978) On the dynamic properties of asymmetric wall-frame structures. *Earthquake Engineering and Structural Dynamics*, **6**, 317–320

Rutenberg, A., Leviathan, I. and Decalo, M. (1988) Stability of shear-wall structures. *Journal of Structural Engineering, ASCE*, **114**, (3), 707–716

Saint-Venant, B. de (1855) Mémoire sur la torsion des prismes. *Mémoires des Savants Étrangers*, **14**, 233–560. Paris

Schueller, W. (1990) *The vertical building structure*. Van Nostrand Reinhold, New York

Singer, J. (1983) Vibrations and buckling of imperfect stiffened shells – Recent developments. In *Collapse: The buckling of structures in theory and practice* (eds J.M.T. Thompson and G.W. Hunt). Cambridge University Press, Cambridge, 443–479

Stafford Smith, B. and Abate, A. (1984) The effects of shear deformations on the shear centre of open section thin walled beams. *Proceedings of ICE*, **77**, Part 2, 57–66

Stafford Smith, B. and Crowe, E. (1986) Estimating periods of vibration of tall buildings. *Journal of Structural Engineering, ASCE*, **112**, (5), 1005–1019 [discussion: (1987), **113**, (9), 2098–2099]

Stafford Smith, B. and Riddington, J.R. (1978) The design of masonry filled steel frames for bracing systems. *The Structural Engineer*, **56B**, (1), 1–6

Stamato, M.C. and Stafford Smith, B. (1969) An approximate method for the three dimensional analysis of tall buildings. *Proceedings of ICE*, **43**, 361–379

Sugiyama, Y. and Kawagoe, H. (1975) Vibration and stability of elastic columns under the combined action of uniformly distributed vertical and tangential forces. *Journal of Sound and Vibration*, **38**, (3), 341–355

Taranath, B.S. (1968) *The torsional behaviour of open section shear wall structures*. PhD Thesis. University of Southampton

Taranath, B.S. (1988) *Structural analysis and design of tall buildings*. McGraw-Hill, London

Tarnai, T. (1980) Generalization of Southwell's and Dunkerley's theorem for quadratic eigenvalue problems. *Acta Technica Hung.*, **91**, 201–221

Terzaghi, K. (1955) Evaluation of coefficients of subgrade reaction. *Geotechnique*, **5**, 297–326

Thompson, J.M.T. and Hunt, G.W. (1973) *A general theory of elastic stability*. John Wiley and Sons, New York

Timoshenko, S.P. (1945) Theory of bending, torsion and buckling of thin walled members of open cross section. *Journal of the Franklin Institute*, **239**, Nos. 3, 4 and 5, 201–219, 249–268 and 343–361

Trahair, N.S. (1993) *Flexural-torsional buckling of structures*. E & FN Spon, London

Williams, F.W. (1977) Simple design procedures for unbraced multi-storey frames. *Proceedings of ICE*, **63**, Part 2, 475–479

Williams, F.W. (1979) Consistent, exact, wind and stability calculation for substitute sway frames with cladding. *Proceedings of ICE*, **67**, Part 2, 355–367

Wong, Y.C. and Coull, A. (1980) Structural behaviour of floor slabs in shear wall buildings. In *Advances in concrete slab technology* (Eds R.K. Dhir and J.G.L. Munday), Pergamon Press, Oxford

Yarimci, E. (1972) The equivalent-beam approach in tall building stability analysis. *Proceedings of the International Conference on Planning and Design of Tall Buildings*, Vol. II-16, 553–563

Name index

Abate, A. 329
Allen, H.G. 204, 315
American Society of Civil Engineers 109
Appeltauer, J. 237, 315
Armer, G.S.T. 19, 121, 177, 193, 205, 324
Asztalos, Z. 193, 315
Atsuta, T. 325

Barbero, E. 248, 315
Barkan, D.D. 37, 38, 315
Barta, T. 325
Basler, K. 16, 126, 129, 319
Beck, H. 8, 10, 81, 87, 245, 315, 325
Becker, H. 327
Beedle, L.S. 326
Beskos, D.E. 153, 322
Beton-Kalender 112, 322
Biot, M.A. 327
Bishop, R.E.D. 61, 315
Biswas, J.K. 325
Blaszkowiak, S. 320
Bleich, F. 325
Blevins, R.D. 325
Bolton, A. 325
Boresi, A.P. 219, 323
Bornscheuer, F.W. 325
Boughton, J. 246, 315
Bowles, J.E. 325
Brohn, D.M. 3, 315, 316
BS5950 176, 177, 316
Building Research Establishment 238

Capuani, D. 61, 316
Cardington Steel Building 182

CEB 176, 316
Chen, W.F. 325
Chilver, A.H. 325
Chwalla, E. 176, 316
Comrie, J. 16, 319
Cook, N.J. 65, 316
Coull, A. 59, 223, 226, 316, 322, 326, 330
Council on Tall Buildings and Urban Habitat 11, 153, 176, 193, 316, 326
Cowan, J. 3, 315
Croll, J.G.A. 248, 316
Crowe, E. 329
Csonka, P. 87, 89, 201, 316, 317

Danay, A. 326
Daniels, B.J. 182, 317
Decalo, M. 329
Den Hartog, J.P. 326
Despeyroux, J. 326
DeWolf, J.T. 219, 317
Dhir, R.K. 330
Dicke, D. 327
DIN 176, 177, 317
Dowrick, D.J. 75, 98, 176, 317
Dulácska, E. 176, 317
Dunkerley, S. 22, 23, 33, 46, 177, 197, 209, 248, 317

Ellis, R.B. 59, 62, 182, 317, 319
Estekanchi, H. 219, 323
Eurocode 42, 43, 64, 65, 176, 177, 317, 327

Fertis, D.G. 327

Name index

Fintel, M. 46, 61, 317
Föppl, L. 27, 37, 39, 40, 52, 57, 58, 198, 200, 221, 327

Gardner, P. 3, 317
Garland, C.F. 43, 317
Gellert, M. 326, 327
Gere, J.M. 19, 21, 28, 43, 54, 195, 221, 318, 323
Gergely, P. 66, 318
Gjelsvik, A. 16, 219, 318
Glück, J. 326, 327, 329
Goldberg, J.E. 62, 318
Goschy, B. 43, 61, 62, 318, 327
Guiqing, L. 328
Griffel, W. 16, 318

Hahn, V. 276, 322
Hall, J.R. 38, 321
Halldorsson, O.P. 176, 318
Harrison, T. 327
Hart, G.C. 61, 318
Hegedűs, I. 29, 35, 36, 199, 205, 215, 237, 318, 327
Hong, C. 328
Hooke, R. 4
Horne, M.R. 193, 318, 320, 322, 323, 327
Hrobst, E.H. 16, 319
Huang, T.C. 61, 319
Hunt, G.W. 329, 330

Iffland, J.S.B. 327
Irwin, A.W. 61, 75, 98, 319, 326
ISE 176, 319

Jeary, A.P. 52, 62, 319
Ji, T. 182, 317
Johnson, D. 152, 153, 321
Jourawski 127
Jubb, J.E.M. 327

Kaliszky, S. 153, 319
Kármán, T. 327
Kawagoe, H. 329
Key, D.E. 36, 40, 67, 319
Knowles, N.C. 3, 319

Kollár, L. 36, 42, 43, 57, 150, 151, 176, 177, 178, 237, 245, 315, 317, 318, 319, 320, 322
Kollár, L.P. 29, 35, 36, 193, 199, 201, 205, 237, 318, 319, 327
Kollbrunner, C.F. 16, 126, 129, 319
König, G. 10, 81, 112, 245, 315, 319, 325
Ku, A.B. 327
Kuzniar, K. 57, 320
Kwok, K.C.S. 52, 324

Lagomarsino, S. 61, 320
Lamb, H. 328
Laudiero, F. 61, 316
Lee, S.L. 57, 321
Le Messurier, A.P. 193, 320
Leviathan, I. 329
Lightfoot, E. 193, 320
Lin, J.K. 43, 54, 318
Liphardt, S. 81, 112, 245, 319
Littler, J.D. 61, 320, 328
London 139
LUSAS 193, 320

Maciag, E. 57, 320
MacLeod, I.A. 3, 61, 81, 177, 193, 210, 245, 320
Madan, A. 226, 320
Mainstone, R.J. 226, 320
Mander, J.B. 320
Manninger, M. 150, 320
Marshall, J. 193, 320
Marshall, W.J. 193, 323
Matthews, D.D. 75, 98, 321
McPharlin, R.D. 193, 320
Mercalli–Sieberg–Cancani 66
Merchant, W. 327
Mofid, M. 219, 323
Morris, L.J. 320, 322, 323
MSC 66
MSZ 176, 320
Mukherjee, P.R. 59, 316
Munday, J.G.L. 330
Murray, N.W. 16, 126, 321

Nadjai, A. 152, 153, 321

Nair, R.S. 328
Newmark, N.M. 67, 321

Papkovich, P.F. 27, 37, 39, 40, 52, 57, 58, 198, 200, 221, 328
Parmelee, R.R. 57, 321
Pearce, D.J. 75, 98, 321
Pearson, C.E. 328
Pelliccione, J.F. 219, 317
Perelman, D.S. 57, 321
Petersson, H. 328
Pflüger, A. 328
Phillips, I.G. 327
Plantema, F.J. 204, 321
Plaut, R.H. 328
Podolsky, D.M. 328
Polyakov, S.V. 226, 321
Póth, L. 36, 150, 151, 245, 319
Price, W.G. 61, 315
Prosec 16, 321
Puri, R.D. 328

Qiusheng, L. 328

Reeh, H. 10, 315
Reinhorn, A.M. 320, 329
Rice, D.B. 326
Richart, F.E. 38, 321
Riddington, J.R. 329
Roark, R.J. 16, 321
Roberts, T.M. 16, 321
Rosenblueth, E. 67, 321
Rosman, R. 43, 245, 321, 328
Rutenberg, A. 328, 329

Saint-Venant, B. de 329
Savoiva, M. 61, 316
Scarlat, A.S. 67, 321
Schäfer, H.G. 8, 81, 245, 315
Schueller, W. 109, 225, 321, 329
Singer, J. 329
Smart, R.A. 3, 321
Southwell Plot 248-252
Southwell, R.V. 41, 43, 56, 200, 209, 232, 248, 282, 322, 328
Spyrakos, C.C. 153, 322
Stafford Smith, B. 150, 223, 226, 322, 326, 329
Stamato, M.C. 329
Steinle, A. 276, 322
Stevens, L.K. 176, 193, 208, 210, 322
Stiller, M. 245, 322
Stoman, S.H. 219, 322
Stüssi, F. 150, 322
Sugiyama, Y. 329
Szittner, A. 248, 322
Szmodits, K. 245, 322

Taranath, B.S. 329
Tarnai, T. 22, 27, 198, 200, 221, 245, 322, 329
Terzaghi, K. 329
Thevendran, V. 219, 323
Thompson, J.M.T. 329, 330
Timoshenko, S.P. 19, 21, 28, 45, 55, 60, 61, 127, 152, 181, 195, 221, 323, 330
Tomblin, J. 248, 315
Trahair, N.S. 330
Tso, W.K. 325

Vafai, A. 219, 323
Valles, R.E. 320
Vasudevan, R. 61, 318
Vértes, Gy. 43, 323
Vézina, S. 150, 322
Virgin, L.N. 328
Vlasov, V.Z. 16, 19, 101, 107, 129, 151, 245, 323

Wahab, A.F.A. 326
Waldron, P. 16, 323
Walker, A.C. 248, 316
Wang, C.K. 176, 318
Wang, C.M. 219, 323
Wang, D.Q. 219, 323
Weaver, W. 61, 323
Weeks, G.A. 226, 320
White, D.S. 14, 238, 248, 323, 324
Williams, F.W. 330
Wolf, J.P. 36, 40, 323
Wong, Y.C. 326, 330
Wood, R.H. 193, 323
Woods, R.D. 38, 321

Zalka, K.A.   14, 16, 19, 20, 42, 43, 47,
    62, 121, 174, 177, 178, 193, 194, 205,
    209, 210, 226, 235, 238, 246, 248, 278,
    319, 320, 323, 324
Zbirohowski-Koscia, K.   126, 129, 324
Zhang, W.J.   52, 324
Zilch, K.   325

Xu, Y.L.   52, 324

Yarimci, E.   330
Young, D.H.   45, 55, 61, 181, 323
Young, W.C.   16, 321

Subject index

3-dimensional analysis/behaviour 63
  see spatial analysis/behaviour

accuracy analysis 236–7, 238, 245
advanced stress analysis 98
analytical procedures 2
approximate methods 2, 18, 27, 43, 52, 59, 62, 66, 98, 192, 193
arrangement of the bracing system 62, 96
  asymmetrical 83
  doubly symmetrical 96
  optimal 173, 259
  symmetrical 82, 175
assumptions 3–4, 63, 82
axial forces 60, 181
axial load of arbitrary distribution 205

basic critical loads 20, 26, 27, 32, 34, 41, 60, 255
basic modes 16, 44
basic natural frequencies 44, 53, 54, 55, 60
beam-columns 41–2, 152
bending moment factor 117, 134
  maximum 119–20, 135–6
bending moments
  due to rotation 117
  in beams 88–9
  in bracing elements 110, 117, 134, 138
  in columns 88–9
  maximum 86, 89, 135, 138
bending stiffness 8, 26, 62, 174
bending torsion 10
bending torsional constant see warping constant

bending type deformation 149, 150, 229, 235
bifurcation 4
bimoment 127
bracing elements 7, 8
  cross-sectional characteristics of 266
bracing system 7
  asymmetrical 83
  balanced 175
  doubly symmetrical 182
  overall deformation of 229
  performance of 259
  symmetrical 82, 175
  under horizontal load 63, 98
bracing walls
  parallel 69
  perpendicular 76
buckled shape of
  frameworks 260, 261, 262, 263
  model 249, 250
buckling
  combined sway-torsional 26
  full-height 219
  storey-height 219
  planar 192–237, 260–63
  spatial 19, 26, 32, 255
built-up columns 218

cantilever developing
  bending and shear deformation 121
centre of external load 12, 16
centre of mass 52
centre of stiffness 8, 71
circular frequencies 44–5, 53
closed cross-section 23, 26, 97, 128, 132

336 Subject index

Code of Practice 64, 65, 66, 67, 176–7
coefficient of elastic uniform compression *see* subgrade reaction
columns in multistorey buildings 14–6, 42, 238–40
combination factor 199
combined
   lateral displacements and torsion 102
   lateral-torsional vibrations 52
   shear and bending situations 35
   sway-torsional buckling 26–9, 30
comparison of
   approximate methods 245
   bending and bending torsion 108
   buckled shapes/deformations/load distributions 121
   horizontal loads 63, 68
   parallel-perpendicular arrangements 94
   performance indicators 179–80
   planar bracing elements 230, 263–4
   shear wall arrangements 52, 169, 173
   torsional resistances 96, 174
compressive forces 60, 181
Computer Aided Disaster 3
computer-based methods
   evaluating 254
computers in design offices 2–3, 192
concentrated
   horizontal force 129, 189
   mass 55
   vertical load 32, 42, 186, 204, 260
conservative forces 4
construction cost 180
construction misalignment 67–8
contact area between foundation and soil 37
continuum model 86, 101, 201–2
convergence problems 193
core 7, 98, 144
   perforated core 276
coupled shear walls 7, 149, 216, 228, 234, 263, 264
coupled vibrations 43, 60
coupling 17, 256
   double 17, 54
   triple 17, 54

coupling of modes 8, 16, 20, 26–9, 42, 44, 52, 54, 255, 256
cracking 10–1
critical load 21, 23, 27, 32, 34, 39, 41, 152, 226
   based of frequency measurements 184
critical load parameter for
   frameworks on fixed support 203, 204
   frameworks on pinned support 204, 211
   pure torsional buckling 23–25
cross-bracing 220
   double 222
   knee 223
   shear stiffness of 220–3
   single 221
cross-section
   open, thin-walled 16–7, 20
cross-sectional characteristics 16, 266–77
   circular sections 275–6
   I-section 266
   L-section 269
   ☐-section 271
   perforated core 276–7
   T-section 268
   TT-section 267
   +-section 272
   Z-section 274
cross-wall system 68, 69, 72, 75, 98, 231

damping 57, 61, 67
deformations
   based on frequency measurements 186
   of the building 75–6, 81, 108, 191, 258
   of the equivalent column 103, 186–90
design formulae for
   coupled shear walls 217–8
   frameworks 208–10, 213, 215, 224
   infilled frameworks 227–8
design guidelines 174–5, 255–65
design office 3, 18
diagonal bar in
   infilled panel 226
   cross-bracing 221
distance between
   shear centre and centre of load 17, 26, 175, 255, 256

Subject index 337

shear centre and line of action of load 100, 102, 258
    see eccentricity
doubly symmetrical systems 20–1, 32, 44, 52, 96, 182
Dunkerley's theorem/formula 22, 33, 46, 177, 197, 209
    graphical interpretation of 22
dynamic coefficient 66
dynamic subgrade reaction 38

earthquake resistant design 67
eccentricity 29, 32, 52, 54, 129, 257, 258
eccentricity parameter 28, 54
economy 1, 257
effect of cracking 10–11
effect of individual columns 14–6
efficiency of bracing system 163, 173, 179, 230
eigenvalue 18, 206, 278
    in the boundary condition 282
elastic foundation 37
elementary
    static considerations 63
    stress analysis 63
'element-based' design process 1, 180
elevator shaft 97
end conditions, different 41
equivalent column 8, 19, 45, 99
    for frameworks 86–7
equivalent diagonal bar
    in cross-bracing 226–7
equivalent floor load 66
equivalent static [seismic] load 65
equivalent static wind pressure 64
equivalent wall 192, 228
    thickness 229, 276–7
Eurocode 42, 64, 65, 176
evaluation of
    computer packages 254
    bracing systems 154, 174, 179–80
    formulae for critical load 26
exact analysis 2, 192
experimental data 182

failure mode of small-scale model 249
fictitious column 110

fictitious (shear) wall 7–8, 13–14, 35, 149
    limitations of 229–30
finite element methods 2, 237
flexible support 36–8, 57–9, 152–3
flexural-torsional buckling 42
floor slabs 4, 7, 10, 72, 83, 110, 121, 182
foundation types 40
Föppl–Papkovich theorem/formula 27, 37, 39, 40, 52–53, 57–8, 198, 200, 221
    graphical interpretation of 40
'frame-like' behaviour 265
frameworks 82, 86, 94, 149, 193, 228, 237
    as bracing elements 82, 149, 210
    characteristic deformations of 194–5 229–30, 260–63
    characteristic stiffnesses of 194–9
    equivalent wall for 192
    evenly spaced 85, 94
    full-height bending of 194–6, 260, 263
    global bending of 196, 260
    in asymmetrical arrangement 83, 94
    in symmetrical arrangement 82, 90, 91
    local buckling of 215, 219
    low-rise 260, 265
    medium-rise 260, 265
    on fixed supports 196, 199–210, 233–5 236, 260, 263, 264
    on pinned supports 196, 210–4, 233–5, 236, 261, 263, 264
    part critical loads of 194–8
    shear deformation of 194, 198, 260
    tall 263
    under concentrated top load 260–1
    under horizontal load 87
    under UDL on the beams 261
    with cross-bracing 218–26, 233–5, 236, 262, 263, 264
    with ground floor beams 213, 233–5, 261, 263, 264
    with ground floor columns of different height 214–5
    without ground floor beams 212, 233–5, 261, 263, 264
frequency analysis 43–62, 256
frequency measurements 182
frequency parameter 47, 49–51

338 Subject index

full-scale test 57
fundamental frequency 47, 54, 55, 62, 66
fundamental mode 46, 49

general safety factor 176
generalized power series method 207, 278–82
geometrical centre 12
geometrical imperfections 4
global
   analysis 2
   approach 1
   behaviour 1
   bending 9, 196, 204
   critical load ratio 176–81, 259
      monitoring 180
   deformations 258
   force 65
   moment of inertia 15
   regularity 225–6
   response 1
   safety (factor) 177, 178, 259, 260
   second moment of area 196
   seismic force 65
   shear 35
   slenderness ratio 225
   stability 255
governing differential equation(s) for
   dynamic problem 43, 44
   frameworks 87, 201, 278
   sandwich model 205
   stability 19, 21
   stress analysis 101–2, 130–1
   torsion 102
   unsymmetrical bending and torsion 101, 130–1
guidelines 174–5, 255–65

height of the building 26, 46, 48, 62, 255
height/width ratio 65, 225, 230
Hooke's law 4
horizontal displacements 103–4, 187, 189
horizontal load 63, 65, 67, 68
   decomposition of 99–100
   direction of 179
   of trapezoidal distribution 98, 187
   uniformly distributed 63, 98

ill-conditioned eigenvalue problem 24, 207, 282
individual
   bracing elements 16
   columns 14–6, 41–2
   structural elements 1–2
infilled frameworks 226–8, 233–5, 263, 264
intensity of horizontal load 67, 68, 90, 111–3
interaction 1, 8, 29, 33, 34, 54
   soil-structure 36–40, 59

laced columns 218
lateral displacements 130, 137
lateral frequencies 45, 46, 56, 57, 62
lateral stiffness 7, 14, 46, 110, 183–4, 256, 257, 258
   based on frequency measurements 183
lateral-torsional vibrations 52
lateral vibration 45, 55, 60
layout of arbitrary shape 11
load
   components 100, 130
   distribution 7, 63, 72, 77, 98, 110
   factor 111
      maximum 112
   function 99
   share 63, 74, 79, 83, 84, 85, 91, 93, 95, 112
   types of 4, 7
load-bearing elements 7, 8, 67
local
   approach 1
   bending 195, 204
   shear 34
   structural analysis 1
location of maximum
   bending moment 89
   Saint-Venant torsional moment 98, 124–5, 138
   shear stresses 129
   translation 76, 81–2
   warping torsional moment 124
low-rise building 13, 69, 82

low-rise frameworks   235, 260

magnification factor   151, 181
masonry in infilled panel   226–8
mass   4, 44, 48, 56, 68, 183
material   4, 71, 75
maximum
  bending moments   86, 118, 135, 138
  displacement   108–9, 130, 137, 187
    ratio   179
  rotation   63, 106–7, 131–2, 137, 179–80
  shear forces   114, 133, 138
  stresses   129, 258
  torsional moments   124–6, 136, 138
  translations   63, 179–80
medium-rise
  buildings   13
  frameworks   89, 235, 260
misalignment   68, 69
mode
  first   18, 57
  second   46, 47, 50, 57
  third   46, 47, 51
mode coupling   29, 42
  see coupling of modes
mode coupling parameter   54
  for double coupling   30–31, 55
  for triple coupling   28, 54, 283–314
mode shapes   45, 46, 48
model
  deformation of bracing elements of
    252
  subjected to forced vibration   246–7
  symmetrical   241
  translations of   242–4
  under horizontal load   241, 243
  under vertical load   248–52
  unsymmetrical   243
modulus of subgrade reaction
  see subgrade reaction
moment of inertia   9, 16
monosymmetrical case   29, 42, 55
MSC seismic scale   66, 69

national codes   176
natural frequencies   45–8, 53, 56–7, 60
natural period   46, 66

non-regular structures   215
non-structural elements   109, 182
non-sway buckling   42
normal stresses   126–127

one-parameter frequency formulae   62
optimum solution   1
overall stability   176

performance indicator   178–80
performance of the bracing system   96,
  174, 176, 178–80, 255–60
planar bracing elements   233, 263
polar moment of inertia   11, 38
power series method   207, 278–82
primary structural elements   7
principal
  axes   9, 12, 17, 29
  directions   21, 102
  planes   8, 12–13, 44–45
product of inertia   9, 12–3, 97, 102, 137–8,
  150, 245
  zero   137–9, 150
pure torsional
  buckling   20, 26
  critical load   23–5, 32, 41
  frequency   47, 56
  vibration   43, 44–8, 54, 55, 56

radius of gyration   11–2, 26, 48, 52
Rayleigh–Ritz method   43
reduction factor for
  axial forces   60
  frequencies   46–7
  soil-structure interaction   39, 59
  stability   21–3, 40, 196–7, 215
regular structures   3
rocking motion   38
rotation   19, 74, 76, 77, 82, 94, 105, 107,
  112, 117, 131, 137–8, 188–9

safety   1, 27, 176, 180, 210, 259, 262
Saint-Venant torsion
  dominant   106
Saint-Venant torsional constant   9–10, 16
Saint-Venant torsional stiffness   26, 48,
  56, 97, 98, 121, 151, 255, 257, 258–9

340 *Subject index*

dominant 112, 114, 117
neglecting 151
zero 112, 119, 245
sandwich model 204, 237
  with thick faces 204, 205
  with thin faces 204–8
second-order effects 109, 151, 178, 181
sectorial
  coordinate 127
  properties 129
  statical moment 129
seismic
  analysis 43, 67
  constant 66
  load 65–7, 99
  zone 66
shear
  full-height 198
  storey-height 198, 212
shear centre 4, 8, 16, 36, 52, 55, 71–2, 77, 121, 137, 150
  coordinates 8–9, 71, 72, 77, 137
  of model 241, 243
shear critical load 197–9
shear deformation 29, 33–6, 61, 150
  global 35
  local 34
shear force factor 113, 133
  maximum 115–6, 133, 138
shear forces
  due to rotation 113
  in the bracing elements 110, 113, 132, 138
shear mode situations 33
shear stiffness 197–9
  for cross bracing 220–3
  full-height 216
  global 199
  local 199
shear stresses 127
  from Saint-Venant torsion 128
  from unsymmetrical bending 127
  from warping torsion 129
  location of 129
shear type deformation 150, 229, 235
shear wall(s) 7, 86, 140, 230–1, 234, 263
  built-up 97, 144

individual 140
sign convention 19, 100, 102, 270
simplified procedures 2
simultaneous loads 22, 33, 177, 210
single-storey building 32, 55, 129, 184, 186, 189
slenderness ratio 65, 225
small deformations 4
small-scale tests/models 1, 238
soil categories 39, 66
soil properties 41, 57, 59
soil-structure interaction 36–40, 57–9, 152–3
Southwell's theorem/formula 41, 56, 200, 209, 232, 282
spatial behaviour 8, 16–7, 20, 27, 98, 255
spatial stiffness 76, 97
spring 36
  coefficient 37–38, 58
  constant 37–8
stability 16–7, 178
  planar 192–237, 260
  spatial 19–42, 255
stable building/structure 26–7, 177–8
staircase shaft 97
static moment of stiffnesses 74
stiffnesses
  based on frequency measurements 183
storey-height columns 42
stress analysis 258
  advanced 98–153
  elementary 63–97
stresses
  in the bracing elements 126
  normal 126
  shear 127
structural
  adequacy 254
  integrity 1
  performance 174, 259, 263
subgrade reaction 38
suitability of structural layouts 254
summation theories 200
  see Dunkerley, Föppl–Papkovich, Southwell
support
  fixed 199, 264

pinned 210, 264
sway 8, 17, 20
  buckling 20, 37, 42
  critical loads 21, 32, 41, 255
sway-torsional
  behaviour 8, 17
  buckling 26, 28

test results 238
tests
  dynamic 246
  for horizontal load 241–44
  large-scale 1
  small-scale 1
  stability 248
third-order effects 4
torque
  total 122, 135
torsion 17, 20, 94, 102, 182, 260
  mixed 126
'torsion arm' 10, 52, 75, 96, 174–5
torsion parameter 24
torsional-flexural buckling 19
torsional moment 100, 121
  distribution 123–4, 136
  in the bracing elements 121–6, 135–6, 138–9
  Saint-Venant 122–6, 135–6, 138
  warping 122–6, 135–6, 138–9
torsional performance 52
torsional resistance 96–97
torsional stiffness 7, 8, 110, 174, 183–4, 256
  based on frequency measurements 183
  zero 63
translation 73, 76
  due to flexible support 153
  maximum 76, 81
translational stiffness 70

uncoupled
  critical loads 29
  equations 102
  frequencies 53, 54
uncoupled case 20, 21, 102
uniformly distributed
  horizontal load 67, 68, 70, 100, 101, 106, 112, 137–8, 245
  vertical load 19, 22, 184
  weight 44
unsymmetrical bending 102, 117, 127, 245

verifying results 3, 254
vibrator on small-scale model 247

wall
  single 83, 94
'wall-like' behaviour 264–5
walls
  built-up 144
  independent 140
  parallel 69, 90
  perpendicular 76, 91
  see shear walls
warping
  dominant 106, 107, 112, 131, 137
warping constant 9–10, 16, 75, 81, 96, 137
  increase the value of 96, 174, 256, 257, 258
warping stiffness 26, 48, 56, 62, 97, 98, 121, 255, 257, 259
  zero 23, 26, 32–3, 48, 52, 106, 121, 132, 137–8, 151, 174, 183, 184, 185, 189
warping torsion 10
weight of building 46, 66, 257
wind 64–5, 99
worked examples 90, 91, 94, 140, 144, 155, 163, 169